# Polymer and Fiber Science: Recent Advances

# Polymer and Fiber Science: Recent Advances

Raymond E. Fornes and Richard D. Gilbert, Editors
Herman Mark,  Honorary Editor

# CHEMISTRY

Raymond E. Fornes
Physics Department
North Carolina State University
Raleigh, North Carolina  27695-8202

Richard D. Gilbert
Fiber and Polymer Science Program
North Carolina State University
Raleigh, North Carolina  27695-8301

Library of Congress Cataloging-in-Publication Data

Polymer and fiber science:  recent advances / Raymond E. Fornes and Richard D. Gilbert, editors ;
Herman F. Mark, honorary editor.
            p.      cm.
      Includes bibliographical references and index.
      ISBN 1-56081-536-1
      1. Polymers.      2. Fibers, Synthetic.      I. Fornes, Raymond E.
      II. Gilbert, Richard D.
      QD381.8.p64  1991
      667'.4--dc20                                                    91-30624
                                                                              CIP


British Library Cataloguing in Publication Data

      Polymer and fiber science : recent advances.
      I. Fornes, Raymond E.  II. Gilbert, Richard D.
      677
      ISBN  1-56081-536-1

Cover Page Figure Caption:    The ends of two cellulose triacetate (CTA) fibers broken under
tension are shown.  The fibers were spun from anisotropic solutions of CTA in $TFA/CH_2Cl_2$.
The fiber at the top appears uniform along its length and has a fracture surface that is highly
fibrillated.  The fiber at the bottom has a distinct banded structure and does not show evidence of
fibrillation at the fracture surface.  The fibers were spun by Q. Wu and C. Zhang under the direction
of R. D. Gilbert and R. E. Fornes.

Printed in the United States of America.

ISBN 1-56081-536-1  VCH Publishers
ISBN 3-527-89536-1  VCH Verlagsgesellschaft

Printing History
10 9 8 7 6 5 4 3 2 1

Published jointly by:

VCH Publishers, Inc.
220 East 23rd Street
Suite 909
New York, New York
10010

VCH Verlagsgesellschaft
P. O. Box 10 11 61
D-6940 Weinheim
Federal Republic of Germany

VCH Publishers (UK) Ltd.
8 Wellington Court
Cambridge CB1 1HZ
United Kingdom

# Preface

This book is the result of a symposium held on March 7-9, 1990 in Raleigh, NC, "In Honor of the Retirement of Richard Dean Gilbert from North Carolina State University and in Celebration of the Twentieth Graduating Class of the Fiber and Polymer Science Program." The book is a fitting tribute to both the program and to Gilbert, a truly outstanding polymer chemist who has contributed immeasurably to its success.

Henry Rutherford was hired by the School of Textiles at North Carolina State University in 1947 to head the Department of Textile Chemistry. At that time he was the only faculty member in the School with scientific research experience or a publication record in recognized journals. He was expected not only to enhance the B.S. degree but also to develop a quality M.S. degree and ultimately a Ph.D. program. A campaign was undertaken in the 1950s to obtain state funding to initiate a basic research program in the School. During this period, the first faculty member with a Ph.D., David Cates, was hired.

Cates, a physical chemistry graduate of Princeton, was asked to undertake the leadership role in the preparation of a proposal for a doctoral degree. The original intent was to offer the degree in textile chemistry, but it soon became evident that, to fill the needs of the School of Textiles, a broader based discipline was necessary. Moreover, there was a void not only at North Carolina State University but also in the southeast region in the field of polymer chemistry and physics. It was thus decided to develop a unique multidisciplinary degree to fill the void with a program that had strong emphasis on fiber forming polymers. The title chosen for the degree was "Fiber and Polymer Science." The program was approved in 1968 by the University of North Carolina upon the recommendation of North Carolina State University. The State of North Carolina had emerged at that point in time as the leading U. S. producer of textiles with manufacturing plants distributed throughout the State. The fiber industry was also thriving, had numerous plants in the region and had need of well-trained personnel with polymer/fiber science backgrounds. The School of Textiles, which enjoyed the support of both the primary textile and the fiber industries, was becoming recognized as one of the leading textile education centers in the world. Therefore, the proposed program was especially timely.

In preparation for the Fiber and Polymer Science (FPS) program, the School of Textiles began to recruit aggressively highly qualified faculty. Synthetic polymer chemists, physical chemists, physicists and engineers were hired as faculty, many of whom had experience in leading industrial research laboratories. In addition, several faculty members in the Engineering School and the Forestry School became members of the FPS program, the most notable of whom was Vivian T. Stannett. As the proposal was nearing its final form, David Chaney, who headed the Chemstrand Division of Monsanto, was hired as dean of the School of Textiles. Throughout his tenure as dean, he provided continuing and strong support for the program.

Professor Herman Mark agreed to serve as Adjunct professor in the School. For more than twenty years he paid regular visits to the School offering special lectures, interacting with faculty and students and providing kindly advice.

A steady array of university, industrial and government research personnel visited the School. An outstanding FPS seminar series was begun in the late 1960s, organized for several years by T. Waller George, in which leading polymer scientists in the world participated. In one year alone, five recipients of the highest award given by the High Polymer Division of the American Physical Society served as seminar speakers. Among these was Paul Flory.

The School had thus attracted the faculty necessary to establish a highly respected doctoral program. Resources were provided to build appropriate laboratories, establish an excellent research library, support research projects and bring distinguished visitors to the program, largely through the seminar series. In the meantime, the faculty developed a challenging curriculum. The program experienced little difficulty attracting high quality students.

Prior to the completion of the FPS proposal, Richard Dean Gilbert was recruited as a faculty member. Gilbert, a native of Winnepeg, Canada, had received his undergraduate degree in physics and chemistry and his masters degree in chemistry at the University of Manitoba. He worked for short periods at the Manitoba Sugar Company and the Polymer Corporation in Canada before undertaking his Ph.D. in synthetic polymer chemistry at the University of Notre Dame where he worked under the direction of Charles Price. Following his Ph.D. in 1950, he worked for a short while with the Department of National Defense of Canada, then took a job with the the American Synthetic Rubber Corporation in Louisville, KY. In 1955 he accepted a position as section manager for the Division of Synthetic Rubber and Latex Chemistry of Uniroyal Chemicals and remained there until he came to North Carolina State University. His first formal interaction with the School of Textiles occurred in 1964, at which time Gilbert negotiated with Rutherford the sponsoring of a research project in the Department of Textile Chemistry. During the course of the project, Rutherford became interested in Gilbert as a faculty member.

Quoting Rutherford: "By this time, we were committed to the Ph. D. Program and began looking for new faculty. We wanted a mature person with a polymer background and with a Ph. D. degree. Gilbert seemed to fit the bill, and after an interview concerning the position, my assessment was as follows: (1) he was interested in a change to an academic atmosphere and in teaching; (2) he would probably be a fair teacher; (3) he would, given freedom, develop a graduate program within the bounds of fiber and polymer science; (4) he was ambitious; (5) and he would work well with other faculty-although he might be short fused at times. So, we made him an offer and he accepted. Turned out I was right on all counts, and I brag about this."

Gilbert came on board in January 1966, two years prior to the approval of the FPS program. He contributed significantly to its final form and was instrumental in implementing improvements in the program as it evolved. Gilbert served on the committee of Joel Williams, the first doctoral student, who graduated in spring, 1971. The spring of 1990, the date of Gilbert's official retirement, thus marked the twentieth graduating class of the FPS program, and at that point, 113 students had completed their Ph.D. degree. Gilbert had chaired or cochaired twenty-five of the students, a number exceeding the output of any other faculty member in the program. The records of those graduates have been truly outstanding.

Gilbert's doctoral work was on copolymerization. His work in industry was mostly of a proprietary nature so that upon entering the faculty ranks at a relatively advanced stage in his career, his publication record was quite limited. Despite the late start, he quickly began to publish his work and had accumulated over 125 publications upon his retirement. While at North Carolina State University, he demonstrated diversified research interests and made major contributions in several different areas. These include fundamental work on the formation ethylene-propylene block copolymers; biodegradable block copolymers using cellulose or amylose as one of the components; development of a new solvent system for cellulose/cellulose derivatives for the preparation of cellulose fibers of the highest strength that has produced to date; basic studies of the liquid crystalline state of cellulose/cellulose derivatives; work on emulsion polymerization; byssinosis research; studies on the effects of high energy radiation on graphite fiber/epoxy composites; development of methods to reduce the moisture sensitivity of epoxies; and environmental effects on polymeric coating materials.

Despite his heavy commitment to research, Gilbert was recognized for outstanding contributions as a teacher. He was the first recipient at North Carolina State of the "Outstanding Graduate Professor Award" and was selected for an "Outstanding Teacher Award". He also found time to serve on numerous committees and boards and was highly active in professional societies. Among this service included: serving four years on the Faculty Senate and eight years on the Administrative Board of the Graduate School of the University; cofounder and two-term chair of the North Carolina ACS Polymer Group; president of the local chapter of Sigma Xi, The Scientific Research Society; and serving as Chair of the Cellulose Division of the ACS.

I first became acquainted with Gilbert in the early spring of 1970, shortly after joining the faculty in the School of Textiles. (Actually, my first real interaction with him was over drinks in a bar in Wilmington, DE, during a visit to du Pont by a small group of faculty from the school.) We began a professional collaboration almost immediately. Our first joint project was with a graduate student who studied the effects of UV light exposure on the structure of a nylon 66 fiber and the relationship to dye uptake. We also prepared our first joint proposal in 1970, which was in the area of byssinosis, a disease associated with exposure to cotton dust, a topic about which neither of us was knowledgeable at the time but one in which we both strongly felt should be a major thrust in the School of Textiles. That work, which was funded in 1972, provided support for graduate students and research associates and also solidified our professional relationship. We discovered that we were able to work together effectively and that our backgrounds were complementary, which led to collaborative work on numerous projects, the writing of many proposals and the sharing of graduate students. My respect grew with time for his professionalism, his ability as an exceptionally talented polymer chemist, his commitment to excellence and honesty in research, his dedication to teaching, his sensitivity, his patience with students (in contrast to his lack of patience with colleagues at times) and his skill at supervising them, his willingness to take on more than his share of responsibility, his creativity, and his knack for getting to the essential elements in any argument. Most of all, I personally value his friendship, and look forward to many more years of friendship as well as continued professional interaction.

R. E. Fornes  
Raleigh, NC  
1991

# Chapter Listings

# Chapter 1

# ENVIRONMENTAL EFFECTS ON POLYMERS

R.E. FORNES AND R. D. GILBERT

*North Carolina State University*
*Raleigh, NC 27695-8202*

## INTRODUCTION

Nearly two decades ago, we initiated a collaborative research program on the effect of various environments on a variety of polymers under the Fiber and Polymer Science Program at North Carolina State University. The first study was concerned with the photodegradation of nylon 66 induced by near ultraviolet radiation particularly in regard to morphological changes.[1-3] Broadline nuclear magnetic resonance spectroscopy was used to measure changes in the molecular mobility as a function of radiation dose, and it was shown that small increases in the crystallinity were induced with exposure in air. This chapter summarizes the more recent investigations that followed.

## HIGH ENERGY RADIATION EFFECTS

Following the initial photdegradation study, we began an extensive set of investigations of the effects of high energy radiation on epoxy resins. Radiation sources included 0.5 MeV electrons, $^{60}$Co γ-rays and 3.0 MeV protons. Emphasis was placed on the epoxy tetraglycidyl-4,4'-diaminodiphenyl methane cured with 4,4'-diaminodiphenyl sulfone. A number of investigative techniques were used, including ESR and infrared spectroscopy, contact angle measurements, x-ray photoelectron spectroscopy (XPS), 3-point bending and interfacial strength measurements, DSC, and dynamic mechanical analysis.[4-15,17] It was demonstrated[4,7] that as the radiation dose was increased, there was a small but significant monotonic increase in ultimate stress and modulus. ESR spectra suggested the epoxy matrix has regions of different crosslink density in which there are significant differences in radial mobility.[8,9] Fast decaying alkyl radicals exist in areas of low crosslink density.[13] XPS and contact angle measurements showed irradiation produces changes in the polarity of the epoxy surface and fracture surfaces of composites.[14,15] The increase in polarity was attributed to chemical changes induced by the ionizing radiation. The interlaminar shear strength of epoxy-graphite fiber increases with high doses (10,000 Mrad) of ionizing radiation.[15] Using a mathematical model based on the finite element method and an extension of Kaelble's theory,[16] a theory that predicts the shear strength of a matrix fiber interface, it was concluded that the interfacial shear strength must increase with radiation dose.[17]

The increase in flexural modulus[4,7] and interlaminar shear strength[5,17] of the composites when irradiated may be accounted for by the interaction of free radicals produced in the irradiated epoxy[8,10,13] at the fiber/epoxy interface and unpaired electrons on the graphite fiber surface. For example, the increases are caused by formation of chemical bonds or chemisorption of radicals.[14,15] An increase in polar-polar interactions may also contribute to the improved interlaminar shear strength.[14,15]

Netravali et al.[11,12,22] confirmed earlier measurements of Gillham et al.[33-35] showing that not all epoxide groups undergo reaction in a typical cure cycle. As the cure proceeds, the Tg increases until it reaches the cure temperature, causing the system to vitrify and reactions of the epoxides to effectively cease. In DSC measurements, TGDDM/DDS cured at 177°C generally exhibited an exothermic peak with an intensity in the range of 10-15% of the exothermic peak for the uncured prepolymer mix. Cured samples exposed to high energy radiation showed a monotonic decrease in the exothermic reaction with radiation dose indicating that the radiation caused continued cure of the epoxy.

Wilson et al.[32] studied the dynamic mechanical behavior as a function of radiation dose. Their data showed that a combination of crosslinking and chain scission was occurring. As the dose increased, the actual Tg increased while the ultimate Tg decreased (the ultimate Tg is defined as the Tg after complete reaction of the epoxide groups). The reaction of the epoxides induced by the radiation increased the crosslinking density and the corresponding vitrification temperature while the chain scission reactions caused a slow decrease in the ultimate Tg. They also showed that exposure of epoxy to high temperature also caused slow degradation of the ultimate Tg. Kent et al. showed that high energy radiation had negligible effect on the graphite fiber.[8]

## WATER EFFECTS

Another aspect of the effect of environment on epoxy/graphite fiber composites is the deterioration of the composite properties following absorption of moisture when the composite is exposed to moist environments. The adverse effect of moisture on mechanical properties is a major drawback of epoxy resins  Water is absorbed into the unoccupied volume of the cured epoxy resin.  The sorbed water plasticizes the resin, lowers the glass transition temperatures, causes swelling, induces stresses, chemical bond cleavage, and debonding of the fiber-matrix interface. Enhanced craze initiation and crack propagation reduce the serviceability of composites.

Yang et al.[18] showed that the amount of water absorbed at room temperature in cured TGDDM/DDS epoxy resins increases as the curing time or temperature increases while the amount of THF-soluble extractables and the density at room temperature decreases. Their data suggest that as the extent of curing increases, the Tg increases and, upon cooling, the density at room temperature is lower because the volume contraction between the cure temperature and room temperature is reduced. This results in an increase in the unoccupied volume and the resins become more accessible to water. Their results suggest that the equilibrium moisture absorption is determined primarily by unoccupied volume of the epoxy resin.

Upon curing, epoxy-amine systems generate hydrophilic groups such as hydroxyl, secondary and tertiary amine groups and residual oxirane and primary amine groups. When DDS is used as the curing agent, sulfone groups are also present. All these polar groups can interact with water and it is believed the interactions provide the driving force for moisture absorption.[18]

It was speculated that if the functional groups in the epoxy were blocked, the moisture absorption could be reduced. In the first of a series of studies using this approach, Fisher et al.[9] reacted thin films of cured TGDDM/DDS (73/27 wt. %) epoxy resin with $\alpha,\alpha,\alpha$ - trifluoro-m-tolyl isocyanate in DMSO at 70°C and reported a 54% reduction in moisture absorption. Thin films were employed so that the extent of reaction could be followed by infrared transmission spectroscopy. Hu et al.[20] blocked the functional groups with a variety of fluorinated aromatic reactants in cured TGDDM/DDS epoxy thin films and obtained reductions in equilibrium moisture absorption by as much as 75%. The blocking reactions were dependent on temperature, time and diffusion. The treated films were stable to hydrolysis even after immersion in distilled water at room temperature for two months as measured by IR spectroscopy.

Kelly et al.[21] studied the effect of cyanoethylation and carbanylation of the residual functional groups on the moisture absorption of thin films of cured TGDDM/DDS epoxy resin. The water uptake at 30, 45, 55 and 70°C of the epoxy resin was monitored gravimetrically. At each temperature, case I or Fickian diffusion was exhibited. The diffusion coefficient increased from 30 to 55°C but decreased at 70°C because of reaction of water with residual oxirane groups. Netravali et al.[22] have shown that absorbed water will react with residual oxirane groups in the epoxy at 70°C, resulting in a decrease in the exothermic energy associated with the completion of the cure in a DSC scan and an increased level of hydroxyl groups, whereas very little reaction occurs at room temperature.

Equilibrium moisture absorptions generally showed a correspondence between the moisture absorption reduction and the number of blocked functional groups, irrespective of the nature of the blocking groups. Moisture absorption reductions as high as 68% were obtained. At 70°C, the values of the diffusion coefficients are significantly lower for films reacted with isocyanate blocking reactants compared with the epoxy resin, which is of significance regarding the effect of hygrothermal history on water absorption. In the above investigations[19-21] involving reactions of either isocyanate or acid chlorides with the epoxy resin, the thin films (ca. 25mm thickness) were preswollen with either dimethyl sulfoxide or dimethyl acetamide to permit uniform diffusion of the reactions into the films. Obviously, this technique is unsatisfactory for thick films ($\geq 150$ mm) because such films would crack a result of excessive swelling; nor would it be feasible for practical applications, as the reaction of blocking reagents would be restricted to functional groups at or near the surface. Of course, it would be preferable for the blocking reactions to occur more or less uniformly throughout the cured epoxy.

To extend this approach of reducing the moisture absorption of epoxy resins to thick films, a method of incorporation of masked isocyanates (blocked isocyanates) into the mixture of epoxy prepolymer and curing agent prior to curing the resin was devised.[23] Several masked isocyanates were synthesized with variations in both the type of isocyanate and the masking group. Those with unblocking temperatures in the range of the cure temperature (150-177°C) of the epoxy resin were selected. IR and DSC analyses showed that residual functional groups in the resin reacted with the masked isocyanates during the cure cycle. Reductions in equilibrium moisture up to ca. 70% were obtained. Sankar et al.[24] used solid state [13]C-CP/MAS NMR data to identify reactions and intermediates involved in the curing reaction.

Rungsimuntakul et al.[25] employed DMA to show that the ultimate Tg of the epoxy is reduced by incorporation of masked isocyanate but that the actual Tg is comparable to the "as cured" Tg of the epoxy. The dynamic moduli up to the Tg are relatively unaffected and, in the case of a number of masked isocyanates, the values of the initial modulus, elongation at break and peak stress are equal to or higher than the corresponding values of unmodified resins at room temperature.

## ACID DEPOSITION EFFECTS ON POLYMERIC COATINGS

Recently, an extensive investigation of the effects of various air pollutants causing acid rain on the photodegradation of polymeric coatings has been undertaken. In one set of experiments, coating samples have been exposed to concentrated $SO_2$ or $NO_x$ atmospheres in the presence of UV. In another set of experiments, clean air has been mixed in a continuously stirred, steady state reaction chamber with pollutants such as $SO_2$, $NO_2$, $O_3$ and hydrocarbons, etc., which are exposed to UV light with a spectral distribution similar to the UV component in sunlight. Acidic deposition products are generated, and the complex mixture is then used to expose coating materials to the pollutants alone or in combination with direct exposure to ultraviolet light and/or moisture. Effects on the microstructure of the base polymers of latex paints have been investigated.[26-28]

Two commercial latex-based polymers have been examined. One is primarily a terpolymer of butyl acrylate, vinyl acetate and vinyl chloride with small amounts of methyl methacrylate and methacrylic acid (I). The other is a equimolar copolymer of butyl acrylate and methyl methacrylate with a minor amount of methacrylic acid (II). A variety of techniques have been used to examine the microstructural changes, including intrinsic viscosity, sol-gel analyses, dynamic mechanical analyses, contact angle measurements, thermal analyses, x-ray photoelectron spectroscopy, elemental analyses, and solution and solid state NMR spectroscopy.

Exposure of the copolymer films to pollutants in the absence of UV light causes virtually no change in film properties. In the presence of UV and clean air (containing $O_2$) there is evidence of both chain scission and crosslinking, with crosslinking predominating. When films of copolymer I are exposed to combinations of either $UV/SO_2$ or $UV/SO_2/H_2O$, there is a dramatic increase in the rate of change in the copolymer properties compared with samples undergoing similar exposures but without the presence of $SO_2$, a result indicating a synergistic interaction between UV and $SO_2$.[26]

Exposure to $UV/SO_2$ or $UV/SO_2/H_2O$ causes loss of acetate groups, incorporation of sulfur in copolymer I and formation of unsaturated carbon-carbon bonds due to acetate group loss and some dehydrohalogenation.[27] Exposure of either copolymer I or II to $UV/NO_2/H_2O$ has only small effects on the microstructure.

[13]C CP/MAS NMR spectroscopy and elemental analyses demonstrated that sulfur incorporation is mainly due to pendant sulfate groups due to $UV/SO_2$ or $UV/SO_2/H_2O$ exposure.[28] It was also suggested that butyl acrylate units exhibit a co-unit effect, primarily between vinyl acetate and butyl acrylate structural units and may be responsible for the observed loss of substituent ($CH_3COO$, $Cl$) and sulfur incorporation.

These studies have been extended to exterior latex and alkyd architectural coatings exposed to different conditions involving simulated sunlight, dew and a photochemical smog in the steady state reactor.[29,30] Changes in surface features after 1370 hours of exposure were characterized by SEM and EDAX. The coatings formulated with $CaCO_3$ as an extender pigment were found to exhibit the greatest changes in surface features and film color change.

The surface topography of exposed films without $CaCO_3$ were almost identical to unexposed control films. EDAX showed that sulfur was present in the exposed films only in trace amounts.

When $CaCO_3$ was present in the films, crystalline material with a variety of morphologies developed. Elemental analyses by EDAX showed the crystalline material is primarily calcium and sulfur. In the presence of dew EDAX showed a significant amount of calcium was removed from the surface.

The dew runoff from the films extended with $CaCO_3$ contained $SO_4^=$. The presence of $SO_4^=$ in the runoff indicate that bisulfite anions $(SO_3^=)$ were oxidized in the chamber smog, most probably by photochemically formed oxidants such as $O_3$ and $H_2O_2$ ($SO_2$ will undergo a series of reversible reactions in the presence of surface moisture and UV light to form $SO_4^=$). It is also possible that some $SO_4^=$ resulted from the $H_2SO_4$ aerosol formed within the photochemical smog.

Polarized microscopy observations suggest that some of the $CaSO_4$ on the film surfaces consisted of hydrates of $CaSO_4$, accounting for the various crystalline morphologies. On several of the films sharp needlelike $CaSO_4$ crystals, identified by x-ray diffraction, were present. Their appearance is very similar to the $CaSO_4$ crystals reported by Wolff et al.[31] to have formed on automotive finish panels exposed outdoors in Florida during the summer of 1988.

## ACKNOWLEDGMENTS

We wish to thank all the students, research associates and colleagues who have contributed to our research programs. We also acknowledge support by the National Aeronautics and Space Administration, the U.S. Army Research Office and the U.S. Environmental Association for work reported here.

## REFERENCES

1. Stowe, B.S., Fornes, R.E., Salvin, V.S., and Gilbert, R.D.,*Textile Res J.* 43, 704 (1973).
2. Stowe, B.S., Fornes, R.E., Salvin, V.S., Gilbert, R.D.,*Textile Res J.*, 43, 714 (1973).
3. Stowe, B.S., Fornes, R.E., Gilbert, R.D., *Polym. Plastics Technolo. and Eng., 3*, 159 (1974).
4. Naranong, N., M.S. Thesis, North Carolina State University, 1980.
5. R.E. Fornes, Memory, J.D., Gilbert, R.D., Long, Jr. E.R., The Effect of Electron and Gamma Radiation on Epoxy-Based Materials, Proc. Large Space Systems Tech., - 1981, *Third Annual Tech. Rev.,* Langley Research Center, Hampton, VA, Nov. 1981.
6. Fornes, R.E., Memory, J.D., Gilbert, R.D., Studies of Epoxy-Resin-Graphite Fiber Composites Using NMR, ESR and DSC, *Proc. Third Annual Army Composites Res.*, Army Research Office, Research Triangle Park, NC; I-7, 1:18(1982).
7. Wolf, K., Fornes, R.E., Memory, J.D., Gilbert, R.D., *J. Appl. Phys., 54(10)*, 5558 (1983).
8. Kent, M., Memory, J.D., Fornes, R.E., Gilbert, R.D., *J. Appl. Polym. Sci., 28*, 3301 (1983).
9. Netravali, A.N., Fornes, R.E., Memory, J.D., Gilbert, R.D., *J. Appl. Polym. Sci., 29*, 311 (1984).
10. Shaffer, K., Fornes, R.E., Memory, J.D., Gilbert, R.D., *Polymer 25(1), 54* (1984).
11. Netravali, A.N., Fornes, R.E., Memory, J.D., Gilbert, R.D., *J. Appl. Polym. Sci., 30*, 1573 (1985).
12. Netravali, A.N., Fornes, R.E., Memory, J.D., Gilbert, R.D., *J. Appl. Polym. Sci., 31*, 1531 (1986).
13. Seo, K.S., Fornes, R.E., Gilbert, R.D., Memory, J.D., *J. Polym. Sci., Polym. Phys. Ed., 26*, 533 (1988).
14. Seo, K.S., Fornes, R.E., Gilbert, R.D., Memory, J.D., *J. Polym. Sci., Polym. Phys. Ed., 26*, 245 (1988).
15. Park, J.S., Seo, K.S., Fornes, R.E., Gilbert, R.D., *Plastics and Rubber Proc. and Applic. 10*, 203 (1988).
16. Dynes, P.J., Kaelble, D.H., *J. Adhesion, 6*, 195 (1974).
17. Park, J.S., Ph.D. thesis, North Carolina State University, 1988.
18. Yang, F., Gilbert, R.D., Fornes, R.E., Memory, J.D., *J. Polym. Sci., Polym. Chem. Ed., 24*, 2609 (1986).
19. Fisher, C.M., Gilbert, R.D., Fornes, R.E., Memory, J.D., J. *Polym. Sci., Polym.Chem. Ed., 23*, 2931 (1985).
20. Hu, H.-P., Gilbert, R.D., Fornes, R.E., J. Polym. Sci., *Polym. Chem. Ed., 25*, 1235 (1987).
21. Kelly, B.K., Gilbert, R.D., Fornes, R.E., Stannett, V.T., J. *Polym. Sci., Polym. Phys. Ed., 26*, 1261 (1988).
22. Netravali, A.N., Fornes, R.E., Memory, J.D., Gilbert, R.D., *J. Appl. Polym. Sci., 31*, 1531 (1986).
23. Lonikar, S.V., Rungsimuntakul, N., Gilbert, R.D., Fornes, R.E.,*J. Polym. Sci., Polym. Chem. Ed., 28,* 759 (1990).
24. Sankar, S.S., Lonikar, S.V., Gilbert, R.D., Fornes, R.E., and Stejskal, E. O., *J. Polym. Sci., Polym Physics Ed., 28*, 293 (1990).

25. Rungsimuntakul, N., Lonikar, S.V., Fornes, R.E., Gilbert, R.D., *Mater. Res. Soc. Proc., 171*, 371 (1990).
26. Patil, D., Gilbert, R.D., Fornes, R.E., *J. Appl. Polym. Sci., 41*, 1641 (1990).
27. Sankar, S.S., Patil D., Schadt, R.J., Fornes, R.E., Gilbert, R.D., *J. Appl. Polym. Sci., 41*, 1251 (1990).
28. Schadt, R.J., Gilbert, R.D., Fornes, R.E., *J. Appl. Polym. Sci.*, in press.
29. Spence, J.W., Lemmons, T.J., Hou, Y., Schadt, R.J., Fornes, R.E., Gilbert, R.D., to be published.
30. Spence, J.W., Lemmons, T.J., Hou, Y., Miller, C., Fornes, R.E., Gilbert, R.D., unpublished results.
31. Wolff, G.T., Collins, D.C., Rodgers, W.R., Verma, M.H., Wong, C.A., J. *Air Waste Manage. Assoc., 40*, 1638(1990).
32. Wilson, T.W., Fornes, R.E., Gilbert, R.D., and Memory, J.D., *J. Polym. Sci., Polym. Phys. Ed., 26*:2029 (1988); 27: 1185 (1989).
33. Lewis, A.F. and Gillham, J.F., *J. Appl. Polym. Sci., 6*: 422 (1962).
34. Lewis, A.F. and Gillham, J.F., *Polym. Eng. Sci., 19*: 683 (1979).
35. Gillham, J.F., *Polym. Eng. Sci., 19*: 676 (1979).

# Chapter 2

# BIOMEDICAL APPLICATIONS OF SYNTHETIC POLYSACCHARIDES

CONRAD SCHUERCH

*Department of Chemistry*
*State University of New York*
*College of Environmental Science and Forestry*
*Syracuse, NY 13210*

The biochemical and biomedical applications of synthetic stereoregular polysaccharides are reviewed.

For nearly thirty years my laboratory was engaged in the chemical synthesis of polysaccharides from simple sugars.[1,2] The methods developed for the synthesis of stereoregular polysaccharides involved, first, converting simple sugars into sugar derivatives with an anhydro ring and with all other substituent hydroxyl groups blocked with benzyl ether functions or other removable protecting groups. Usually an eight or nine step synthesis was required to prepare suitable pyranose sugar derivatives with 1,2-, 1,3-, 1,4- or 1,6-anhydro rings. Under appropriate conditions, these polymerize with backside attack and inversion on C-1 to form linear polysaccharide derivatives 95 to 99% stereoregular in structure. Debenzylation gives polysaccharides of defined structure. Various modifications of this approach, developed in our laboratory and elsewhere, have given polysaccharides that bear other substituent groups, or that are branched, or have more than one sugar distributed randomly in the chain. Participation of a neighboring ester function on C-2 in the polymerization process has, at least in one case, allowed the formation of a regular oligosaccharide with *retention* of configuration on C-1.[3]

Now, with the great variety of polysaccharides available in nature, the question arises, "What utility do synthetic polysaccharides have?". In point of fact, they have been useful in investigating the structure of natural polysaccharides, the course of allergic and immunological reactions, and the action of plant lectins and synthetic and hydrolytic enzymes. There has also been exploratory research on their use as components of liposomes and in a few other therapeutic applications. Some examples follow.

The stereoregular structures of these polysaccharides have been established by their method of synthesis and by physical measurements, primarily proton and [13]C nuclear magnetic resonance spectroscopy, and secondarily by optical rotation. The results have been confirmed in a number of cases by enzymic degradation.  E.T. Reese and F.W. Parrish[4] first showed that a synthetic α-1,6-linked glucan was degraded by an endoenzyme, that is, one that degrades by random specific attack on α-1,6-linkages along the chain.  The mixture of monomers and low oligomers resulting was interpreted to show a frequency of only 2% of linkages other that α-1,6.  More extensive studies on a variety of synthetic glucans by G. Walker, to be discussed later, confirmed this level of stereoregularity.  The corresponding α-1,6-linked mannan was treated with an exoenzyme by J.S. Tkacz and J.O. Lampen.[5] Exoenzymes attack only at the nonreducing chain end and degrade the chain down to the first structural flaw.  This more sensitive analysis indicated that the mannan had more than 99% stereoregularity, while demonstrating the activity of the enzyme on a high polymeric substrate.

These two polysaccharides have been used by I.J. Goldstein[6,7] to investigate the structural selectivity of concanavalin A, a plant protein or lectin that forms precipitates on binding with various α-linked glucans and mannans.  His group demonstrated that no precipitate formed with these linear polysaccharides and that consequently the lectin could be used to identify branching in related natural polysaccharides, with which precipitation did occur.

Several of these polysaccharides are synthetic analogues of polysaccharides produced by microorganisms.  The parasitic fungus, *Trichophytum rubrun*, a dermatophyte, is known to produce a linear (1-->6)-α-D-mannopyranan and to cause skin reactions.  A synthetic mannan of the same structure was shown to show the same delayed-type skin reactions on inoculation of guinea pigs and delayed-type cross-reactivity against natural mannan antiserum.[8]  The same synthetic mannan has been used in the investigation of the structure of haptenic groups in the yeast *Candida stellatoidea*,[9] and in identifying (1-->6)-α-D-mannopyranan haptenic groups in mutant A strain of the yeast *Saccharomyces cerevisiae*.[10]  Intravenous injection of a neutral mannan from a wild-type strain of the same yeast has been shown to cause anaphylaxis and a rapid lethal effect in mice.  Synthetic (1-->2)- and (1-->3)-α-linked mannans have been used to investigate the mechanism of action and structural features inducing the reaction.[11]

More extensive studies have been carried out using synthetic glucans.  These serve as model substances for the extracellular polysaccharides produced by two genera of lactic acid bacilli, the genus *Leuconostoc* and genus *Streptococcus*.  One strain of *Leuconostoc mesenteroides* (NRRLB512) produces a soluble dextran that is used as a blood plasma extender, "Dextran", and for various Sephadex products by Pharmacia.  *Streptococcus mutans* and related species produce insoluble glucans that promote the deposition of dental plaque and are also involved in various heart infections.

The soluble dextran used as a blood plasma extender has a chain of (1-->6)-α-linked glucose units with about 5% branch points. It has one disadvantage as a blood plasma extender, for in very rare cases there is a serious, sometimes fatal allergic reaction with anaphylactic shock. The immunology of this reaction has been studied extensively.

Many years ago, Professor Elvin Kabat investigated the binding energy of antidextran antibodies to glucose and a series of (1-->6) α-linked glucose oligosaccharides. He found that the monosaccharide was bound most strongly, and each additional glucose unit in the chain was bound less and less strongly. This result was interpreted by him to mean that the antibody binds to terminal units preferentially. But it could equally well mean that the first sugar unit entering the active site of the antibody binds most strongly. Dr. Wolfgang Richter later tested our linear (1-->6) α-glucan in the precipitin reaction. This is a crosslinking reaction between polysaccharide and protein and like gelation reactions in other polymer systems, requires multiple functional groups, or reacting sites, on the components. If Kabat's interpretation were correct, our linear dextran with only one nonreducing end group would not precipitate, but in point of fact, it acted exactly like the natural dextran, which has a number of nonreducing end groups, one on each branch. Richter's conclusion was that the dextran-antidextran reaction involved predominantly oligosaccharide sequences along the chain, not just on terminal units on the polysaccharide.[12] He also demonstrated that antibodies to B512 were specific for (1-->6)-α-D-glucosyl linkages and were unreactive to the corresponding synthetic mannan.[13] Sometime later Kabat's group was able to obtain monoclonal antibodies to dextran and by also using our linear dextran was able to find individual antibodies that acted by each mechanism.[14] L.G. Bennett and C.P.J. Glaudemans[15] have determined the binding constant of one of these monoclonal antibodies with synthetic linear (1-->6)-α-glucan and found it essentially identical to that obtained with a corresponding pentasaccharide, thus confirming that in this case the active site reacts with terminal sequences. Research on the combining sites of antibodies to (1-->6)-α-D-glucans or dextrans continues, and the presence of groove-type sites for linear sequences and cavity-type sites for nonreducing terminal units is now established.[16]

There are several different immunological and allergic reactions to dextrans, and their relative importance has been shown to differ between species, specifically humans and rats. Presumably these differences relate to the varying proportion of antibodies that react with terminal units and chain units on the polysaccharides. The differences have been demonstrated by the use of both natural and linear synthetic dextran.[17]

The specificity of antibodies raised to dextrans produced by other strains of *Leuconostoc mesenteroides* has been investigated with synthetic linear dextran and a series of synthetic dextrans with 3-O-α-D-glucopyranosyl side chains. Antibodies to the N-4 strain dextran were shown to be specific for the $\alpha(1-->6)$ linkage and inhibited by 3-O-$\alpha$-D-glucopyranosyl or - mannosyl side chains.[18] Antidextran B-1375 antibodies were of three types specific for $\alpha(1-->6)$ linkages, $\alpha(1-3)$ branch points and an unidentified structural feature.[19]

A.W. Richter and R. Eby prepared a series of antigens by coupling seventeen di- and trisaccharides of glucose and mannose to protein. In this way immunogenic analogues of branched natural and synthetic dextrans and linear synthetic mannans and glucomannans were obtained. Antibodies to the oligosaccharides cross-reacted and precipitated natural or synthetic polymers carrying the same structural patterns, and various aspects of the cross-reactivity were studied.[20]

Dental plaque contains an extracellular or capsular polysaccharide that adheres to tooth enamel and is produced by the lactic acid bacterium *Streptococcus mutans* inter alia. The organism generates lactic acid and the capsular plaque retains it near the enamel, slowing down its diffusion and thus causing caries. The structure of this material and the enzymes that synthesize it and degrade it have been studied in great detail by G.J. Walker. In this extensive research, she has used the water-soluble synthetic linear $(1-->6)$-$\alpha$-D-glucan, a series of soluble synthetic $(1-->6)$-$\alpha$-D-glucans with varying numbers of 3-O-$\alpha$-D-glucopyranosyl side chains, and water-insoluble synthetic $(1-->3)$-$\alpha$-D-glucans.

There are two classes of enzymes that synthesize the dextran of dental plaque. They are called glucosyl transferases, and they transfer a glucose unit from sucrose to the nonreducing end of glucan molecule. One class called GTF-S produces a soluble glucan, or dextran, and the other class, GTF-I, produces an insoluble glucan. When both enzymes are operating together, an insoluble plaque matrix is formed.

There are also a number of different enzymes that can be obtained from many different sources that degrade dextrans. They belong to two general classes, the exoenzymes that degrade polysaccharides from the non-reducing end group, usually one mono- or disaccharide unit at a time, and continue the degradation down the chain until they reach a flaw in the structure which prevents their proceeding. The second class of enzyme, endoenzymes, attack the polysaccharide along the chain and fragment it into oligosaccharides. The enzymes have an active site that associates with several monosaccharide units, often six to nine glucose units, along the chain during the cleavage reaction and are most active when the site is full. In general, the final products are mono-, di-, and perhaps trisaccharides from degradation by means of an endoenzyme. Even among enzymes that act on a single type of bond, there can be different specificities. For example, two different enzymes may attack only sequential $(1-->6)$-$\alpha$-bonds and yet stop at different distances away from a $\beta$-linkage or a branch point so that the ultimate di- or trimeric products obtained from the action of two different endoenzymes acting on the same polysaccharide linkage will not necessarily be the same. Therefore, the characterization of the enzymes used is not a simple matter.

As mentioned previously, exoenzymes are very sensitive indicators of structural irregularities in polymers. For example, in polymers with DP>1000, random irregularities of 5 mol % will limit the degradation to less than 2 %. One percent of irregularity in a polymer of DP 100 allows only 50 % of the polymer to degrade and of course, less for large DPs. Thus, if a synthetic polysaccharide of high regularity is degraded by an exoenzyme, the results can both confirm the activity of the enzyme and the percent irregularities estimated from other measurements.

Walker isolated a pure exoenzyme and a pure endoenzyme from two Streptococcus species and tested their action on the synthetic dextran of known (1-->6)-α-D-glucan structure. She found that the enzymes were specific for this linkage and the *exo*enzymes degraded the polysaccharide to the extent of 35%.[21] This corresponded to results expected for a polymer with 2% flaws in the chain, and confirmed previous estimates by Reese and Parrish and from NMR measurements. She then showed that the *endo*enzyme degraded the synthetic dextran first to α-(1-->6) tetramers and pentamers and then more slowly to the glucose monomer, dimer, and trimer in the α-(1-->6) or isomaltose series.[22] She also used these enzymes to investigate the structure of a number of different natural soluble and insoluble dextrans and the effect of branching and linkages other than α-(1-->6) on enzyme action. The action of the *exo*enzyme was halted by α-(1-->3) or α-(1-->4) linkages and by branch points, and the extent of the polysaccharide degradation was determined by the proportion of these flaws.[23]

She next investigated the mechanism of synthesis of dental plaque by glucosyl transferases and demonstrated that the synthesis of insoluble glucan by GTFI was enhanced in rate by a factor of 20-23 times when traces of synthetic (1-->6)-α-dextran or slightly branched B512 dextran were added to the medium. Hence these acted as primers, but a dextran with alternating (1-->3)- and (1-->6)-linkages did not act as a primer.[23]

Structural analysis by both chemical and enzymic methods showed that the insoluble dextran was formed by synthesis of (1-->3)-α-D-glucosan chains attached to the primer. Enzymic degradation of a *number* of the insoluble dextrans from various live *Streptococci* showed that they consist of chains of 1-->3 linkages, linked to chains of 1-->6 linkages with varying degrees of branching. Presumably the soluble (1-->6)-α-dextran is first formed by GTF-S and acts as a primer for the action of GTF-I, which forms the insoluble adhesive matrix of dental plaque. The degree of branching and crosslinking was shown to be dependent on experimental conditions and bacterial strain.[23,24]

A later article discussed the mechanism of branching in more detail.[25] For this Walker used a series of *synthetic* dextrans with different percents of 3-0 glucosyl units as branches, prepared by Dr. Hiroshi Ito.[26] The dextrans were put in a suitable medium with radioactive sucrose and GTF-I from Streptococcus mutans. She isolated the dextrans formed at an *early stage of* branching and was able to show that the amount of radioactivity attached to the dextrans correlated inversely with the amount of branching on the original dextran. As a result she was able to conclude that the (1-->3)-α-glucan chains were attached to unbranched (1-->6)-glucose units in the soluble dextran and were not attached to the single glucose side chains.

Walker also used one of the same synthetic branched dextrans to investigate the structure of branches in two natural *dextrans* NRRL B512 with 5% branches and NRRL B1351 with 10% side chains.[27] She degraded the three dextrans with an *endodextranase* (cleaving α-1-->6 links) down to glucose, isomaltose and a series of branched oligosaccharides. These were separated by chromatography and she was able to show that B1351 and the synthetic dextran had one homologous series of oligosaccharides with single 3-O-α-D-glucosyl units attached to a linear α-(1-->6) chain, while there were two homologous series of oligosaccharides derived from B512. One series was identical with the above, and the second series had a dimeric 3[3]-isomaltosyl side chain attached to the linear fragment of the main chain. The latter demonstrated the presence of side chains longer than a single glucosyl unit in B-512, and their absence in B-1351.

Walker's most recent article[28] describes the analysis of about *thirty* different dextrans by methylation and by enzymic hydrolysis. Some of the dextrans are soluble and some insoluble. Most are *Leuconostoc* and *Streptococcus* genera of the *Lactobacillaceae*, one from *Aspergillus niger* [100% (1-->3)-α-D] and one from *Polyporus* ([(1-->3; 1-->4)-α-D]. She determines by enzymatic analysis the amount of linear consecutive (1-->3)-α-D-linkages, branching features, etc. For this analysis comparison is made with two synthetic linear (1-->3)-α-D-glucans prepared by F.J. Good in our laboratory.[29] In addition, degradation by means of endoglucanase that cleaves (1-->3)-α-D-glucosyl linkages specifically demonstrates that the better of the two synthetic products is 97% stereoregular, and the flaws are identified as (1-->3)-β-D-glucosyl linkages.

Sulfated polysaccharides have a number of important physiological functions. Heparin, for example, is a natural sulfated polysaccharide that has important uses as an anticoagulant. Because of its high cost, a great deal of effort has been expended in finding a suitable semisynthetic replacement. Sulfated polysaccharides also have significant antiviral effects and effects on the immune system that complicate their use. It has been possible to monitor these various effects and select polysaccharides of varying structure that are much more effective in one sense than another. A number of natural and synthetic polysaccharides have been tested by Professor T. Uryu and colleagues. His group has found one natural polysaccharide, after sulfation, to be strongly antiviral against the AIDS virus with essentially no anticoagulant activity.[30]

Professor K. Kobayashi has developed methods for preparing a kind of regular glycolipid by polymerizing or copolymerized octadecyl derivatives of anhydrosugars.[31] At low degrees of substitution these have been used to coat and strengthen liposomes in an attempt to make them more stable drug delivery systems.

The polysaccharides discussed above have all been simple homopolymers or polymers with random distribution of branches on, or units in the chains. The much more difficult task of synthesizing polysaccharides containing a regular sequence of sugars and linkages along the chain has been studied most extensively in the Zelinsky Institute under the leadership of K. Kochetkov. Their approach has been to couple dimeric or trimeric saccharide derivatives by condensation polymerization. The products after deprotection in principle might correspond to capsoular polysaccharides of *Pneumococcus, Meningococcus, Salmonella,* etc. Their most notable achievement to date has been the synthesis of a polysaccharide with a trimeric repeating unit corresponding to the O-factor of *Salmonella newington.* A second synthetic polysaccharide differing only in the configuration of the mannose linkage was not active against the antiserum to the O-factor.[32,33]

In summary, synthetic polysaccharides have proven to be of use as model compounds in basic studies on allergy, immunity, the action of enzymes and lectins, and for structure proof of natural polysaccharides of biomedical interest. They are beginning to be tested as components of pharmaceuticals and probably will be of some interest at least as a novel tour de force as long as people study polysaccharides.

## REFERENCES

1.  C. Schuerch, *Adv. Carbohydr. Chem. Biochem.*, 39, 157-212 (1981). Other contributors to this field are given in refs. 1 and 2.
2.  T. Uryu, "Polysaccharides" in Models of Biopolymers by Ring-Opening Polymerization, S. Penczek, Ed., CRC Press, Boca Raton, FL 1990.
3.  H. Ichikawa, K. Kobayashi, H. Sumitomo and C. Schuerch, *Carbohydr. Res.,* 179, 315-320 (1988).
4.  E.T. Reese and F.W. Parrish, *Biopolymers,* 4, 1043 (1966).
5.  J.S. Tkacz, J.O. Lampen, and C. Schuerch, *Carbohydr. Res.,* 21, 465-472 (1972).
6.  I.J. Goldstein, R.D. Poretz, L.L. So, and Y. Yang, *Arch. Biochem. Biophys.*, 127, 787-794 (1968).
7.  R. Robinson and I.J. Goldstein, *Carbohydr. Res.*, 13, 425-531 (1970).
8.  S.F. Grappel, *Experimentia,* 27, 329-330 (1971).
9.  W.O. Mitchell and H.F. Hasenclever, *Infect. Immunol.,* 1, 61-63 (1970).

10. Y. Obuko, N. Shibata, T. Matsumoto, M. Suzuki, C. Schuerch and S. Suzuki, *J. Bact.*, 144, 92-96 (1980).
11. T. Nagase, T. Mikami, S. Suzuki, C. Schuerch, and M. Suzuki, *Microbiol. Immunol.*, 28(9), 997-1007 (1984).
12. W. Richter, *Int. Arch. Allergy*, 46, 438-447 (1974).
13. W. Richter, *Int. Archs. Allergy Appl. Immunol.*, 48, 505-512 (1975).
14. J. Cisar, E.A. Kabat, M.M. Dorner, and J. Liao, *J. Exp. Med.*, 142, 435-459 (1975).
15. L.G. Bennett and C.P.J. Glaudemans, *Carbohydr. Res.*, 72, 315-319 (1979).
16. E.A. Kabat, Abstract No. 8, Div. Carbohydr. Chem., 200th Amer. Chem. Soc. national meeting, Washington, DC, Aug. 1990.
17. A.K. Delitheos, T.H.P. Hanahoe, and G.B. West, *Int. Archs. Allergy Appl. Immunol.*, 50, 436-445 (1976).
18. M. Torii, S. Ogawa, K. Watabe, T. Koshikawa, M. Yamazoe, T. Uryu, and C. Schuerch, *J. Biochem.*, 99, 263-267 (1986).
19. M. Torii, M. Yamazae, S. Ogawa, K. Watabe, T. Koshikawa, T. Uryu and C. Schuerch, *Microbiol. Immunol.*, 30(3), 261-268 (1986).
20. A.W. Richter and R. Eby, *Molec. Immunol.*, 22(1), 29-36 (1985).
21. G.J. Walker and A. Pulkownik, *Carbohydr. Res.*, 29, 1-14 (1973).
22. A. Pulkownik and G.J. Walker, *Carbohydr. Res.*, 54, 237-251 (1977).
23. M.D. Hare, S. Svensson and G.J. Walker, *Carbohydr. Res.*, 66, 245-264 (1978).
24. G.J. Walker, R.A. Brown and C. Taylor, *J. Dental Res.*, 63(3), 397-400 (1984).
25. G. Walker and C. Schuerch, *Carbohydr. Res.*, 146, 259-270 (1986).
26. H. Ito and C. Schuerch, *J. Am. Chem. Soc.*, 101, 5797-5806 (1979).
27. C. Taylor, N.W.H. Cheetham and G.J. Walker, *Carbohydr. Res.*, 137, 1-12 (1985).
28. B.J. Pearce, G.J. Walker, M.E. Slodki, and C. Schuerch, *Carbohydr. Res.*, 203, 229-246 (1990).
29. F.J. Good and C. Schuerch, *Macromolecules,* 18, 595-599 91985).
30. H. Nakashima, O. Yoshida, T.S. Tochikura, T. Yoshida, T. Mimura, Y. Kido, Y. Motoki, Y. Kaneko, T. Uryu, and N. Yamamoto, *Jpn. J. Cancer Res. (Gann),* 78, 1164-1168 (1987) and related papers.
31. K. Kobayashi, H. Sumitomo and H. Ichikawa, *Macromolecules,* 19, 529-535 (1986).
32. N.K. Kochetkov, *Sov. Sci. Rev. Sect. B.* 4, 1 (1982).
33. N.K. Kochetkov, V.I. Betaneli, M.V. Ovchinnikov and L.V. Backinowsky, *Tetrahedron Suppl.*, (9) 149 (1981).

# Chapter 3

# KINETICS OF HYDROLYSIS AND THERMAL DEGRADATION OF POLYESTER MELTS

KAB S. SEO AND JAMES D. CLOYD

*Eastman Kodak Company*
*Eastman Chemical Company,*
*Research Laboratories*
*Kingsport, TN 37662*

## INTRODUCTION

The degradation rate of polymer melts, in the opposite sense of thermal stability, is often defined as the rate of chain breakage as represented in the following equation[1]:

$$1/[\overline{DP}] = 1/[\overline{DP_0}] + k_d t \qquad (1)$$

where $\overline{DP_0}$ and $\overline{DP}$ are the initial and final number average degree of polymerization, respectively and $k_d$ is the degradation rate constant. The degree of polymerization can be monitored by various methods including end-group analysis, gel permeation chromatography (GPC), light scattering, osmometry, ebulliometry or viscometry. The solution methods can be employed only for soluble polymers in an appropriate solvent and usually require a periodic measurement of the chain length at different degradation stages. On the other hand, the melt viscometry continuously monitors the chain length and is useful for insoluble polymers. When only one degradation rate is involved, the thermal degradation rate constant for a linear polymer may be determined from the melt viscosities measured at different times:[2]

$$1/[\eta^a] = 1/[\eta_0^a] + kt \qquad (2)$$

where $\eta_0$ is the initial melt viscosity of the polymer and $\eta$ is the viscosity at time t, k is the thermal degradation rate constant and a is a constant equal to 1/3.4 from the 3.4th power relationship of molecular weight and melt viscosity. Equation (2) predicts a straight line for $1/[\eta_0^a]$ versus time plot for a single degradation rate. However, polyester melts usually show dual slopes consisting of an initial fast rate and a later slow rate. For polyethylene terephthalate (PET) the initial rate is very sensitive to the moisture content in the sample while the later slow rate is much less sensitive to the moisture content. Therefore, the initial fast rate is attributed to hydrolysis driven mainly by the residual water in the sample, while the later slow rate is attributed to thermooxidative chain breakdown initiated by the thermal energy.[3] We refer to the latter as "thermal degradation" hereafter. The rate of hydrolysis of ester links has been reported to be several orders of magnitude higher than thermal breakdown rate.[4,5] Thus, the initial hydrolysis contribution is often expressed as a constant term in a degradation kinetic equation for molten polyesters, or even neglected for the dry sample.[6] As industrial processes show trends of moving to shorter extrusion melt residence times, the simplification of treating hydrolysis in this manner will not be appropriate. In the present study we derive a combined kinetic equation which represents both hydrolysis and thermal degradation of polyester melts, and we discuss the experimentally determined degradation rate constants in view of the kinetic equation.

## EXPERIMENTAL

### Materials
The polyethylene terephthalate (PET) samples from Eastman Chemical Company with weight average molecular weights ($M_w$) ranging from 40,000 to 60,000 were used in this study. The moisture content of sample was controlled by drying time and measured by Karl Fischer titration method. Samples were dried at 100°C under the vacuum.

### Rheology
The melt viscosity change with time of the sample at 285°C was monitored with a Rheometrics Mechanical Spectrometer (RMS 7220) with eccentric rotational disks (ERD) at 25 rad/s. The sample was heated for 3 minutes before measurements.

## DEGRADATION KINETICS

Typical plots of melt viscosity versus time for dried and undried PET samples are shown in Figure 1. Two degradation rates are clearly seen especially for the undried sample. The initial degradation rates between dried and undried samples are substantially different, while the degradation rates at the later stage are close to each other.

In view of the experimental results, the degradation of the polyester melt is considered to have two contributions, namely hydrolysis and thermal degradation. The oxidative reaction and alcoholysis, which will affect the degradation rate in any stage of degradation, is not considered separately from the two contributions in this study. A random scission of ester links is assumed in any case.

### Hydrolysis
The rate of breakdown of ester links by hydrolysis may be expressed as[5,7]:

$$[d\phi/dt]_H = -k_H(\phi_0 - r)(x_0 - r) \qquad (3)$$

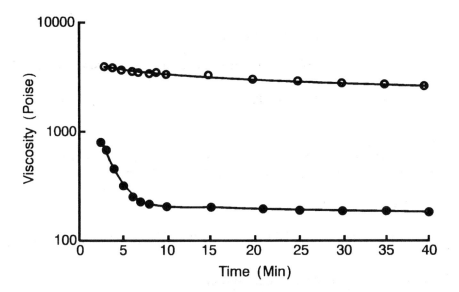

Figure 1. Melt viscosity of PET at 285°C. (o) dried for 24 hours; (•) undried.

where $\phi$ is the number of available ester links at time t, $\phi_0$ is the initial number of ester links, r is the number of moles of water reacted, $x_0$ is the initial number of moles of water, and $k_H$ is the rate constant. The subscript H represents hydrolysis. All the parameters here are in number density (e.g. number per unit volume). Rewriting equation (3), we have

$$[d\phi/dt]_H = -k_H \cdot \phi \cdot x \qquad (4)$$

where x is the number of moles of unreacted water. We assume that all the water molecules react with ester links without drying, i.e., the rate of water consumption is same as the rate of ester link breakdown by hydrolysis. Then the rate of water consumption can be written as:

$$dx/dt = [d\phi/dt]_H = -k_H(\phi_0 - x_0 + x)x \qquad (5)$$

Since $\phi_0 - x_0 \gg x$ for a moisture content less than 0.5 % (by weight) and the number average degree of polymerization DP much greater than unity, equation (5) becomes

$$dx/dt = -k_H( \phi_0 - x_0 )x \tag{6}$$

The solution for equation (6) gives the concentration of unreacted water at time t as:

$$x = x_0 \exp[ -\xi t ] \tag{7}$$

where $\quad \xi = k_H ( \phi_0 - x_0 )$

By combining equations (4) and (7), the rate of hydrolysis can be written as:

$$[ d\phi/dt ]_H = -k_H x_0 \phi \exp[ -\xi t ] \tag{8}$$

**Thermal Degradation**

If the thermal breakdown reaction is assumed to be a first-order reaction the concentration of ester links can be written as[3,8]:

$$[ d\phi/dt ]_T = -k_T \phi \tag{9}$$

where $k_T$ is the thermal breakdown rate constant. The subscript T represents thermal degradation.

**Hydrolysis and Thermal Degradation**

The overall breakdown rate of ester links may be represented as the sum of hydrolysis and thermal breakdown contributions. Combining equations (8) and (9) we have:

$$[ d\phi/dt ]_H = -k_H x_0 \phi \exp[ -\xi t ] - k_T \phi \tag{10}$$

Solving the differential equation (10) we get

$$\ln[ \phi/\phi_0 ] = -x_0/[\phi - x_0 ]\cdot\{ 1 - \exp[ -\xi t ] \} - k_T t \tag{11}$$

The number of ester links $\phi$ in equation (11) can be replaced by the number average degree of polymerization DP through the following relationship[8]:

$$\phi = N \{ 1 - 1/[ \overline{DP} ] \} \tag{12}$$

where N is the total number of monomer units. Using the following approximation when DP>>1:

$$\ln(\{ 1 - 1/[ \overline{DP} ] \}) \cong -1/[ \overline{DP} ]$$

equation (11) can be written as:

$$1/[\overline{DP}] - 1/[\overline{DP_0}] = x_0/[\phi - x_0] \cdot \{1 - \exp[-\xi t]\} + k_T t \quad (13)$$

Using the following relationship[9]:

$$\overline{DP} = \overline{M_n}/M_0$$
$$\eta = KM^{3.4}$$

where M = molecular weight of polymer $M_0$ = molecular weight of monomer, e.g. 96 for poly(ethylene terephthalate) (PET) and K is a front factor,

equation (13) can be expressed as:

$$1/[\eta^a] = 1/[\eta_0^a] + A\{1 - \exp[-\xi t]\} + k_2 t \quad (14)$$

$$\text{where} \quad a = 1/3.4$$
$$A = x_0/[M_0 K^a(\phi_0 - x_0)]$$
$$k_2 = k_T/M_0 K^a$$

Here we assume that the polydispersity ($\overline{M_w}/\overline{M_n}$)does not change during degradation. Equation (14) represents the viscosity as a function time and includes both hydrolysis and thermal degradation contributions. For practical purposes equation (14) can be separated into two components. When time t is small, more strictly $\xi t \ll 1$, equation (14) can be approximated into an equation which represents the hydrolysis only. Neglecting the thermal degradation term ($k_2 t$) and expanding the exponential term, we get:

$$1/[\eta^a] = 1/[\eta_0^a] + A\xi t \quad (15)$$

$$\text{or} \quad 1/[\eta^a] = 1/[\eta_0^a] + k_1 t \quad (16)$$

$$\text{where} \quad k_1 = A\xi = k_H x_0/(M_0 K^a)$$

The apparent hydrolysis rate constant $k_1$ is determined from the slope of a $1/[\eta^a]$ versus t plot when t is small. It is noted that $k_1$ is proportional to $k_H$ and the initial water content $x_0$ but independent of the molecular weight. On the other hand, at the later stage when t is long enough, equation (14) can be reduced to:

$$1/[\eta^a] = 1/[\eta_o^a] + A + k_2t \qquad (17)$$

Equation (17) represents the thermal degradation after completion of hydrolysis. The thermal degradation rate constant $k_2$ depends only on the degradation rate constant $k_T$ as defined in equation (14). It is noted that the term A in equation (17) is the difference in intercepts of equations (16) and (17). If there is no initial water in the sample, A is zero and equation (17) represents thermal degradation only as expressed in equation (2).

## RESULTS AND DISCUSSION

In Figure 2 is shown a curve fitting result of equation (14) to experimental data for a dried PET sample. The model equation agrees well with the experimental data for the dried PET. In Figure 3 a curve fitting for the undried sample was made using the values of $\eta_o$, $\xi$ and $k_2$ estimated for the dried sample as in Figure 2. Although the lack of fitting of the model equation to the experimental data is noticeable for the undried sample the qualitative agreement is very good. The discrepancy can be attributed to many things such as evaporation of water, generation of small bubbles during melt viscosity measurements resulting in erroneous viscosity values, and self-catalytic effect of end-groups produced by hydrolysis on the subsequent hydrolysis reaction.

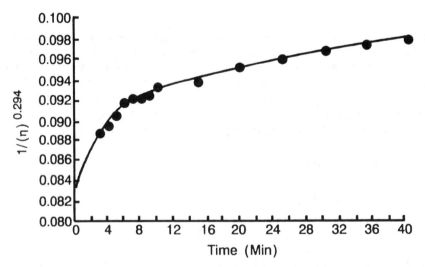

Figure 2.    Curve fitting of equation (14) for dried PET. (•) experiment; (—) 0.0920 - 0.0095 exp (-0.32t) + 0.000181t.

Although curve fitting of equation (14) will provide $\eta_o$, $\xi$, A, $k_1$ and $k_2$, one may want to use simpler forms as in equations (16) and (17) to estimate the parameters. In Figures 4 and 5, the effects of moisture on the hydrolysis rate constant ($k_1$) and the thermal degradation rate constant ($k_2$) determined by equation (16) and equation (17), respectively, are plotted for various molecular weights. The estimation of $k_1$ was based on the data when $t \leq 5$ minutes and $k_2$ was estimated when $t \geq 15$ minutes. It is seen that the hydrolysis rate constant is approximately proportional to the moisture content while the thermal degradation rate constant is relatively insensitive to the moisture as predicted in the present kinetic equations. It is also noticed that the hydrolysis rate constant is at least an order of magnitude higher than the thermal degradation rate constant depending on the moisture content. The effect of molecular weight on both hydrolysis rate constant and thermal degradation rate constant is small except at 0.06% moisture content where the low molecular weight sample shows higher hydrolysis rate constant. In Figure 6 is shown the effect of temperature on the hydrolysis rate constant ($k_1$) for dried samples but still containing a small amount of residual water. The activation energies of the hydrolysis determined from Figure 6 are 6.4 and 16.7 kcal/mole for the low and high molecular weights, respectively. These values are somewhat lower than the literature values of 20-30 kcal/moles for hydrolysis which were determined at 60-200°C.

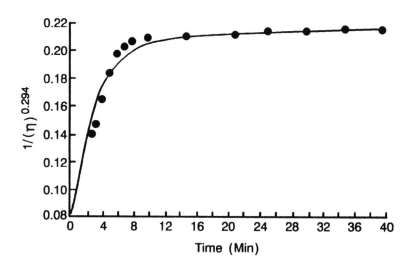

Figure. 3.    Curve fitting of equation (14) for undried PET. (o) experiment; (—) $0.20875 - 0.12626\exp(-0.32t) + 0.000181t$.

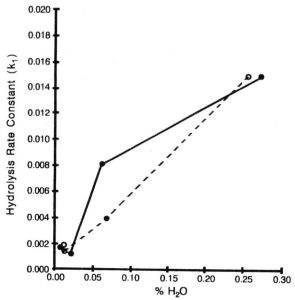

Figure 4.    Moisture effect on hydrolysis rate constant ($k_1$) of PET at 285°C

$M_w$ = 38,500 (·) and  59,000 (o).

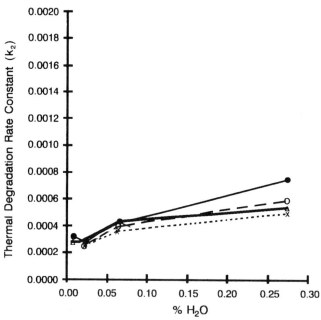

Figure  5.  Moisture effect on thermal degradation rate constant $k_2$ of PET at

285°C; $M_w$ = 38,500(·), 44,600($\Delta$), 52,500(x)  and 59,000 (o).

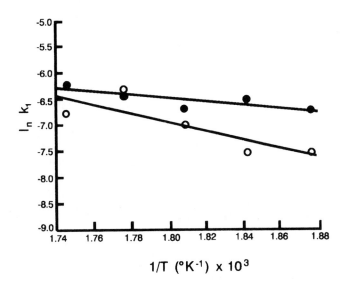

Figure 6.    Arrehnius plot of hydrolysis rate constant (k1): $M_w$ = 38,500 (•) and = 59,000 (o).

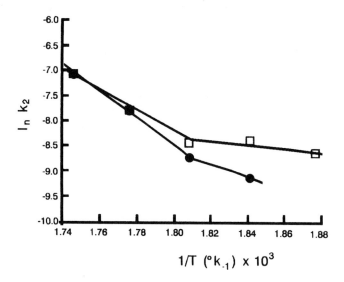

Figure 7.    Arrehnius plot of thermal degradation rate constant (k2): $M_w$ = 38,500 (•) and 59,000 (o).

temperature range.[5,7,10] The observed higher hydrolysis rate constant for the lower molecular weight in Figures 4 and 6 is probably due to the effect of end groups that catalyze the hydrolysis reaction in the melt.[10] The effects of temperature on the thermal degradation rate constant ($k_2$) are more prominent than on the hydrolysis rate constant ($k_1$) as illustrated in Figure 7. The activation energies of the thermal degradation at the temperature range of 280°C to 300°C are 41.6 kcal/mole and 52.7 kcal/mol for low and high molecular weight samples, respectively, and agree with the literature values.[6,11] The effect of molecular weight on the thermal degradation rate constant (k2) is not significant at higher temperature (> 280°C). However, at lower temperatures (< 280°C), the Arrhenius plot breaks down and the activation energy decreases significantly especially for the high molecular weight sample. The thermal degradation rate constant of the lower molecular weight sample at 260°C shows a negative value and is not included in Figure 7. At this lower temperature (260°C), polymerization may predominate in the thermal degradation of low molecular weight chains. The thermal degradation rate constant of the high molecular weight at the lower temperature is higher than that of the lower molecular weight. This is probably due to the decreased mobility (or increased viscosity) of the longer polymer molecules at the lower temperature, where the chemical reaction rate is greatly influenced by the polymer chain dynamics. As the temperature decreases, the two competing reactions (depolymerization or scission and polymerization or crosslinking) may change their kinetical balance so that the chance of recombination of two radicals, for example, becomes smaller, resulting in more degradation.[12] It is also possible that the higher viscosity at lower temperature causes more shear degradation between parallel plates during melt viscosity measurements. A wider range of molecular weights should be examined to observe more detailed effects of end groups or molecular weights on the degradation rate constants.

## CONCLUSION

The hydrolysis and thermal degradation rates of molten polyesters can be expressed in a single kinetic equation. The suggested model equation describes well the melt viscosity change of polyesters as a function of time and can be used to determine hydrolysis rate constant, thermal degradation rate constant, initial moisture content and molecular weight of polyesters. The hydrolysis rate strongly depends on the moisture content in the solid state while the thermal degradation rate is relatively insensitive to the moisture. Although the effect of molecular weight on both hydrolysis and thermal degradation rate constants is not significant, the lower molecular weight sample shows a slightly higher hydrolysis rate constant probably due to end group effects. The activation energy of hydrolysis is in the range 6.4 to 16.7 kcal/mole and the activation energy of thermal degradation is in the range of 41.6 to 52.7 kcal/mol depending on the molecular weight of PET.

## ACKNOWLEDGEMENT

The authors wish to thank Eastman Chemical Company for the support of this work. They are also grateful to Mr. J. E. Briddell, Ms. G. A. Williams and Ms. A. E. Morelock who made the viscosity measurements, and Dr. G. B. Caflisch for his helpful comments on the kinetic equation.

## REFERENCES

1. H. Mark and A. V. Tobolsky, Physical Chemistry of High Polymeric Systems, High Polymers, Vol. II, 2nd ed., Chapt. XIII, Wiley-Interscience Pub., Inc. New York, 1950.
2. D. R. Gregory, Paper presented at 2nd World Congress of Chemical Engineers.-Polymer Engineering, Montreal, Canada, October (1981).
3. E. F. Cassasa, J. Polym. Sci. 4,405-407 (1949).
4. L. Marshall, A. Todd, Trans Faraday Soc., 49,67-78 (1953).
5. W. McMahon, H. A. Birdsall, G. R. Johnson and C. T. Camilli, J. Chem. Eng. Data, Vol. 4, No. 1, 57-79 (1959).
6. F. C. Wampler and D. R. Gregory, J. Appl. Polym. Sci., 16,3253-3263 (1972).
7. R. C. Golike and S. W. Lasoski, Jr., J. Phys. Chem., 64, 895 (1960).
8. D. R. Gregory and M. T. Watson, Polym. Eng. Sci., Vol. 12, No. 6, 454-458 (1972).
9. J. D. Ferry, Viscoelastic Properties of Polymers, 3rd.ed., Wiley, New York (1980).
10. D. A. Ravens and I. M. Ward, Trans. Faraday. Soc., 57, 150 (1960).
11. W. L. Hergenrother, J. Appl. Polym. Sci., Polym. Chem. ed., Vol. 12., 875-883 (1974).
12. N. M. Emanuel and A. L. Buchachenko, Chemical Physics of Polymer Degradation and Stabilization, in New Concepts in Polymer Science (ed. C.R.H.I. de Jonge), 1st. ed., VNU Science Press, CH13, (1987).

# Chapter 4

# SOLVENT ABSORPTION IN FREE-STANDING POLYIMIDE FILMS

WALTER J. PAWLOWSKI AND MARK I. JACOBSON*

*IBM Systems Technology Division
Endicott, New York 13760
*Department of Materials Science and Engineering
Ohio State University
Columbus, Ohio 43210*

## INTRODUCTION

Because of their excellent properties, polyimides are used extensively in the electronics industry.[1] Kapton is prepared via the polycondensation of pyromellitic dianhydride (PMDA) and oxydianiline (ODA) in N,N-dimethylacetamide (DMAc), resulting in the corresponding polyamic acid. Although the actual processing of the film is proprietary and cannot be determined, a review of the literature[2,3] indicates that the polyamic acid is chemically imidized using the appropriate chemicals with a subsequent heat treatment to remove residual chemicals and complete the imidization. Recently, several new polyimide films have been introduced including Upilex (S and R). Upilex S is made via the polycondensation of biphenyldianhydride (BPDA) and p-phenylenediamine (PDA) in DMAc providing the corresponding polyamic acid. Upilex R is formed by the polymerization of BPDA and ODA to give the appropriate polyamic acid. The literature[4] suggests that these films are wholly thermally imidized. Table 1 shows the chemical structures and some selected properties of Kapton H and Upilex (S and R).

Since these films can be used for similar applications, it is important to understand the interaction between various processing chemicals and the polyimide film being used. Studies have been performed on the sorption and transport of various vapors/gases in Kapton film.[5-8] Yang et al.[5,6] found that water vapor sorption is Fickian in both 0.3 and 2.0 mil thick Kapton film, although the diffusion coefficient for the 2.0 mil film is roughly three times larger than the 0.3 mil film. Iler et al.[7] demonstrated that sulfur dioxide sorption in 0.5 and 1.0 mil thick Kapton is Fickian. However, ammonia sorption exhibited significant deviations from Fickian behavior as a result of chemical interaction with the polyimide. Tong et al.[9] and Gattiglia and

Russell[10] have studied the solvent/polymer interaction of thermally imidized PMDA-ODA with dimethylsulfoxide (DMSO) and N-methylpyrollidone (NMP) via laser interferometry (thin films - less than 10μm) and gravimetric analysis (thick films - 40 to 50μm), respectively. In both cases, it was found that DMSO sorption and swelling was considerably faster than NMP. Gattaglia and Russell[10] proposed that this difference most likely resulted from the smaller molecular volume of DMSO (71.3 cm³/mol) compared with NMP (96.6 cm³/mol) and/or to a difference in the interaction parameter between PMDA-ODA and the solvents. Both studies reported case II behavior for the PMDA-ODA/DMSO system. Tong et al.[9] described the transport of NMP as intermediate between Fickian and case II. Gattiglia and Russell[10] reported that PMDA-ODA swells via a case II mechanism for NMP.

The intent of this study was to observe the absorption of various typical processing chemicals and some additional model solvents in Kapton H, Upilex S, and Upilex R.

## EXPERIMENTAL

### Gravimetric Solvent Absorption
Kapton H, Upilex S, and Upilex R samples (2 mils thick) were cut into 0.75" X 1.25" pieces and stored at ambient until use. Reagent grade solvents were used as received. Samples of the polyimide films were weighed prior to solvent exposure using a Mettler AE240 Analytical Balance (accuracy = +/- 0.01 mg). The samples were then suspended in quiescent solvent at the specified temperature which was controlled to +/- 1°C. Samples were removed from the solvent bath at specified intervals, wiped dry, and reweighed on the analytical balance.

### Rutherford Backscattering Spectrometry (RBS) Measurements
Ion beam analysis is well suited to the problem of obtaining both qualitative as well as quantitative information regarding the penetration of solvents into polymers.[11,12] Samples of polyimide film were attached to an aluminum substrate with an epoxy adhesive. Separate samples were mounted with the rough (drum) and smooth (air) side of the Kapton exposed to allow for analysis of each side of the film. The samples were then immersed in the appropriate solvent at the specified temperature for various time intervals. On removal from the solution, the samples were blown dry with nitrogen gas for 2-3 seconds to remove the surface layer of solvent and submerged in liquid nitrogen (LN₂) until analysis.

Incident He⁺ ions at an energy of 2.4 MeV were used to irradiate the samples. The incoming beam was at normal incidence and focused to a spot size of approximately 1 mm². The accumulated charge ranged from 20 to 40 μC. Beam current was held constant at 30 nA. The vacuum in the end chamber during the collection of data was maintained at 10⁻⁷ torr. The substage platform in the end chamber was cooled by $LN_2$ maintaining the samples at low temperature ($100^\circ K$) throughout the analysis. Backscatter spectra were analyzed using a simulation program named RUMP.[13,14]

## RESULTS AND DISCUSSION

### Gravimetric Solvent Absorption: Effect of Film Type

The fractional increase in weight (weight at a specified time, $Wt$, divided by the weight at time zero, $W_0$) as a function of exposure time is shown in Figure 1 for Kapton H, Upilex S and Upilex R immersed in methylene chloride (MC) at $25^\circ C$. It is apparent that the Kapton H film absorbs MC at a faster rate than Upilex S and Upilex R. For Upilex S, no detectable weight gain was observed even after 24 hours of exposure. A review of Table 1 shows that Upilex S has the lowest reported moisture regain value. The increased rigidity in the polymer backbone, due to replacement of ODA with PDA, should result in tighter packing of the molecules. This is further demonstrated by the higher density of Upilex S compared with both Upilex R and Kapton H. In addition, thermal imidization could play an important role. Gattiglia and Russell[10] showed that the rate of weight gain of NMP decreased as the imidization temperature of PMDA-ODA increased. The fact that Upilex S absorbs 1.2% moisture after a 24 hour exposure at $23^\circ C$ and no measurable MC, is confusing. One would expect that the Upilex S would absorb some small portion of MC, and subsequent RBS data confirmed that there was absorption of low concentrations after 1.5 hours of exposure (see RBS section). One reason for the low absorption could be that the interaction parameter between Upilex S and MC is prohibitive.

The chemical/thermal imidization of Kapton H film and increased flexibility of the polymer chain due to free rotation around the ether linkage of the ODA may explain the rapid absorption of MC. The transport of the chemical imidization by-products out of the film most likely results in microvoid formation. Russell has shown the presence of microvoids even in wholly thermally imidized PMDA-ODA.[15]

Upilex R contains the flexible ODA portion of Kapton H in addition to the biphenyl structure of the Upilex S, which results in its inherently lower density. One might anticipate that a less dense material of similar chemical structure would exhibit faster and possibly greater solvent uptake. This is not the case for MC absorption in Upilex R compared to Kapton H (see Figure 1). Further evidence of this difference can be seen by comparing the moisture regain of Kapton H (2.8%) with Upilex R (1.3%). Since the two polyimides have similar chemical structures (i.e., same functional groups), it appears that in addition to the differences in method of imidization there may be other processing factors that influence the individual film properties (bulk or surface). Another possibility may be that the BPDA portion of the polyimide plays an important role in molecular ordering and/or solvent uptake.

| NAME | CHEMICAL FORMULA | DENSITY (G/ML) | Tg(°C) | MOISTURE[a] REGAIN (%) | TCE[b] (PPM/°C) | DIELECTRIC CONSTANT |
|---|---|---|---|---|---|---|
| KAPTON H | | 1.42 | 360–410 | 2.8 | 20 | 3.5 |
| UPILEX S | | 1.47 | >500 | 1.2 | 8 | 3.5 |
| UPILEX R | | 1.39 | 285 | 1.3 | 15 | 3.5 |

a – IMMERSION IN WATER AT 23°C FOR 24 HOURS.
b – THERMAL COEFFICIENT OF EXPANSION ($\times 10^{-6}$ CM/CM/°C).

TABLE 1. CHEMICAL STRUCTURES AND PROPERTIES OF SELECTED POLYIMIDE FILMS

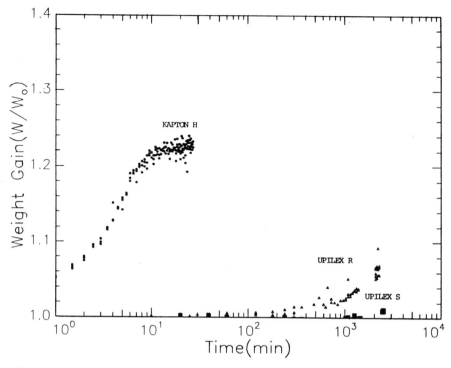

Figure 1.    Weight gain as a function of time in MC at 25°C.

## Effect of Solvent Type

The chemical structures of the solvents used and their associated molecular volumes are shown in Table 2. Kapton H showed a decreasing absorption rate with increasing molecular volume until the molecular volume reached 78.73 cm$^3$/mol (DCE). At molecular volumes equal to or greater than this, Kapton H showed no measurable absorption after exposures of up to 1 hour (e.g., no weight gain was observed for DCE, MCF, and DCP). Gattiglia and Russell[10] showed a similar trend for PMDA-ODA swelling with DMSO and NMP at long exposure times. Although the time scale of these experiments was relatively short, typical exposures to these types of chemicals in a processing environment would not exceed 15 minutes. It is also recognized that this study has not taken into account the interaction parameter of the solvent/polymer system. The intent is only to show that a general trend exists.

As mentioned previously, no measurable absorption of MC was detected gravimetrically in Upilex S, even after exposure for 24 hours. No measurable absorption was found for any of the other processing chemicals after exposure for 1 hour.

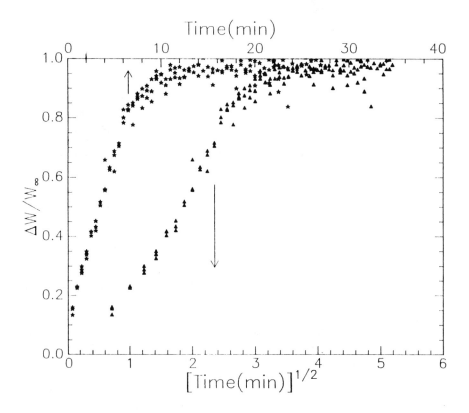

Figure 2.    Weight gain as a function of time and the square root of time for MC absorption in Kapton H.

Table 2.    Chemical structures and molecular volumes of solvents

| CHEMICAL NAME | CHEMICAL FORMULA | BOILING POINT ($^{\circ}$C) | VOLUME (CM$^3$/MOL) |
|---|---|---|---|
| METHYLENE CHLORIDE | Cl-CH$_2$-Cl | 40 | 64.08 |
| METHYLCHLOROFORM (1,1,1-TRICHLOROETHANE) | Cl-C-CH$_3$ with Cl above and Cl below | 75 | 99.72 |
| GAMMA-BUTYROLACTONE | | 205 | 76.28 |
| DIBROMOMETHANE | Br-CH$_2$-Br | 95 | 69.67 |
| 1,2 - DICHLOROETHANE | Cl-CH$_2$CH$_2$-Cl | 83 | 78.73 |
| 1,2 - DICHLOROPROPANE | Cl-CH$_2$-CH-CH$_3$ with Cl below | 95 | 97.49 |

## Mode of Transport (Gravimetric Analysis)

Normalized sorption curves plotted against the square root of time tend be initially linear for Fickian diffusion.[5] When the initial weight gain is linear with time, the diffusion is termed case II.[16] A typical plot of MC absorption in Kapton H versus both time and the square root of time at 25$^{\circ}$C is shown in Figure 2. It is evident that the weight uptake is linear with time initially. As stated before, similar observations were made for swelling measurements of thermally imidized PMDA-ODA when immersed in DMSO and NMP.[10] The characteristic linearity with time at short times was also observed for the absorption of dibromomethane (DBRO) and gamma-butyrolactone (BLO) in Kapton H. Low absorption of the solvents in Upilex S and Upilex R prevented the making of a gravimetric determination of the diffusion characteristics.

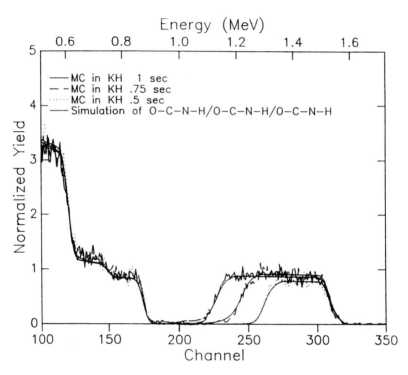

Figure 3.      RBS spectra depicting MC absorption in Kapton H

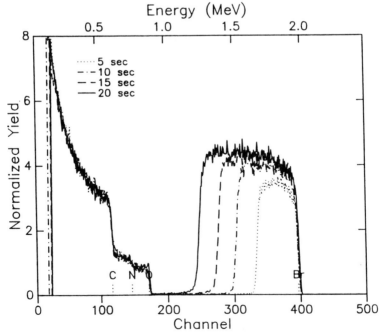

Figure 4.      RBS spectra of DBRO absorption in Kapton H

## Rutherford Backscattering Spectrometry (RBS) Measurements

Spectra of MC uptake in Kapton H are shown in Figure 3. The relatively flat Cl peak associated with the steep solvent front is indicative of case II diffusion. The slight slope associated with the MC is attributed to partial evaporation of solvent during sample preparation and/or transport to the sample holder. The absorption of DBRO in Kapton H was also found to be Case II using RBS (Figure 4). The steeper solvent front for DBRO compared to MC was attributed to the higher boiling point of DBRO.

Kapton film has a rough (drum) and smooth (air) side because of the film-forming process (Figure 6). A comparison of front depth versus time for DBRO in Kapton H showed a very slight orientation effect for Kapton H (Figure 5). A greater depth of the case II front at similar times was observed for the rough side indicating that a very slight "head start" occurs for solvent penetrating from the rough surface of the Kapton H. Both sides of the Upilex films are smooth in appearance when viewed microscopically.

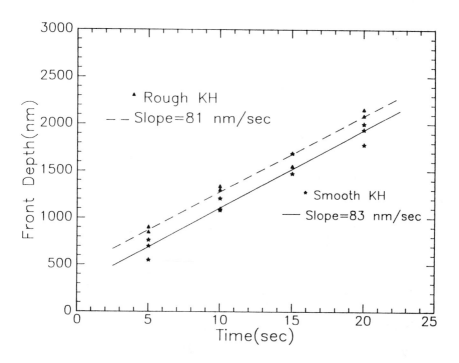

Figure 5.     Front velocity versus time for DBRO absorption into the the rough (drum) and smooth (air) side of Kapton H

Figure 6.      Optical interference photomicrographs (200x) of the rough (top) and smooth (bottom) side of Kapton H

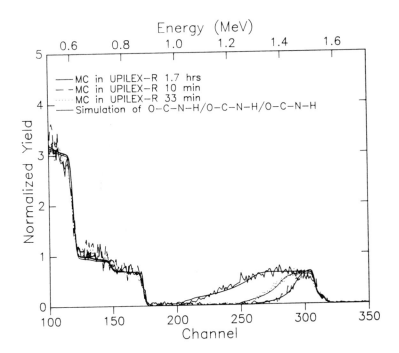

Figure 7.    RBS spectra depicting MC absorption in Upilex R

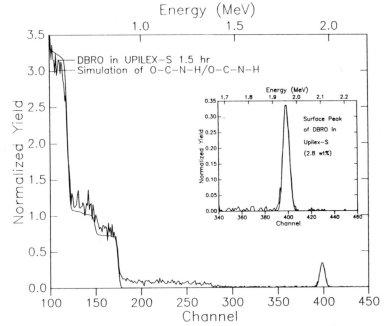

Figure 8. RBS spectrum of MC absorption in Upilex S after 1.5h. exposure

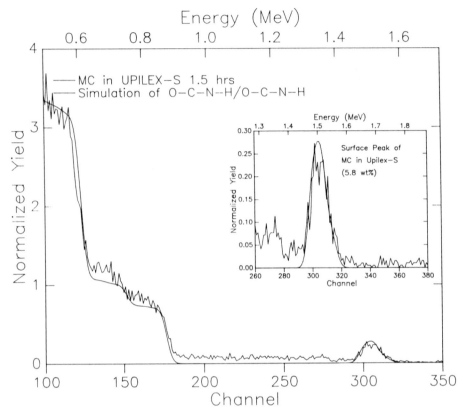

Figure 9.     RBS spectrum of DBRO absorption in Upilex S after a
1.5-hour exposure.

The absorption of MC in Upilex R is shown in Figure 7. The plot shows an ill-defined boundary and the long lead time associated with relatively slow solvent absorption. Despite the slower penetration, the MC absorption appears to be Case II. The very slow absorption of MC and DBRO in Upilex S hindered determination of the mode of transport. However, on exposure of Upilex S to MC and DBRO for 1.5 hours it was found there was 5.8 wt% of MC through 180 nm (Figure 8) and 2.8 wt% of DBRO through 90 nm. (Figure 9).

## CONCLUSIONS

Kapton H absorbs MC and DBRO at a faster rate and to a greater extent than Upilex S and Upilex R. Upilex S showed the least absorption on exposure at the same conditions for the same amount of time. This may be attributed to the differences in chemical structure, method of imidization, and/or film formation. A general trend exists where solvents with molecular volumes greater than or equal to 78.73 $cm^3$/mol are not absorbed in Kapton H after exposures of 1 hour. Both MC and DBRO diffusion in Kapton H were shown to be case II by gravimetric and RBS analysis. The diffusion of BLO in Kapton H was also determined to be case II gravimetrically. In addition, RBS measurements of DBRO diffusion into the rough (drum) and smooth (air) side of Kapton H film showed that solvent penetrating from the rough side initially penetrates slightly more rapidly. The diffusion of MC in Upilex R was shown to be case II using RBS. Gravimetric analysis proved unsuccessful because of the slow diffusion rate. RBS measurements showed a small weight gain for MC and DBRO in Upilex S after a 1.5 hour exposure.

## ACKNOWLEDGMENTS

Special thanks are extended to Dan Van Hart for the Optical interference photomicrographs.

## REFERENCES

1.    E. Sugimoto, IEEE Electrical Insulation Magazine, Vol. 5, No. 1, 15, January/February, 1989.
2.    W.M. Edwards, U.S. Patent, No. 3,179,614 (1965).
3.    R.J. Angelo, R.C. Golike, W.E. Tatum, and J.A. Kreuz, Proceedings of the Second International Conference on Polyimides, 67, 1985.
4.    Y. Sasaki, U.S. Patent No. 4,473,523 (1984). 0 K. Kumagawa, U.S. Patent No. 4,725,484 (1988).
5.    D.K. Yang, W.J. Koros, H.B. Hopfenberg, and V.T. Stannett, J. Appl. Polym. Sci. 31, 1619 (1986).
6.    D.K. Yang, W.J. Koros, H.B. Hopfenberg, and V.T. Stannett, J. Appl. Polym. Sci. 30, 1035 (1985).
7.    L.R. Iler, R.C. Laundon, and W.J. Koros, J. Appl. Polym. Sci. 27, 1163 (1982).
8.    E. Sacher and J.R. Susko, J. Appl. Polym. Sci. 23, 2355 (1979).
9.    H.M. Tong, K.L. Saenger, and C.J. During, J. Polym. Sci., Polym. Phys. Ed., 27, 689 (1989).
10.   E. Gattiglia and T.P. Russell, J. Polym. Sci., Polym. Phys. Ed., 27, 2131 (1989). P.J. Mills, C.J. Palmstrom, and E.J. Kramer, J. Materials Sci. 21, 1479 (1986).
11.   J.W. Mayer, W-K. Chu, and M. Nicolet, Backscattering Spectrometry, Academic Press, New York, 1978.
12.   L.R. Doolittle, Rapid Simulation of RBS Spectra, Nucl. Inst. Methods B9, 344 (1985).
13.   J.F. Zeigler, Helium Stopping Powers and Ranges in all the Elements, Pergamon Press, New York, 1977.
15.   T.P. Russell, Polym. Eng. Sci. 24, 345 (1984).
16.   R.C. Lasky, T.P. Gall, and E.J. Kramer, Case II Diffusion, in Principles of Electronic Packaging, edited by D.P. Seraphim, R.C. Lasky, and C.Y. Li (McGraw-Hill, New York, 1989), p. 796.

*Kapton is a registered trademark of E.I.du Pont de Nemours & Co.

**Upilex is a registered trademark of UBE Industries, Ltd., Japan.

# Chapter 5

# ANISOTROPIC DEFORMATION AND CRYSTALLIZATION IN POLYMER MELTS (WITH REFERENCE TO FIBER FORMATION )

UDAY S. AGARWAL, P. ASHER, W.W. CARR, F. PINAUD, P. DESAI AND A.S. ABHIRAMAN

*Polymer Education and Research Center, Georgia Institute of Technology Atlanta, Georgia 30332-0100*

## INTRODUCTION

Formulating a rational basis for the prediction of process-morphology relations in the fabrication of products from crystallizable polymer melts requires full comprehension of the interactive roles of stress (or the consequent deformation-) and crystallization-induced order. The interaction between these two primary phenomena in governing morphogenesis arises primarily from the following aspects:

(1) Onset of crystallization during fabrication would effect a major transition in the rheological behavior of the polymer and thus influence significantly the kinematics of the process.

(2) Any order obtained through the stress field prior to crystallization in the process, especially if it is anisotropic, would influence strongly the kinetics of crystallization *and* the morphological features arising from it. Thus the stress field can play a dual role in dictating the kinematics of deformation, through its direct influence via the appropriate rheological constitutive relationships in the precrystallized polymer and its indirect influence through its role in initiating crystallization and the consequent changes in the rheological characteristics of the polymer.

Although many questions still remain regarding the interrelationships between temperature, stress field, deformation and phase transformations, and the consequent morphology and properties obtained in polymers (Figure 1), much progress has been made in identifying many of the fundamental phenomena pertaining to these interactions in polymer processing. We discuss in the following some of these features, with examples drawn from melt-based fiber formation. Besides its obvious technological significance, melt spinning offers an excellent vehicle for examining the validity of the conceptual postulates and the analyses that might be used in establishing the links between thermorheological behavior and phase transformations of polymers.

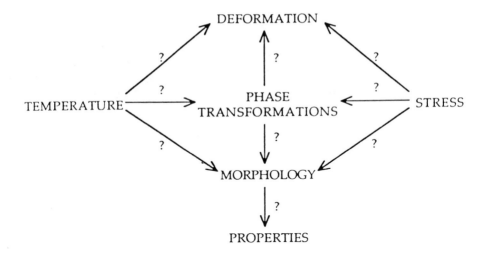

Figure 1. Material–Process interactions in the evolution of polymer morphology.

The development of structure in melt-processing of a crystallizable polymer can be viewed as a combination of the effects of flow in the melt state and of crystallization, if the latter occurs. For example, we consider a melt spinning threadline as shown in Figure 2. Flow-induced orientation of polymer chains occurs in the extensional and shear flow at the entrance to the capillary, in the shear flow inside the capillary, and in the extensional flow of the threadline. The effects of this orientation in the melt, introduced predominantly in the threadline where simultaneous deformation and cooling occur, persists in the as-spun fiber. Essentially all the deformation in the threadline occurs prior to glass transition, in the case of noncrystallizing polymers, or crystallization, in the case of crystallizing polymers.

Classification of the different zones in the evolution of structure as shown in Figure 2 can be used effectively to develop a generalized scheme for melt spinning of different polymers.[1] Earlier schemes, while incorporating appropriately the role of stress-induced order in the melt, had implied a monotonic functional dependence of the order in the as-spun fiber on the orientation that evolves in the melt.[2] The inconsistencies that can arise in the absence of incorporation of the nature and extent of reordering caused by crystallization in the process have been shown by Abhiraman and Hagler.[1]

The analysis and discussion which follow are presented in the context of a two-phase model of bulk polymers, consisting of crystalline regions in which the segments of molecules are packed in a periodic array (generally three-dimensional although sometimes only in two dimensions) and non-crystalline regions where such repetitive order is absent. Such a scheme, although recognized as a simplification of the real situation,[3,4] is necessary at the present time in the development of a rational framework for the evolution of order in polymer processing.

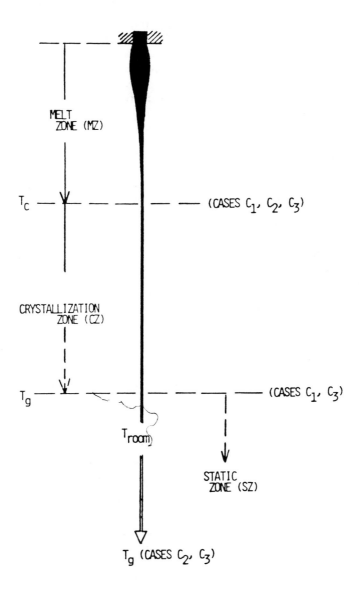

Figure 2.   Zones of structure development in a melt spinning threadline. $C_1$, crystallizable polymer with $T_g > T_{room}$; $C_2$, crystallizable polymer with $T_g < T_{room}$; $C_3$, crystallizable polymer with $T_g > T_{room}$ when dry but $T_g < T_{room}$ in the presence of moisture. (From ref. 1.)

## CRYSTALLIZATION, ORIENTATION DISTRIBUTIONS AND CONSEQUENCES

If a noncrystalline polymer that has been oriented through deformation should crystallize, the orientation distributions (OD) of the two phases that evolve will be determined by crystallization-induced reorientations and possible orientational relaxations in the two phases. A kinetic model has been formulated[5] to follow the evolution of orientation with the extent of crystallization, based on the following assumptions and postulates:

(i)   Primary crystalline nuclei, once formed, do not undergo any reorientation during their growth phase.

(ii)  Changes in the OD of the segments in the non-crystalline phase during crystallization are dictated completely by the segments that are transformed to the crystalline state (incorporated in the crystals), i.e., a segment that remains in the non-crystalline phase retains the orientation it had in the oriented precursor.  The model neglects, therefore, any orientational relaxations in uncrystallized polymer chain segments.

(iii) Growth of a crystal from its primary nucleus occurs through incorporation of segments from its neighborhood, with the rate of such incorporation dependent on the orientation of the segments relative to the crystal.  The *relative* rates of transformation of uncrystallized segments at different inclinations to the crystal in their neighborhood are assumed to remain unchanged with the progression of crystallization. With the following nomenclature,

$$\vec{\eta} = \text{vector in orientation space}$$

$$f_p\{\vec{\eta}\} = \text{precursor orientation density function ( i.e., ODF}$$
$$\text{of the polymer just prior to the onset of crystallization)}$$

$$f_n\{\vec{\eta}\} = \text{ODF of primary crystalline nuclei}$$

$$f_t\{\vec{\eta}; x_c\} = \text{ODF of segments in the precursor that have}$$
$$\text{been incorporated in the crystalline phase at an extent of crystallization of } x_c$$

$$f_a\{\vec{\eta}; x_c\} = \text{ODF of noncrystalline segments at } x_c$$

$$f_c\{\vec{\eta}; x_c\} = \text{ODF of crystalline segments at } x_c$$

assumptions (i) and (ii) may be represented as follows:

$$f_c\left\{\vec{\eta};\, x_c\right\} = f_n\left\{\vec{\eta}\right\}$$

$$(1)$$

$$f_p\left\{\vec{\eta}\right\} = x_c f_t\left\{\vec{\eta};\, x_c\right\} + (1-x_c)\, f_a\left\{\vec{\eta};\, x_c\right\} \quad (2)$$

The analysis to determine the ODF's has been described in detail by Abhiraman.[5] Briefly, it consists of the following steps in sequence.

The OD of primary crystalline nuclei is related to the OD of the precursor polymer by following procedures such as those of Krigbaum and Roe[6] or Ziabicki and Jarecki.[7] For example, the analysis of Krigbaum and Roe yields

$$f_n\left\{\vec{\eta}\right\} = c\int\left[f_p\left\{\vec{\eta}\right\}\right]^{N} d^3\vec{\eta}$$

$$(3)$$

where
   $N$ = number of segments in a critical size nucleus
   $c$ = constant, obtained from

$$\int f_n\left\{\vec{\eta}\right\} d^3\vec{\eta} = 1$$

$$(4)$$

Transformation in the neighborhood of a primary nucleus with its orientation along $\vec{\eta}_c$ is obtained from postulate (iii) above as

$$-\frac{dm}{dt}\left\{\vec{\eta};\, \vec{\eta}_c;\, x_c\right\} = \overline{K}\{\beta;\, t;\, x_c\}\, m\left\{\vec{\eta};\, \vec{\eta}_c;\, x_c\right\} \quad (5)$$

where

   $\beta$ = angle between $\vec{\eta}$ and $\vec{\eta}_c$

   $t$ = time

   $\overline{K}\{\beta;\, t;\, x_c\}$ = rate function, with

where $\overline{K}\{\beta_1;\, t;\, x_c\}/\overline{K}\{\beta_2;\, t;\, x_c\}$ is a function of $\beta_1$ and $\beta_2$ and *not* t or $x_c$

that is       $$\overline{K}\{\beta;\, t;\, x_c\} = K\{\beta\}\, K_1\{t;\, x_c\}.$$

$$(6)$$

Using a transformation from the time domain to the extent of crystallization domain,[5] the ODF of uncrystallized segments in the neighborhood of a growing crystal, $g_a\{\vec{\eta}; \vec{\eta}_c; x_c\}$ , is obtained as

$$g_a\{\vec{\eta}; \vec{\eta}_c; x_c\} = \frac{f_p\{\vec{\eta}\}}{1-x_c} \exp\left[-K\{\beta\}J\right] \tag{7}$$

where J has to satisfy the condition

$$\int g_a\{\vec{\eta}; \vec{\eta}_c; x_c\} d^3\vec{\eta} = 1 \tag{8}$$

The ODF of all the non-crystalline segments is obtained by integrating $g_a$ (from equations 7) with respect to the ODF of primary nuclei, which yields

$$f_a\{\vec{\eta}; x_c\} = \frac{f_p\{\vec{\eta}\}}{1-x_c} \int f_n\{\vec{\eta}\} \exp\left[-K\{\beta\}J\right] d^3\vec{\eta} \tag{9}$$

The overall ODF is then given by

$$f\{\vec{\eta}; x_c\} = x_c f_c\{\vec{\eta}\} + (1-x_c) f_a\{\vec{\eta}; x_c\} \tag{10}$$

Significance of the analysis presented here lies in its implications regarding the consequences of crystallization in an oriented precursor which can be summarized as follows.

— Whether one uses theories of crystal orientation based on *a priori* probabilistic considerations[6] or on a thermodynamic analysis that assigns an orientation-dependent free energy change for the incorporation of an amorphous segment into a crystal,[7] the following conclusion would be reached regarding crystallization in an initially anisotropic amorphous precursor. "Crystallization in an oriented precursor would lead to a crystal orientation distribution possessing a much greater degree of anisotropy than that of the precursor itself." This is shown schematically in Figure 3.

— The degree of reorientation required in the process of incorporating a precursor segment into a crystal increases with increasing inclination of the precursor segment with respect to the crystal. This would cause the rate of transfer of segments into a crystal to decrease monotonically with increasing inclination to the crystal, i.e., $K\{\beta\}$ is a monotonically decreasing function of $\beta$. The consequence of this preferential transfer with respect to the orientation distribution of amorphous segments remaining at any extent of transformation is straightforward. The high anisotropy of crystalline orientation distribution (Figure 3) leads to more rapid transfer into crystals of precursor segments that are closer to the preferred direction. Its effect on the OD of amorphous segments is also shown schematically in Figure 3.

– Since the average orientation of the crystals with respect to the preferred direction is much greater than that of the precursor segments that undergo the amorphous $\rightarrow$ crystal transformation, the overall orientation should increase significantly with crystallization in an initially amorphous, oriented precursor (Figure 3).

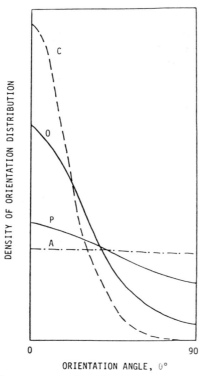

Figure 3. Schematic representation of orientation distributions resulting from crystallization. Orientation distributions of all segments prior to crystallization (P), crystallized segments (C), and uncrystallized segments (A); the weighted combination of C and A is the orientation distribution of all segments following crystallization (O). (From ref. 5.)

## Examples from Melt Spinning

Numerous examples exist to demonstrate the high crystalline orientation and the consequent large increase in overall orientation arising from crystallization in an oriented precursor.[8-12] Preferential transformation of oriented segments in the precursor to the crystalline phase is seen most dramatically in the thermorheological behavior, especially thermal shrinkage, of fibers.  An interesting demonstration of this aspect has been provided by Vasillatos[13] in melt spinning of poly(ethylene terephthalate), with the threadline quenched rapidly at different points around $T_c$ (case $C_1$ in Figure 2) at constant output and winding speed.  We have followed a similar procedure to produce fibers quenched at different extents of crystallization.[14] Rapid quenching in these experiments was accomplished with cold water pumped to the filament bundle through a large capacity metered finish applicator.  Figures 4 and 5 show, respectively, the equatorial WAXS scans and the densities (optical and bulk) of these fibers.  The consequences of progression of crystallization to different extents in this process can be seen clearly in the orientations of the as-spun fibers and in the shrinkages obtained subsequently in boiling water (100°C, 1 min) under free conditions (Figure 6).  The dramatic drop in shrinkage obtained here with the progression of crystallization is much higher than that which might be attributed simply to the reduction in the noncrystalline fraction in the material that undergoes this shrinkage.  For example, the difference in densities between Q9 and Q18  would imply an increase in the extent of crystallization by only ~11% (based on crystalline and noncrystalline densities of 1.529 and 1.336 g/cm$^3$, respectively) but the shrinkage decreases between the two samples from > 45% to ~7%.

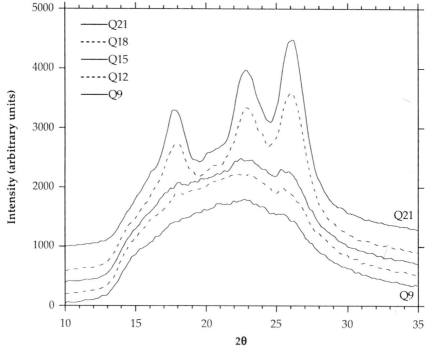

Figure 4. WAXD equatorial scans of high speed spun (5300 m/min), quenched PET fibers. Quenching distance from spinneret: (Q9) 9 inches; (Q15) 15 inches; (Q18) 18 inches.

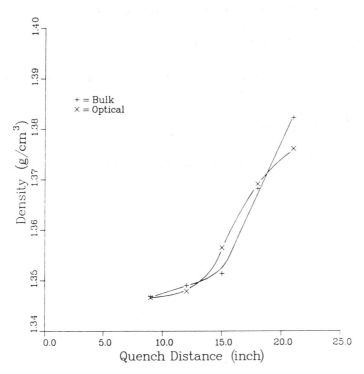

Figure 5. Bulk and optical (Lorentz) density of high speed spun, quenched PET fibers as a function of quenching distance from the spinneret.

Another interesting feature is the monotonic increase in sonic modulus with quench distance (Figure 6), which indicates that significant entropic orientational relaxation does not occur during crystallization in the threadline. Even with preferred transformation of segments with higher orientations in the precursor to the crystals, the very high orientation of the primary nuclei from which the crystals grow contributes to this increase in sonic modulus. If the precursor segments are separated into those which are destined to be in the crystals at an extent of crystallization of $x_c$ and the remaining segments, the following would be obtained in the absence of orientational relaxation.

<u>In the precursor:</u>

$$\left\langle f_p\left\{\vec{\eta}\right\}\right\rangle = x_c\left\langle f_t\left\{\vec{\eta}; x_c\right\}\right\rangle + \left(1-x_c\right)\left\langle f_a\left\{\vec{\eta}; x_c\right\}\right\rangle \quad (11)$$

<u>After crystallization</u>

$$\left\langle f\left\{\vec{\eta}; x_c\right\}\right\rangle = x_c\left\langle f_c\left\{\vec{\eta}\right\}\right\rangle + \left(1-x_c\right)\left\langle f_a\left\{\vec{\eta}; x_c\right\}\right\rangle$$

If only moderate orientation exists in the precursor,

$$\left\langle f_c\left\{\vec{\eta}\right\}\right\rangle \gg \left\langle f_t\left\{\vec{\eta};\, X_c\right\}\right\rangle \tag{13}$$

contributes to the large increase in overall orientation and sonic modulus, and

$$\left\langle f_a\left\{\vec{\eta};\, X_c\right\}\right\rangle \ll \left\langle f_t\left\{\vec{\eta};\, X_c\right\}\right\rangle \tag{14}$$

contributes to the drop in shrinkage being more than in proportion to $(1\text{-}X_c)$.

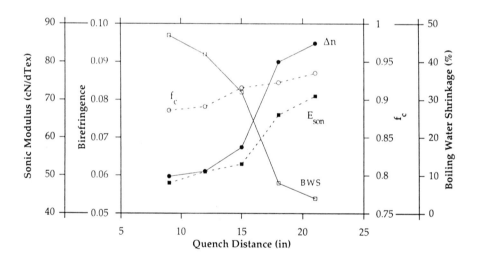

Figure 6. Birefringence, sonic modulus, Hermans orientation function for the crystalline regions ($f_c$) and boiling water shrinkage (100°C, 1 min) of high speed spun, quenched PET fibers as a function of quenching distance from the spinneret.

## KINETICS OF DEFORMATION AND ONSET OF
## CRYSTALLIZATION IN AN ANISOTROPIC STRESS FIELD

In fabrication processes from polymer melts, the dynamics of the nonisothermal deformation process is governed by, among other things, the strong influence of the stress field in dictating the onset of crystallization and the consequent gross changes in the rheological characteristics of the material. For example, the deformation and stress profiles in a process such as conventional melt spinning, which operates at an imposed cumulative deformation (winding speed/extrusion velocity), is dictated by the position at which crystallization occurs. The latter, in turn, is determined by the stress-induced orientation in the melt. Theories of Ziabicki and Jarecki,[7] and Krigbaum and Roe[6] reveal clearly the enormous increase in the rate of nucleation which can result from induced anisotropy in a polymer (see, e.g., Figure 7).

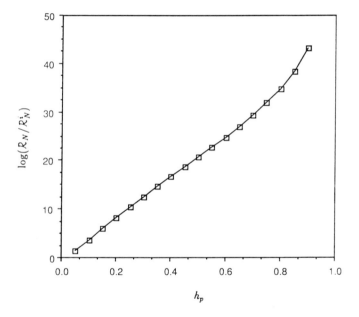

Figure 7.   Enhancement in rate of nucleation of a crystal, $\log\left(R_N/R_N^i\right)$, as a function of precursor Hermans orientation function $h_p$. Number of segments in critical size nucleus, $N = 25$. (From ref. 26.)

A dramatic example of the rheological consequences of stress-induced anisotropic crystallization is seen in stress field effects on the deformation of initially uncrystallized, but crystallizable polymers in the temperature range where they can crystallize. We have demonstrated[15,16] it by following the deformation of an uncrystallized filament of PET ( intrinsic viscosity = 0.62

dL/g; produced by melt spinning at 1600 m/min) when it is heated rapidly to a temperature between $T_g$ and $T_m$ while maintaining a constant applied tensile force.  The deformation profiles seen in the examples of Figure 8 are consequences of the following mutually interacting physical processes.

1. The well-known retractive force that develops because of the tendency of oriented molecules to coil, this force diminishing with increased extent of coiling in the molecules. The retractive process occurs until the retractive force is negated by the applied tensile force. If the external tensile force is higher than the internal retractive force, the orientation of the molecules tends to increase.

2. Fluidlike deformation of the filaments under the applied tensile force, with the rate of this deformation process increasing with the tensile force.

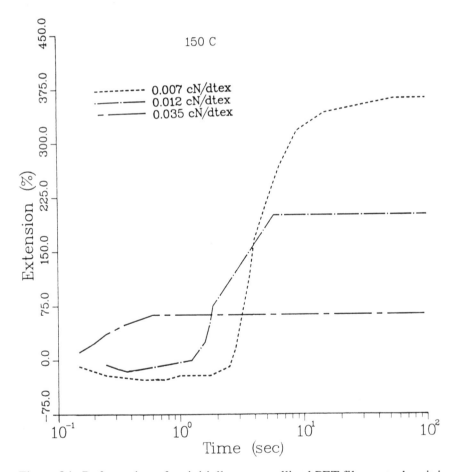

Figure 8A. Deformation of an initially uncrystallized PET filament when it is heated rapidly, at a constant force, to (A) 150°C and (B) 180°C. The stresses shown are the initial stresses; sample melt spun at 1600 m/min; intrinsic viscosity = 0.62 dL/g.

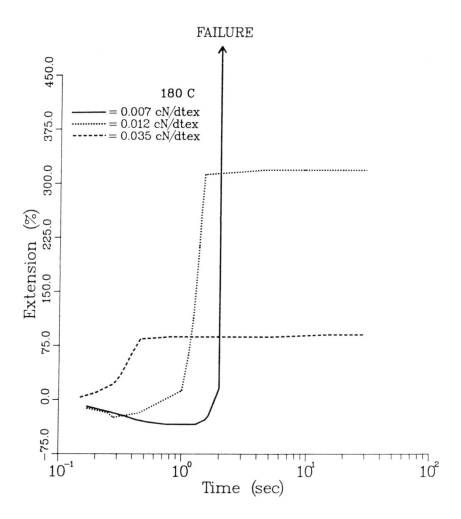

Figure 8B. Deformation of an initially uncrystallized PET filament when it is heated rapidly, at a constant force, to (A) 150°C and (B) 180°C. The stresses shown are the initial stresses; sample melt spun at 1600 m/min; intrinsic viscosity = 0.62 dL/g. (8B from ref. 16.)

    3. The initiation of crystallization in the filament that would transform it into a solidlike material. The rate of initiation of this phase transition increases significantly with the level of initial orientation and the degree to which it is maintained or enhanced by the applied stress. Thus, although the effect of higher tensile stress is to increase the rate of deformation in the fluidlike state of the polymer, it also leads to an earlier arrest of this deformation process by promoting a more rapid phase transition to the crystalline state.

The following are readily seen from the experimental data.

1. At a low external tensile force, there is an initial contraction in length prior to the onset of fluidlike tensile deformation. As expected, the initial retraction is reduced or eliminated with an increase in the applied stress level.

2. The fluidlike deformation continues until the onset of crystallization arrests this process completely. The induction period for this deformation decreases with increasing applied stress level.

3. An interesting observation is that the net ultimate deformation, which is dictated by the extent of initial retraction as well as the rate and the duration of tensile deformation in the fluidlike state, decreases when the tensile force is increased from 0.007 cN/dTex to 0.035 cN/dTex. This apparently strange behavior is a consequence of the role of stress in maintaining or promoting orientation, and thus crystallization, which is unique to polymers. As is shown in Figure 8B, the net effect of stress-induced deformation and crystallization can be even more dramatic in the sense that failure occurs at the lowest stress level but not at the higher stress levels shown. Rapid onset of crystallization in the latter cases arrests the deformation before ductile failure can occur.

An even more dramatic example than the preceding experiments with PET is seen in similar experiments with a random copolymer of ethylene terephthalate and ethylene isophthalate (ET/EI) at a 70:30 molar ratio.[17] Fibers obtained by melt spinning this copolymer at 2000 m/min show no evidence of crystallization even after prolonged annealing above $T_g$ without any applied stress.[17] If this as-spun fiber is heated to above $T_g$ while subjecting it to a moderate tensile force, the consequent deformation can be arrested through crystallization in less than a second (Figure 9). As with the homopolymer, the random copolymer also fails in this experiment at a low applied stress due to ductile failure before crystallization. Further evidence of rapid anisotropic stress-induced crystallization of this copolymer, which is extremely difficult to crystallize under quiescent conditions, can be seen in the WAXS patterns (Figure 10) obtained in fibers melt-spun at high speeds (eg., 6000m/min) to produce a significant orientation and then annealed for a short time (2 min at 100°C) at constant length. Additional experiments pertaining to thermorheological behavior and oriented crystallization in melt spinning of this copolymer and a PET-based blend are described in ref. 17.

## KINETICS OF CRYSTALLIZATION IN ORIENTED POLYMERS

It has been well known from studies with initially uncrystallized polymers that any induced anisotropy in the precursor causes a large increase in the overall rate of crystallization.[6,7,18-24] The thermorheological experiments described in the preceding section and theoretical predictions, such as those from Krigbaum and Roe (Figure 7), demonstrate clearly the large influence of an anisotropic stress field, and the consequent orientation, on *initiation* of crystallization in polymers. Experiments in kinetics of crystallization indicate, however, that the effect of precursor orientation and stress field diminishes

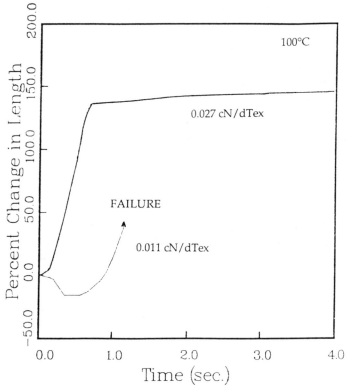

Figure 9. Deformation of a melt spun (2000 m/min) 70:30 random poly(ethylene terephthalate-co-ethylene isophthalate) copolymer fiber when it is heated rapidly, at constant force, to 100°C. The initial stress is shown in each case. (From ref. 17.)

(A)    (B)

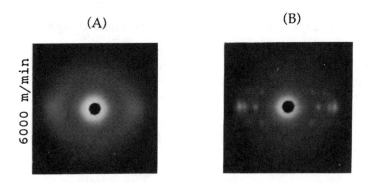

Figure 10. Flat plate WAXD photographs of a 70:30 random poly(ethylene terephthalate-co-ethylene isophthalate) copolymer fiber melt spun at 6000 m/min (A) and annealed at 100°C at constant length (B).

drastically beyond the initiation of crystallization (i.e., beyond primary nucleation) provided the stress level is not high enough to cause deformation during the progression of crystallization.[16,18,19,25] Data from an experiment with partially precrystallized PET fibers are shown in Figure 11 to exemplify this aspect. Here a PET fiber, melt-spun at 4500 m/min to effect the initiation of crystallization in the threadline, has been heated for different durations by passing it at different velocities through a tubular heater at 150° C and quenching it rapidly at the exit. The experiments have been conducted under two conditions, (i) without a net change in length while monitoring the force generated (CCLC) and (ii) at a constant force (CCFC), significantly lower than those generated in (i). The relative insensitivity of rate of crystallization at this stage to the level of anisotropic stress is seen clearly in these data.

The theoretical framework, described earlier in the context of orientational changes caused by crystallization in a polymer, can be used also to explore the relative rates of primary nucleation as well as growth of crystals in anisotropic and isotropic precursors. Following a similar procedure, it can be shown[26] that the ratio of crystal growth rates in initially anisotropic and isotropic polymers, $R_g$ and $R^i_g$, at any extent of crystallization, $x_c$, is given by

$$\left(R_g/R^i_g\right)\{X_c\} = \frac{\int K\{\beta\} f_a\left\{\vec{\eta}; X_c\right\} d^3\vec{\eta}}{\int K\{\beta\} f^i_a\left\{\vec{\eta}; X_c\right\} d^3\vec{\eta}} \qquad (15)$$

where $K(\beta)$ is the inclination-dependent component of the crystallization rate function, $\beta$ is the angle between $\vec{\eta}_c$ and $\vec{\eta}$, and $f_a$ and $f^i_a$ are the ODFs of non-crystalline segments in the neighborhood of the growing crystal.

If we examine crystal growth in a polymer with axially symmetric ODF around a preferred direction, the maximum of this ratio will be obtained when $x_c = 0$ and the crystal direction $\vec{\eta}_c$ is along the preferred direction. The results obtained for this case with different levels of sensitivity of $K\{\beta\}$ to $\beta$ (Figure 12) show that enhancement in kinetics of crystal growth through anisotropy in its neighborhood is lower by several orders of magnitude than the corresponding enhancement in primary nucleation (Figure 7).

## STRESS FIELD IN CRYSTALLIZATION AND GENERATION OF MECHANICAL PROPERTIES

The development of stress-induced orientational order during deformation of bulk polymer fluids and the consequent evolution of morphological order and mechanical properties in fabricated products are, if not well understood, at least well recognized. The following examples from melt spinning serve to illustrate two critical aspects in this regard, viz., the influence of the strength of anisotropic stress during crystallization in the generation of mechanical properties and the consequences of crystallization in oriented polymers vis-a-vis subsequent deformation.

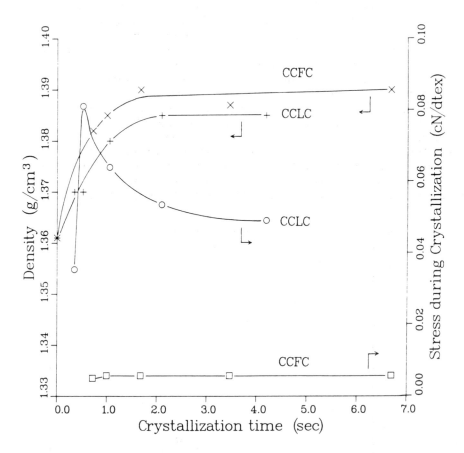

Figure 11. Progression of crystallization in an initially partially crystallized PET fiber at 150°C. CCFC: continuous, constant force crystallization; CCLC: continuous, constant length crystallization. (From ref. 16.)

**Example 1**
     PET fibers that had been melt spun under conditions that yield uncrystallized fibers were heated rapidly in a tubular furnace at 150° C while being subjected to a predetermined tensile force. The experiment and the material used were the same as those pertaining to Figure 8, except that the heating process was terminated after 5 seconds. The applied stress levels in these experiments were above the critical stress level required to cause crystallization and thus arrest the deformation process before failure. The samples thus obtained were tested for mechanical properties. The sonic moduli and stress-strain curves of these fibers are shown in Figures 13 and

14, respectively. ( <u>Note</u>: The data are shown as a function of initial stress levels instead of stress at the *initiation* of crystallization. The latter might be a more appropriate quantity but it is difficult to determine exactly the point at which crystallization is actually initiated. The critical extent of deformation required for fluid→solid transformation and the rheological characteristics of the material in the transition zone are still quantitative challenges to be resolved in polymer rheology.) The net deformation undergone by the material in the stress-induced crystallization experiment is also shown in Figure 13. It is seen clearly from the data in Figure 13 that stress, not the net deformation, provides a single valued monotonic relationship with the mechanical properties that are generated in the fiber. An important feature in these data pertains to the properties obtained at the highest stress level (0.18 cN/dTex) used in these experiments. The sonic modulus (> 120 cN/dTex) and the stress-strain behavior (Figure 14) of the fibers obtained through crystallization at this stress level are within the realm of drawn fibers that might be used for textile applications. The stress applied here (initial stress = 0.18 cN/dTex; final stress = 0.50 cN/dTex) is well within the range of stresses that evolve in intermediate and high speed melt spinning processes.

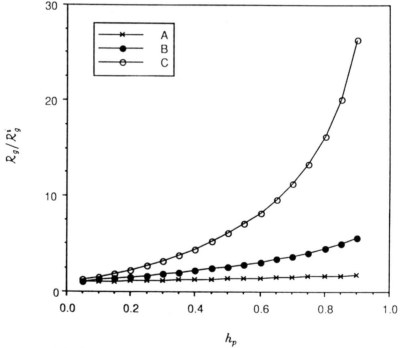

Figure 12. Maximum enhancement in rate of crystal growth of a crystal, $R_g/R_g^i$, as a function of precursor Hermans orientation function $h_p$, for linear (A) and exponential (B, C) angular dependences of $K(\beta)$. $K(p/2)/K(0) = 0.2146$ (A); $10^{-2}$ (B); $10^{-8}$(C); (From ref. 26.)

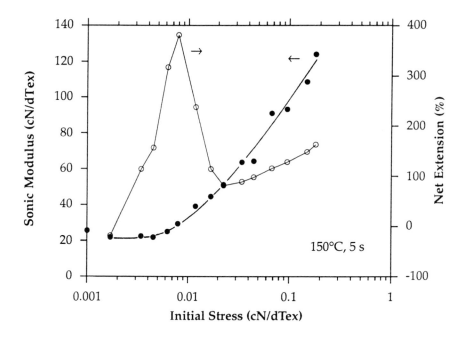

Figure 13. Sonic modulus of PET fibers after thermal deformation/crystallization for 5 seconds at 150°C as a function of initial stress (initial fiber is the same as in Figure 8). The net thermal deformation undergone by the fiber is also shown. (From ref. 16.)

A clear implication from these data pertains to the inferences that can be drawn regarding production of useful fibers directly through dynamically induced drawing in melt spinning. If one conducts a gedanken experiment in this regard by considering the "draw ratio" obtained in the stress-induced crystallization experiment (DR = 2.6), the speed of melt spinning used in the formation of the initial uncrystallized fiber (1600 m/min), the low stress levels at which deformation and crystallization can be induced in this fiber and the properties obtained as a consequence (Figures 13 and 14), the inference would be that such drawing could be induced in high speed melt spinning by heating the normally cooling threadline to a temperature above $T_g$ at an appropriate position along its path. Brody[27] has clearly demonstrated this possibility in practical melt spinning. It should be noted here that the data shown in Figures 13 and 14 do not extend to the limits of stress levels that can be applied before cohesive failure might occur during the stress-induced "deformation-crystallization" experiment. It is necessary to determine these limits and the corresponding evolution of properties as functions of temperature and intrinsic material characteristics so that the maximum potential for "single step" melt spinning processes with thermorheologically induced drawing can be inferred.

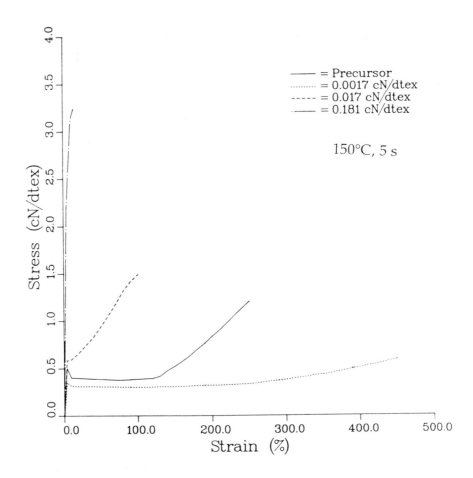

Figure 14. Stress-strain curves of PET fibers after thermal deformation/crystallization: details in Figure 13. (From ref. 16.)

**Example 2**

Two important aspects pertaining to crystallization in an oriented polymer that have not been addressed theoretically are the orientational relaxations that might occur prior to, and during, crystallization and the combined influence of orientational relaxations in the noncrystalline fraction and oriented crystallization on the mechanical properties of the polymer.    An experimentally observed feature of crystallization in an oriented polymer with important practical consequences is that the ductility of the precursor is often retained after crystallization, provided gross orientational relaxation does not occur prior to the onset of crystallization.[28]  If the latter should occur, subsequent crystallization produces often a brittle polymer.  It is a consequence of poor connectivity among the crystalline domains in the material.  This aspect can be demonstrated by annealing PET fibers that

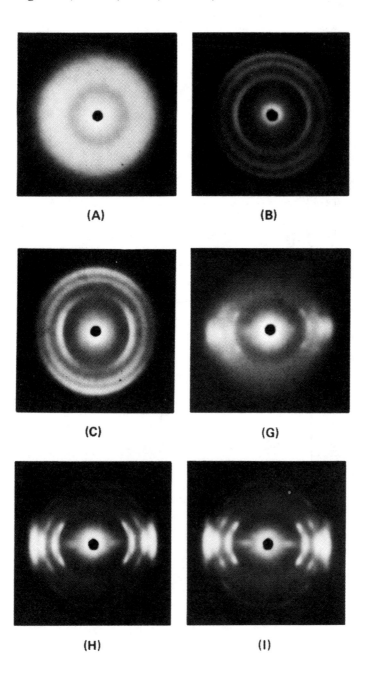

Figure 15. Flat plate WAXD photographs of PET fibers. A: melt spun at 1600 m/min B: after free length annealing (FLA), and C: after constant length annealing (CLA); G: melt spun at 4500 m/min, H: after FLA, and I: after CLA. All the annealing was carried out at 150°C for 2 minutes (From ref. 28.)

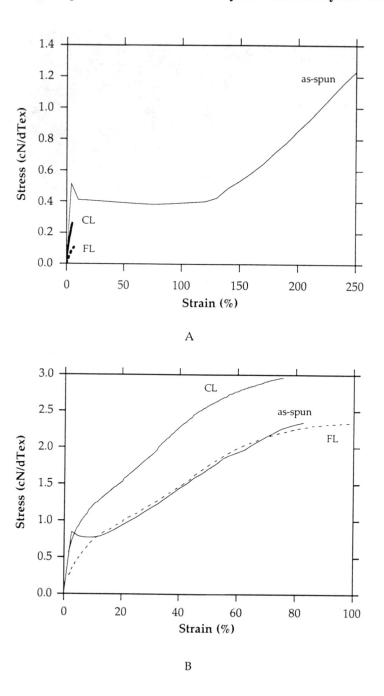

Figure 16. Typical stress-strain curves of as-spun PET filaments and after CLA and FLA at 150°C for 2 minutes; melt spun at (A) 1600 m/min and (B) 4500 m/min.

had been melt-spun at low (e.g., 1600 m/min) and intermediate or high (e.g., 4500 m/min) speeds at a temperature above their $T_g$. Wide angle x-ray scattering patterns and the stress-strain behavior of these fibers which are obtained before and after annealing are shown in Figures 15 and 16. Almost complete orientational relaxation is seen to occur prior to crystallization in the initially uncrystallized fiber of low orientation (Figures 15B and 15C), irrespective of whether a constraint against shrinkage is imposed or not during annealing. Consequently, crystallization in this material leads to an extremely brittle mechanical behavior (Figure 16A). The breaking elongation in this instance decreases from more than 350% to less than 10% upon crystallization. On the other hand, the fiber that had undergone partial oriented crystallization in the melt spinning process is seen to retain its ductility with significant additional crystallization (Figures 15B and 15C). These phenomena have been shown to result in unusual behavior in the drawing processes that might follow melt spinning, if an annealing step is incorporated prior to drawing.[29] Also the high degree of overall orientation without loss of ductility which can result from oriented crystallization (Figures 15G and 16B) has been used advantageously to produce PET fibers of extremely high order by the combination of annealing at a temperature T with $T_g < T << T_m$ followed by drawing at $T \sim T_m$.[29,30]

## REMARKS

Anisotropic crystallization constitutes the most important phase transformation in the morphological evolution of many polymeric materials, ranging from the partially ordered structures of flexible and semi-rigid polymers to the highly ordered morphologies of rigid aromatic polymers with high mechanical performance characteristics. The morphology obtained in the fabrication of most crystallizable bulk polymers is governed by the interactive influences of deformation-induced anisotropic order and crystallization-induced reordering of the anisotropic precursor. Interactions between the two phenomena extend from their combined effects on the dynamics of deformation during fabrication to the evolution of the solid-state structural features that determine the properties of the fabricated product. This presentation should serve to illustrate that much progress has indeed been made, both experimentally and theoretically, in elucidating the fundamental aspects of these phenomena. It should be recognized, however, that a quantitative unified link from processes and materials to the resulting morphologies and properties has yet to be developed and serves as a challenge for future pursuits in this field.

## ACKNOWLEDGMENTS

This chapter honors the contributions made to fiber and polymer science by Professor Richard Gilbert and is dedicated to the memory of Professor Waller George. The authors are grateful to Mrs. Donna Brown for her patient help in the preparation of the manuscript.

## REFERENCES

1. Abhiraman, A.S. and Hagler, G.E., *J. Appl. Polym. Sci., 33,* 809 (1987).
2. Ziabicki, A., *Fundamentals of Fiber Formation,* Wiley-Interscience, London, 1976.
3. Tzou, D. L., Desai, P., Abhiraman, A. S., and Huang, T. H., *Polym. Preprints, 31(1),* 157 (1990).
4. Tzou, D. L., Huang, T. H., Desai, P. and Abhiraman, A. S., *J. Polym. Sci. Part B: Polym. Phys. , 29,* 49 (1991).
5. Abhiraman, A.S., *J. Polym. Sci. B: Polym. Phys., 21* , 583 (1983).
6. Krigbaum, W.R., and Roe, R. J., *J. Polym. Sci. A, 2,* 4391 (1964).
7. Ziabicki, A. and Jarecki, L., *Colloid Polym. Sci., 256,* 332 (1978).
8. Ziabicki, A. and Jarecki, L., *High Speed Fiber Spinning,* A. Ziabicki and H. Kawai, Eds, John Wiley, New York, 1985, p. 225.
9. Hashiyama, M., Gaylord, R., and Stein, R. S., *Die Makromol. Chem. Suppl., 1,* 579 (1975).
10. Krigbaum, W. R., and Taga, T., *J. Polym. Sci. Part B: Polym. Phys. , 17,* 393 (1979).
11. Wasiak, A.,*Colloid Polym. Sci., 259,* 135 (1981).
12. Nadella, H. P., Henson, H. M., Spruiell, J. E., and White,J. L., *J. Appl. Polym. Sci., 21,* 3003 (1977).
13. Vassilatos, G., Knox, B. H., and Frankfort, H. R. E., *High Speed Fiber Spinning,* A. Ziabicki and H. Kawai, Eds, John Wiley, New York, 1985, p. 383.
14. Desai, P. Ph.D. thesis, GA Institute of Technology, Atlanta, GA, 1988.
15. Desai, P., and Abhiraman, A. S., *J. Polym. Sci. Polym. Lett. Ed., 24,* 135 (1986).
16. Desai, P., and Abhiraman, A. S., *J. Polym. Sci. B: Polym. Phys. , 26,* 1657 (1988).
17. Agarwal, U. S., M.S. Thesis, GA Inst. of Tech., Atlanta, GA, 1987.
18. Andrews, E. H., *Proc. Royal Soc. London A, 277,* 562 (1964).
19. Andrews, E. H., Owen, P. J., and Singh, A., *Proc. Royal Soc. London A, 324,* 79 (1971).
20. Alfonso, G.C., Verdona,M.P., and Wasiak,A., *Polym. 19,* 711 (1978).
21. Wasiak, A., *Colloid Polym. Sci., 259,* 135 (1981).
22. Smith, F. S., and Steward, D. R. *Polymer, 15,* 283 (1974).
23. Nakamura, K., Watanabe, T., Katayama, K., and Amano, T., *J. Appl. Polym. Sci., 16,* 1077 (1972).
24. Nakamura, K., Watanabe, T., Amano, T., and Katayama, K. *J. Appl. Polym. Sci., 18,* 615 (1974).
25. Desai, P., and Abhiraman, A., S., *J. Polym. Sci. Polym. Lett. Ed.., 23,* 213 (1985).
26. Desai, P., and Abhiraman, A. S., *J. Polym. Sci. B: Polym. Phys. , 27,* 2469 (1989).
27. Brody, H., European Patent Appl. EP 0,041,327 A1 (September) 1981.
28. Hamidi, A., Abhiraman, A. S., and Asher,P., *J. Appl. Polym. Sci.., 28,* 567 (1983).
29. Yoon, K. J., Desai, P., and Abhiraman, A S., *J. Polym. Sci. B: Polym. Phys., 24,* 1665 (1986).
30. Abhiraman, A.S. and Song,J. *J. Applied Polym. Sci., 27,* 2369 (1982).

# Chapter 6

# SURFACE MODIFICATION OF ULTRAHIGH STRENGTH POLYETHYLENE FIBERS THROUGH ALLYLAMINE PLASMA POLYMER DEPOSITION FOR IMPROVED INTERFACIAL STRENGTH

A. N. NETRAVALI AND Z.-F. Li

*Fiber Science Program*
*Cornell University*
*Ithaca, NY 14853*

## INTRODUCTION

In the past couple of decades there has been significant progress toward making stronger and lighter fibrous composites for aerospace applications. Until recently, glass, graphite, and aramid fibers were the main fibers used to make advanced composites. However, recently introduced ultra-high strength polyethylene (UHSPE) fibers, which combine both high strength and modulus with low density are being used increasingly in some composites. UHSPE fibers have good impact toughness, abrasion resistance and chemical resistance.[1] These fibers, however, have the lowest melting point and the poorest adhesion to any matrix material compared to other reinforcing fibers.

Performance of the composites depends not only on the properties of the fibers and matrices but also on the interface between them. Good interfacial adhesion provides composites with structural integrity and good load transfer from fibers to matrix and vice versa. It is generally acknowledged that good adhesion is achieved through chemical bonding, favorable wettability, large specific area and mechanical interlocking.[2] The fundamental cause of poor adhesion of UHSPE fibers to any matrix lies in their chemical composition consisting of methylene groups. The nonpolar nature makes them difficult to wet and impossible to chemically bond to matrices. Besides, these fibers have smooth surfaces, which eliminate mechanical interlocking, and relatively high diameters that reduce the specific contact area.

One of the best ways to improve interfacial bond strength is to introduce chemical reactive groups onto fiber surfaces that can react with matrices as well as increase the surface energy for improved wetting. Approaches to achieve this include chemical reaction or grafting, coating and oxidation. Chemical reaction or grafting can be easily carried out if a surface contains some reactive groups. For example, surface modification of aramid fibers by grafting small chains end-capped with reactive groups to increase the reactivity of the main chains was done by Takayanagi et al.[3] Coating of graphite fibers with epoxy or polyimide resins is done many times to improve adhesion.[4] Carbon fibers, being electrically conductive, can also be coated with vinyl polymers through electropolymerization.[5] A potential problem with coating is that the coating itself may not adhere effectively to the fiber. Chemical modification by oxidation may be the least preferred way, although it is used for graphite, because in most cases it is accompanied by a decrease in fiber strength.

Most of the surface modification methods available for other fibers may not be employed for UHSPE fibers because of their inert nature. In this chapter we report on the effect of allylamine plasma polymer coating on the strength of UHSPE fibers and their interfacial shear strength in epoxy matrix.

A gas plasma can be produced by introducing the desired gas into a vacuum chamber, typically 0.1-10 torrs, and subsequently exciting the gas using radiofrequency (RF) energy. The energy applied dissociates the gas into electrons, ions, free radicals and other metastable excited species.[6] The free radicals and electrons created in the plasma collide with the exposed material surface, rupturing covalent bonds and creating free radicals. The activated material surface can then readily combine with the excited gas species and provide chemically reactive groups. Since plasma surface treatment causes changes to a limited depth (several molecular layers), bulk properties of the fibers do not change significantly.[7] Plasma process can be easily controlled by several independent variables such as flow rate, pressure, power input and time.

Plasma processes can be divided into two groups: polymer-forming and non-polymer-forming. Plasmas of gases such as oxygen, nitrogen, hydrogen, ammonia and argon etc. are non-polymer-forming. They modify fiber surface chemistry by reacting with fibers through abstraction of hydrogens from polymer chains and by creating free radicals, which later may be oxidized into polar groups when exposed to air. Polymer forming plasmas are formed by most organic gases, vapors of many liquids or mixtures of some non-polymer forming gases. They deposit a polymeric film onto fiber surface containing reactive/functional groups. These films generally adhere well to the substrate because of the covalent chemical bonding and the polar nature. They also have superior thermal stability and low solubility as a result of their highly crosslinked nature.

Several researchers have used plasma treatment for fiber surface modification. Ladizesky and Ward[8] found that oxygen plasma treatment increase interfacial bond strength significantly but at the expence of fiber strength. Recently, Holmes and Schwartz treated UHSPE fabrics with ammonia plasma[9]. The adhesion of the fabrics to epoxy resin was found to be improved without affecting the fiber strength. Kaplan et al.[6] and Nguyen et al.[10] treated fibers with unspecified gas plasmas. Their results showed a two-fold increase in the composite interlaminar shear stress (ILSS) without fiber strength loss.

Although non-polymer-forming plasmas have been popular for surface modification of fibers, polymerforming plasmas have also been used by some.  Sung et al.[11] treated graphite fibers with acrylonitrile plasma and found a substantial improvement in ILSS.  Wertheimer and Schreiber[12] treated aramid fabrics with allylamine and hexamethyldisiloxane plasma. Using peel tests, they found that the bond strength decreased after treatment with allylamine plasma alone.  However, when fabrics were first activated in argon plasma before introducing allylamine plasma, bond strength improved measurably.   Krishnamurthy and Kamel[13] reported allylamine plasma polymer grafting onto glass fibers.  They observed a fairly thick coating  with a particulate texture but did not measure the interfacial strength.

Several authors have studied and reviewed the mechanisms of plasma polymerization.[14-16] Because of the complex nature of the plasma, the exact mechanisms of plasma polymerization have not been clearly understood. However, it is generally accepted that free radical and ionic initiation and propagation coexist in plasma polymerization and that polymerization may take place in gas as well as on the substrate surface.  Plasma formed polymers are characteristically amorphous, irregular and highly crosslinked, and often contain various functional groups that are not present in monomers.[15]

## EXPERIMENTAL  PROCEDURES

### Materials
Ultrahigh strength polyethylene (UHSPE) fibers, Spectra® 900, were obtained from Allied Fiber Company in the form of yarn containing 118 filaments.  Allylamine ($H_2C=CHCH_2NH_2$), 98% purity, was obtained from Aldrich Chemical Corporation and distilled before using. Epoxy resin, DER 331, and curing agent, DEH 26, used in the study were obtained from Dow Chemical Company.  DER 331 is a diglycidyl ether of bisphenol A (DGEBA) with an average epoxy equivalent weight of 187.  DER 26 is a tetraethylene-pentamine (TEPA) with amine hydrogen equivalent weight 27.1.

Spectra® yarns were wound on an aluminum frame (150 mm x 50 mm) and the ends were glued to the frame using epoxy.   The plasma polymerization of allylamine was carried out using a plasma discharge System, Model 504, manufactured by LFE Corporation.  The aluminum frame with the mounted fibers was kept in the reaction chamber.  Prior to introducing the allylamine vapor, the chamber was vacuumed to 60 mtorr. Allylamine vapor was then introduced and the pressure was maintained at 500 mtorr by controlling the vapor flow.  The power was turned on when a steady pressure was established.

Single filaments were tensile tested, before and after plasma treatment, on an Instron testing machine Model 1122 using small capstan jaws as described by Schwartz et al.[17]  Before each tensile test, the cross-sectional area of each fiber was measured using a vibroscope according to ASTM D1577 -79.  The fiber density was assumed to be 0.97 g/mm$^3$.  Jaw spacing of 50 mm and crosshead speed of 5 mm/min were used.

To measure the interfacial shear strength, single fiber pull out tests were performed.  Specimens were made using silicone rubber molds, having a slit up to half the depth, as shown in Figure 1.[18]  The slit in the silicone mold opened to form a V-shaped crack, when bent, to mount a single filament. When the mold was released, the crack closed.  The embedment length was kept constant at 5 mm for all experiments.  DER 331 resin and DEH 26 curing agent were mixed in stoichiometric proportion and poured carefully in the mold, after degassing, without disturbing the fiber.  The epoxy resin was cured in the mold for 3 hours at 80°C, post cured for an additional hour at 110°C and then allowed to cool slowly by turning the oven off.  The lower part of the protruding fiber was cut using a sharp razor blade.  The two legs of the specimen were held in the lower Instron clamps while the upper part of the fiber was wound on the same capstan jaw, used in the tension tests, and held in the upper Instron clamp.  The fiber was pulled out at the rate of 1 mm/min.  Before mounting the fiber in the mold, the cross-sectional area of each fiber was measured using the vibroscope technique.  Average interfacial shear strength (t) was obtained using the following formula:

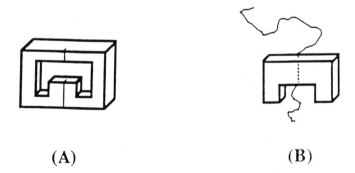

(A)                                    (B)

Figure 1.      (A) Silicone rubber mold and (B) epoxy resin specimen for single-fiber pull-out test.

$$t = F_p/\pi dL \qquad\qquad (1)$$

where $F_p$ is the fiber pull out load obtained from the load vs displacement plots (Instron), d is the fiber diameter and L is the embedded fiber length. Typical plots for control and treated fibers are shown in Figure 2. Scanning electron microscopy was used to observe any change in the surface topography before and after the treatment as well as after the pull-out tests.

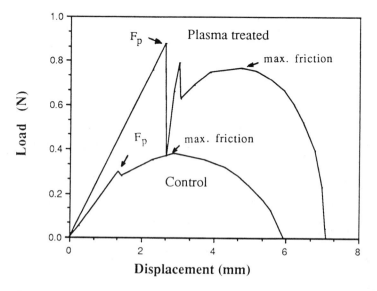

Figure 2. Typical load vs. displacement plots of single fiber pull out tests. $F_p$ is the fiber pull-out load.

## RESULTS AND DISCUSSION

### Single-Fiber Strength

The mean fiber strength and the corrected mean fiber strength are shown in Table 1. For calculating the corrected fiber strength, the polymer coating thickness was subtracted from the vibroscope-measured fiber diameter. In doing so it was assumed that the contribution of the polymer coating to the strength is negligible. It can be seen from Table 1 that for a fixed 10 minute treatment time, the fiber strength decreases monotonically with the plasma power. Klein et al.[19] have shown that for ultra high modulus polyethylene (UHMPE) fibers, chain scission occurred more easily than crosslinking when the samples were irradiated under vacuum. Higher levels of UV radiation and electron density exist at higher power levels[14]. It is therefore probable that the strength reduction was due to increased chain scission by UV and electrons at higher power levels.

Table 1.
Effect of Allylamine Plasma Treatments on Fiber Strength

| Fiber Treatment | Mean Strength (MPa) | Corrected Mean Strength (MPa) |
|---|---|---|
| Control | 3530 | 3530 |
| 30W 10 min | 3441 | 3453 |
| 45W 10min | 3274 | 3289 |
| 70W 10min | 3152 | 3176 |
| 45W 5min | 3375 | 3383 |
| 45W 10min | 3274 | 3289 |
| 45W 20min | 3478 | 3502 |
| 45W 40min | 3319 | 3384 |

Table 1 also shows the effect of treatment time on fiber strength at a constant power input of 45 W. It can be seen that the strength drops for the 5-minute treatment time but remains essentially unchanged for any further increase in time. This suggests that the plasma degradation or chain scission occurs only during the initial period and possibly before the coating starts to build up or when it is very thin. However, once the deposition is a few monolayers thick, it can absorb the radiation and thus protect the fiber from further chain scission. Li[18] reported that, under the experimental conditions used, at 45 W power level, 4.25 nm layer of polymer deposited per minute. It is, therefore, possible that the strength degradation occurred within the initial first minute.

**Interfacial Shear Strength**
    The single fiber pull out load vs. displacement plots are shown in Figure 2. It can be seen from the plots that load increases linearly in the initial part of the plot. Some specimens, observed under a polarized optical microscope, indicated that the interface failed progressively from the top, where maximum shear stress exists, to the bottom during this period. When the debonding is complete, the fiber slips relieving the stress instantaneously. The load at the first peak $F_p$ when the fiber slips was used in equation (1) to calculate average IFSS.    The load then rises slowly, until the fiber starts gradually slipping and then falls as the embedded length decreases. In Table 2 we show the average interfacial shear stress, improvement in IFSS and maximum friction for all treatment conditions. Maximum friction values refer to the maximum load after the first peak as shown in Figure 2. Significant improvement in IFSS, between 220 and 325%, is obtained for all treatment conditions. However, the IFSS does not change significantly with power input. A slight decrease in IFSS at higher power levels may be due to reduction in primary amines, which are converted to secondary and tertiary amines or imines as observed by Li.[18] Tertiary amines and imines do not react with epoxy resins to provide chemical bonding, while the secondary amines have only one reactive hydrogen. The presence of secondary amines and imines has also

been reported by Krishnamurthy and Kamel.[13]  They also reported the presence of nitrile groups.  Quantitative measurements of primary amine concentration of ammonia plasma done by Allred et al.[20] showed that the concentration increased rapidly in the first minute, remained constant till 10 minutes and decreased for longer exposures.  At constant power input of 45 W the IFSS increased from 1.54 to 1.73 MPa with the increase of exposure time from 5 to 10 minutes.  However, further increase in exposure time showed no clear trend.

Table 2.
Interfacial Shear Strength and Maximum Friction Values from the Single-Fiber Pull-out Tests.

| Fiber Treatment | IFSS (MPa) | Improvement (%) | Maximum Friction (g) |
|---|---|---|---|
| Control | 0.56 | --- | 39.6 |
| 30W 10min | 1.82 | 225 | 79.5 |
| 45W 10min | 1.73 | 209 | 58.7 |
| 70W 10min | 1.64 | 193 | 76.9 |
| 45W 5min | 1.54 | 175 | 67.6 |
| 45W 10min | 1.73 | 209 | 58.7 |
| 45W 20min | 1.25 | 123 | 61.1 |
| 45W 40min | 1.68 | 200 | 81.6 |

Most pulled out fiber surfaces when observed under SEM were clean, without any epoxy adhering to them, suggesting that the failure occurred mainly at the interface.  This is obvious from the small values of the IFSS.  All these IFSS values, even after the plasma treatment, are smaller by more than an order of magnitude compared to other fibers such as glass,[21] graphite[22] and Kevlar.[22,23]  In a very few cases fibrils were pulled out.  It is possible that fibrillation may have been initiated at the preexisting fiber defects or while the specimens were being made.

## CONCLUSIONS

* Fiber strength decreases slightly after the allylamine plasma treatment. Higher decrease in strength is observed with increase in power level. However, at constant power level, effect of time on strength is insignificant.

* Allylamine plasma polymer deposition increases the IFSS to between 120 and 225% of the control fibers. However, in the single-fiber pull-out test, failure occurs mainly at the interface.

## DEDICATION

This chapter is dedicated to Professor Richard D. Gilbert: teacher, scholar, researcher and sportsman.

## REFERENCES

1.  Lemstra, P. J. , Kirschbaum. R. , Ohta, T. , and Yasuda, H. , High-Strength/High-Modulus Structures Based on Flexible Macromolecules: Gel Spinning and Related processes, in *Developments in Oriented Polymers-2*, I. M. Ward (Ed.), Elsevier Applied Science, London and New York (1987).
2.  Scola, D. , High Modulus Fibers and the Fiber-Resin Interface in Resin Composites, in Composite Materials, Vol. 6, *Interface in Polymer and Matrix Composites*, E. Pluddemann, (Ed.), Academic Press, New York (1974).
3.  Takayanagi, M. , Lei, W.-Y. , and Koga, K. , Polymer J. , 19, 467 (1987).
4.  Lubin, G. , *Handbook of Composites*  Van Nostrand Reinhold Co., New York (1982).
5.  Subramanian, A. V. and Crasto, A. S. , *Polym. Comp.*, 7, 210 (1986).
6.  Kaplan, S. K. , Rose, P. W. , Nguyen, N. X. , and Chang, N. W. , Gas plasma Treatment of Spectra Fiber, 33rd Intl. SAMPE Symp. , p. 551 (1988).
7.  Kolluri, O. S. , Kaplan, S. L. , and Rose, P. W. , Gas Plasma and The Treatment of Advanced Fibers. SPE/Advanced Polymer Composites '88 - Technical Conference Los Angeles, CA November 1988.
8.  Ladizesky, N. H. and Ward, I. M. , *J. Mat. Sci., 18*, 533 (1983).
9.  Holmes, S. and Schwartz, P. , Amination of Ultra-High Strength Polyethylene using Ammonia Plasma, *Comp. Sci. Tech. 38*, 1 (1990).
10.  Nguyen, N. X. , Riahi, G. , Wood, G. , and Poursartip, A., Optimization of Polyethylene Fiber Reinforced Composites Using a Plasma Surface Treatment, 33rd Intl. SAMPE Symp. , (1988).
11.  Sung, N. H. , Dagli, G. , and Ying, L. , Surface Modification of Graphite Fibers via Plasma Treatment, 37th Annual Conference, RP/C Inst. , Soc. Plastics Ind. Inc. , Session 23-B, Jan. 11-15 (1982).
12.  Wertheimer, M. R. and Schreiber, H. P. , Surface Property Modification of Aromatic Polyamides by Microwave PLasmas, *J. Appl. Polym. Sci., 26*, 2087 (1981).
13.  Krishnamurthy, V. and Kamel, I. K. , Plasma Grafting of Allylamine onto Glass Fibers, 33rd Intl. SAMPE Symp. , p. 560 (1988).
14.  Yasuda, H. , *Plasma Polymerization*, Academic Press, Inc. , New York, 1985.
15.  Millard, M. , Synthesis of Organic Polymer Films, in *Techniques and Applications of Plasma Chemistry*, J. R. Hollahan and A. T. Bell (Eds.) John Wiley & Sons, New York (1974).
16.  Shen, M. , and Bell A. T. , A Review of Recent Advances in Plasma Polymerization, in *Plasma Polymerization*, ACS Symposium Series No. 108, ACS, Washington, D.C. (1979).
17.  Schwartz, P. , Netravali, A. N. , and Sembach, S. , Text. Res. J. , 56, 502 (1986).
18.  Li, Z.-F. , Surface Modification of Ultra-High Strength Polyethylene Fibers Through Allylamine Plasma Polymer Deposition, M.S. Thesis, Fiber Science Program, Cornell University, May 1989.
19.  Klein, P. G. , Woods, D. W. , and Ward, I. M. , *J. Polym. Sci. , B - Polym. Phys. , 25*, 1359 (1987).

20. Allred, R. E. , Merill, E. W. , and Roylance, D. K. , in *Molecular Characterization of Composite Interface*, H. Ishida and G. Kumar (Eds.), Plenum Press, New York (1985).
21. Netravali, A. N. , Schwartz, P. , and Phoenix, S. L. , *Polym. Comp., 10*, 385 (1989).
22. Netravali, A. N. , Henstenburg, R. B. , Phoenix, S. L. , and Schwartz, P. , *Polym. Comp., 10*, 226 (1989).
23. Netravali, A. N. , Li, Z.-F. , Sachse, W. H. , and Wu, H. F. , Acoustic Emission Coupled Single-Fiber-Composite Technique for the Measurement of Interfacial Shear Strength of High Performance Fibers in Epoxy, presented at the Fiber-Tex 1989, The III Conference on Advanced Engineering Fibers and Textile Structures for Composites, Greenville, S. C., Oct. 30-Nov. 1, 1989.
24. Penn, L. , Bystry, F. , Karp, W. , and Lee, S. , in *Molecular Characterization of Composite Interface*, H. Ishida and G. Kumar, (Eds.), Plenum Press, New York (1985).

# Chapter 7

# STATISTICS FOR THE SHORT TERM STRENGTH AND CREEP RUPTURE OF *p*-ARAMID FIBERS

PETER SCHWARTZ

*Fiber Science Program*
*Cornell University*
*Ithaca, New York 14853-4401*

## INTRODUCTION

Brittle fibers, such as poly(*p*-phenylene terephthalamide), are often used in applications that require them to sustain loads that are a significant fraction of their ultimate tensile strength. In products such as ropes and fiber-reinforced composite pressure vessels, the inherent variability in these materials often has a deleterious effect on the efficiency of the structure as a whole. Reporting on lifetime studies of pressure vessels, Gerstle and Kuntz[1] noted that much of the lifetime variability could be attributed to the diameter variability found in the filaments on a spool.

Using the coefficient of variation (CV% = standard deviation/mean x 100%) as a measure, Wagner *et al.*[2] noted that the filament diameter in spools of Kevlar®49 could vary by as much as 25%, although 10% was more typical. Filaments from spools with different variabilities showed virtually no difference in ultimate *stress* although differences in creep lifetime at constant stress at 21 °C yielded order of magnitude differences.

The effect of temperature, too, affects the lifetime for these fibers. Activation energy models such as those suggested by Tobolsky and Eyring,[3] Coleman[4], Zhurkov[5] and Wilfong and Zimmerman[6] all indicate a temperature effect on lifetime, although there is disagreement as to the mechanism. Both slippage of chains as well as chain scission have been mentioned.

The activation energy associated with creep rupture was studied experimentally, and the results are cast within a theoretical framework following that of Tobolsky and Eyring[3] and Zhurkov[5] .

## THEORETICAL FRAMEWORK

### Lifetme Statistics

The distribution for lifetimes of single aramid filaments under constant stress $S$ has been shown to follow a two-parameter Weibull distribution[7] of the form

$$F(t) = 1 - \exp [ - (t/r)^s] \qquad t \geq 0 \qquad (1)$$

where $r$ and $s$ are the Weibull scale and shape parameters, respectively. Recasting Equation 1 in a slightly different format, we may write

$$F(t) = 1 - \exp \{ - [ k(\Sigma)t]^s \} \qquad (2)$$

where $k(\Sigma)$ is termed the 'breakdown' rule, and we assume an exponential rule of the form

$$k(S) = a \exp (\beta \Sigma). \qquad (3)$$

We can connect the constants $a$ and $b$ to molecular events by taking

$$a = \tau_0^{-1} \exp [ - (U_0/kT)] \qquad (4)$$

and

$$b = \gamma /kT \qquad (5)$$

where $U_0$ is the stress-free activation energy, $\gamma$ is an associated activation volume, $\tau_0$ is related to the period of bond vibration, $k$ is Boltzmann's constant, and $T$ is the absolute temperature. For this, the Weibull scale parameter $r$ in Equation (1) can be seen to be

$$r = \tau_0 \exp [(U_0 - \gamma \Sigma)/kT] \qquad (6)$$

and is identical to the form given by Zhurkov.[5]

### Strength Statistics

The statistical distribution for short term strength of $p$-aramid fibers has been shown[2] to be well-fitted by the two-parameter Weibull distribution given by

$$W(\sigma) = 1 - \exp [ - (\sigma /a)^b] \qquad \sigma \geq 0 \qquad (7)$$

where $a$ and $b$ are the scale and shape parameters, respectively. Coleman[4] has pointed out that, under the exponential breakdown rule, the strength distribution is not Weibull in form but rather (for exp $(\beta\sigma) \gg 1$) is approximately the double exponential

$$W(s) \cong 1 - \exp \{ - [ \ \alpha(\beta R)^{-1} \exp (\beta\sigma)]^s \} \tag{8}$$

where $R$ is the loading rate. Fortunately the two distributions given in Equations (7) and (8) give similar results for suitably chosen values of the various parameters, and therefore the more easily used Weibull form will provide the tool for the analysis.

## EXPERIMENTAL PROCEDURES

### Materials and Conditions

The $p$-aramid fibers used in this work were Kevlar®49 supplied by E. I. du Pont de Nemours & Co., Inc. in the form of 23 tex yarns containing 134 filaments. Fibers from two different production spools, hereafter termed S1 and S2, were used.

Both short- and long term tests were carried out at three different temperatures—21, 80 and 130°C. Prior to mounting each filament, an adjacent length was measured for linear density using the vibroscope method (ASTM D1577-79), and the assumption was made that the linear density was unchanged for adjacent filaments. Table 1 contains the CVs of the tests of filament linear density from vibroscope measurements.

Table 1.
Average percent coefficient of variation (%CV) in linear density for p-aramid filaments[9] measured at 21°C, 65% relative humidity.

| Spool | Average  CV% |
|-------|--------------|
| S1 | 8.68 |
| S2 | 27.26 |

### Tensile Tests

All filaments were tested at a gauge length of 25 mm and a strain rate of 0.5 min$^{-1}$ using an Instron Model TM tensile testing machine. For tests above room temperature, the Instron was fitted with a heat chamber.

### Creep Rupture Tests

Creep rupture tests were run on a specially designed, computer-monitored frame first described by Wagner *et al.*[8] At elevated temperatures, the frame was modified to fit inside a large convection oven, as described in Wu *et al.*[9]

The filaments were tabbed in a similar manner to those used in tensile tests and were the same gauge length (25 mm). The loading was by dead-weights, and each load was adjusted by taking the filament diameter into account so as to maintain equal *stress* at each level. Different stress levels were used, each a fraction of the Weibull scale parameter for strength obtained from tensile tests at that level and temperature.

## RESULTS AND DISCUSSION

In Table 2 is presented the maximum likelihood estimators of the Weibull parameters for short term strength for the aramid filaments from S1 and S2, tested at three different temperatures. The data indicate that there is significant reduction in strength (Weibull scale parameter, $a$)as a function of temperature, as has been noted earlier[6]. But note the variability, as indicated by the Weibull shape parameter, $b$, remains remarkably constant and similar for both spools. This is in spite of the greater variability in filament diameter in S2, as seen in Table 1, and reinforces the caution given in Wagner *et al.*[6] that each filament's size should be taken into account in order to get the true failure stress distribution.

Table 2
Maximum likelihood estimators (MLE) of the Weibull shape (b)
and scale (a) parameters for strength at 21, 80, and 130°C[9].

| Spool | Temperature | $b$ | $a$ |
|-------|-------------|-----|-----|
|       | [°C]        |     | [MPa] |
| S1    | 21          | 10.2 | 3538 |
|       | 80          | 10.1 | 3304 |
|       | 130         | 10.5 | 3019 |
| S2    | 21          | 10.4 | 3594 |
|       | 80          | 10.2 | 3385 |
|       | 130         | 10.1 | 3041 |

The effect of size variability is, however seen in the results for creep rupture lifetime. At every stress level and temperature, the median lifetime of filaments from S1 (smaller diameter variability) is approximately an order of magnitude longer than those from S2 (greater diameter variability).

In order to calculate the model parameters in equation (6), we plot, in Figure 1 all of the lifetime data at each temperature. From these we determine the following parameter values:

$$U_0 \cong 3.35 \times 10^5 \text{ J mol}^{-1} = 80 \text{ kcal mol}^{-1}$$

$$\tau_0 \cong 0.0607 \text{ s}$$

$$\gamma \cong 0.208 \text{ nm}^3$$

The value for the stress-free activation energy $U_0$ is consistent with that quoted for the rupture of the C-N bond in aramids.[10] The activation volume is approximately that of the unit repeat in Kevlar®, although there are some indications of stress effects.

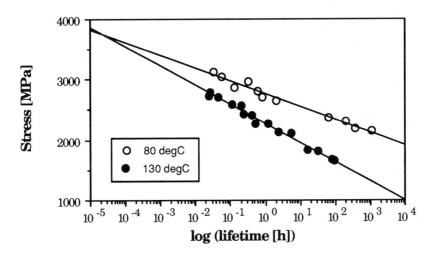

Figure 1. Creep-stress level plotted against ln(lifetime) for filaments from spools S1 and S2: (o) 80°C; (•) 130°C (from ref. 9). $U_0$ @ 3.35 ¥ $10^5$ J mol$^{-1}$, $t_0$ = 0.067 s, g = 0.208 nm$^3$.

In many models, $t_0$ is often approximated as the molecular bond vibration period, which for the C-N bond is on the order of $10^{-12}$ sec. However in a sample loaded to a significant fraction of its breaking stress, one would expect large variations from the stress-free value, and this seems to be the case in our data.

## CONCLUSIONS

The major conclusions are as follows:
- Both the strength and creep rupture of Kevlar®49 filaments follow a two-parameter Weibull distribution.
- Filament diameter variability plays no role in filament stress variability (assuming it is properly accounted for), but does play a significant role in creep rupture lifetime.
- From an exponential breakdown model, the data yield an activation energy for the creep failure in the range of $3.35 ¥ 10^5$ J mol$^{-1}$, consistent with the energy of scission of the C-N amide bond.

## ACKNOWLEDGMENTS

This work was supported in part by funding from the Cornell University Agricultural Experiment Station, in part by the Cornell Materials Science Center (an NSF DMR-MRL), and in part by E. I. du Pont de Nemours & Co., Inc.

## DEDICATION

This paper is dedicated to Professor Richard D. Gilbert—teacher, colleague, and friend.

## REFERENCES

1. Gerstle, F. P. and Kuntz, S. C. in Proceedings of ASTM Symposium on Long term Behavior of Composites, ASTM STP 813. Philadelphia: ASTM, p. 263, 1983.
2. Wagner, H. D., Phoenix, S. L., and Schwartz, P., *J. Comp. Mater.* **18** (1984) 312.
3. Tobolsky, A. and Eyring, H., *J. Chem Phys.* **11** (1943) 125.
4. Coleman, B. D., *J. Appl. Phys.* **27** (1956) 862.
5. Zhurkov, S. N., *Int. J. Fract. Mech.* **1** (1965) 311.
6. Wilfong, R. E. and Zimmerman, J., *J. Appl. Polym. Sci., Appl. Polym. Symp.* **31** (1977) 1.
7. Schwartz, P., *Polym. Engr. Sci.* **27** (1987) 842.
8. Wagner, H. D., Schwartz, P., and Phoenix, S. L., *J. Mater. Sci.* **21** (1986) 1868.
9. Wu, H. F., Phoenix, S. L., and Schwartz, P., *J. Mater. Sci.* **23** (1988) 1851.
10. Phoenix, S. L., in Proceedings of 9th US National Congress of Applied Mechanics, Book N⁰ H00228, Cornell University, Ithaca, New York, 21-25 June 1982. New York: American Society of Mechanical Engineers, p. 219, 1982.

# Chapter 8

# THERMAL PROPERTIES OF NYLON 6/ POLY(ETHYLENE-CO-ACRYLIC ACID) BLENDS

K. J. YOON*, R. E. FORNES AND R. D. GILBERT

*Fiber and Polymer Science Program*
*North Carolina State University*
*Raleigh, NC 27695-8302*
**Department of Textile Engineering*
*Dankook University*
*Seoul 140-714, Korea*

## INTRODUCTION

Polymer blends have been studied extensively and progress has been made in understanding the phenomena of polymer mixing. However, the interactions occurring between polymers and their effects on blend miscibility, compatibility, the resulting structure and blend properties are not completely understood.

Miscibility can be defined as the spontaneous mixing over the whole composition range of the component polymers. For spontaneous mixing, the change in Gibbs free energy, $\Delta G_m$, must be negative at temperature T. Since the entropy change $\Delta S_m$ for mixing of high molecular weight molecules is negligible, the enthalpy change on mixing, $\Delta H_m$, is the primary criterion for spontaneous mixing.

$$\Delta G_m = \Delta H_m - T\Delta S_m < 0$$

$$\Delta S_m \approx 0$$

therefore, $\quad \Delta H_m < 0.$

The requirement of an exothermic interaction is a necessary condition for complete miscibility, although it is not a sufficient one. Another criterion which must also be met for complete miscibility[1] is

$$\left| \frac{\partial^2 \Delta Gm}{\partial \phi^2} \right|_{T,P} > 0$$

where $\phi$ is the volume fraction of a component and P is the pressure.

Immiscible polymer blends tend to separate into their respective phases. When the interfacial adhesion between the two phases is poor, the blends exhibit poor mechanical properties. Interaction between polymers is favorable not only for miscibility but also for good interfacial adhesion between immiscible components. An excellent review of the interactions possible in polymers, as well as the basic thermodynamics and other general aspects of blending, is presented by Olabisi.[1] Proper control of the processing parameters and/or addition of "compatibilizing agents" (block or graft copolymers of the two components) may enhance the interfacial adhesion between the respective phases and prevent the deterioration of physical properties. Systems that are not miscible but exhibit superior properties are termed compatible.[2] There are many more examples of compatible polymers than miscible polymers.

Based on the approach that interaction is favorable in obtaining miscible or compatible blends, blends of nylon 6 with a variety of polymers containing functional groups capable of interaction with the nylon 6 amide group have been studied by previous workers of this group.[3,4] Based on end group analysis and insolubility of the blend in formic acid, Venkatesh et al. suggested that the reaction between nylon 6 and poly(acrylic acid) may result in a small amount of crosslinking[3]. Solution blended samples containing greater than 30% poly(acrylic acid) did not show the melting endotherm of nylon 6 in the thermal analysis, indicating the restriction of nylon 6 chains by the presence of poly(acrylic acid). Because of this extensive interaction or reaction only limited amounts of poly(acrylic acid) could be added to nylon 6 in the melt processing of the blends.

Since the amide group is basic, polymers having an acidic functionality may be capable of reacting with nylon 6. Reactions that may be possible are described in Figure 1. Interchange reactions[5-10] and condensation reactions may be possible. Interchange reactions are generally considered to occur slowly and in the presence of catalysts. Polymers containing carboxylic acid groups or anhydride groups have been suggested to undergo chemical reactions with nylon 6 to form block, graft or crosslinked polymers.[3,11-23] Poly(acrylic acid) has been shown to form inter and/or intramolecular anhydrides with the evolution of water above 420°K and subsequently undergo decarboxylation above 493°C.[24,25] Ester groups that have lower activity compared to anhydrides have been shown to react with primary amines in a model compound study with dibutyl succinate and tridecylamine[26]. Consequently any anhydride formed may be capable of undergoing reaction with the end amino groups of nylon 6.

Mechanochemical reaction of the methyl groups via radical formation has also been suggested to be possible.[16] The reactions may also produce crosslinked polymers.

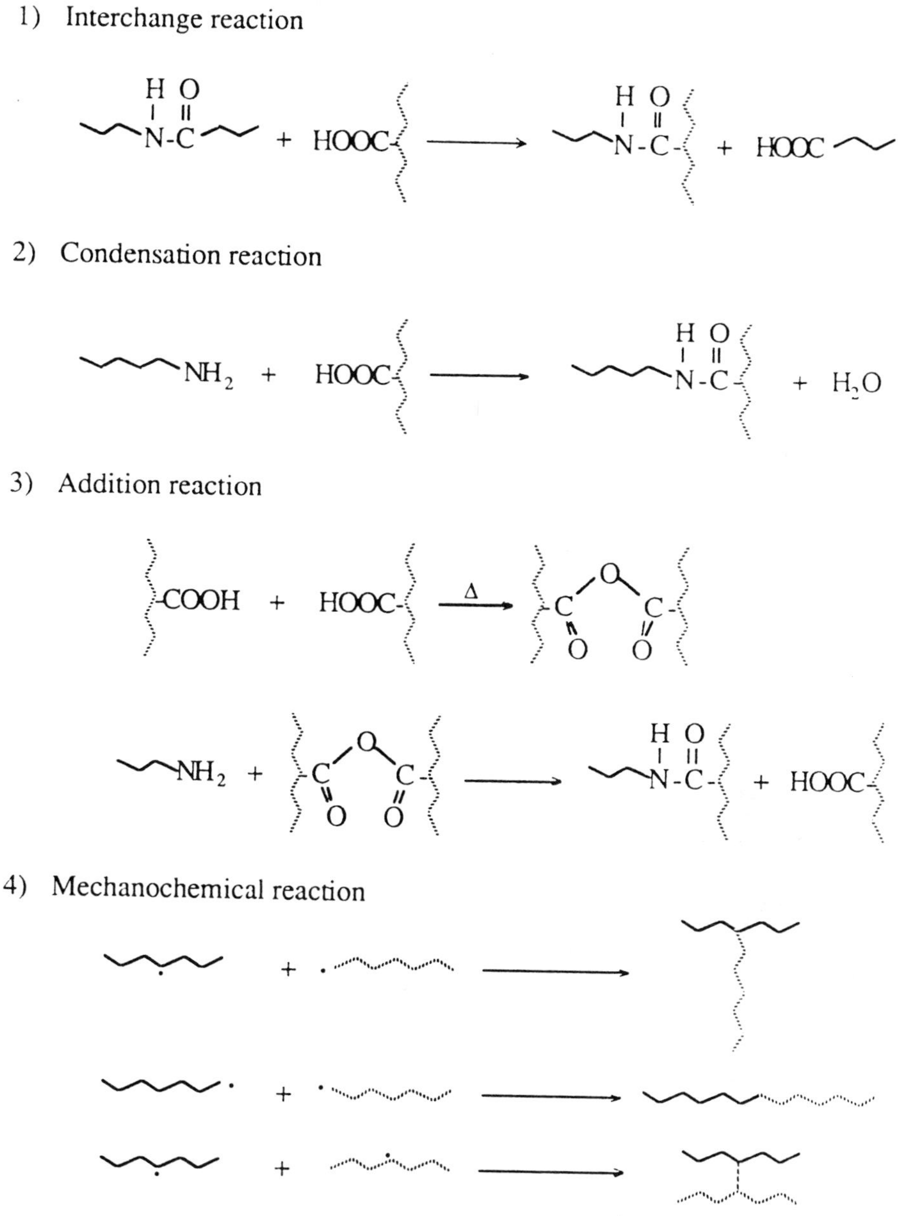

1) Interchange reaction

2) Condensation reaction

3) Addition reaction

4) Mechanochemical reaction

Figure 1.    Schematic of the possible reactions between nylon 6 and poly(ethylene-co-acrylic acid).

If such reactions do occur, they may produce block or graft copolymers which are known to enhance the interfacial adhesion by acting as "compatibilizers" in immiscible blend systems.[27] It is generally accepted that such block or graft copolymers can preferentially locate at the interface between the two phases to provide enhanced interfacial adhesion. Ide et al. had used this approach to obtain a better dispersion of polypropylene in nylon 6 employing poly(propylene-g-maleic anhydride) as an interfacial agent[11-14]. They suggest that reaction between the anhydride and the end amino groups had produced graft copolymers that function as interfacial agents. Many researchers have investigated this approach recently, showing a renewed interest in nylon 6 blends.[3,15-17,22]

To control the excess interaction and/or reaction present in nylon 6/poly(acrylic acid) blends, blends of nylon 6 with poly(ethylene-co-acrylic acid) (PEAA) containing ~9 mole % acrylic acid was studied. In previous studies on the blend of nylon 6 with PEAA containing ~8 mole % acrylic acid,[15-17] mechanical loss data have shown that this system has a $T_g$ intermediate between those of the two constituents. However the loss peak was broader than the loss peak of nylon 6 indicating a somewhat heterogeneous system.[15] Illing[16] inferred that some nylon 6 had reacted with the acidic groups of PEAA from extraction experiments and a positive Molau test. The Molau test, a test for the presence of graft copolymers of nylon 6, is said to be positive when the nylon 6 blend forms an emulsion in formic acid.[16] The infrared spectrum of the toluene extract, which should have been basically PEAA, exhibited characteristic bands of the amides.[16] The blend displays higher impact strengths due to an enhanced dispersibility and interfacial adhesion which was inferred from the photomicrographs.[17]

If small amounts of block or graft copolymers are formed by the reaction of nylon 6 with PEAA, they would undoubtedly retard the crystallization of nylon 6 component in the blend. This effect should apply to sufficiently annealed samples, as well as freshly prepared samples. In the absence of such copolymers, the two components should phase separate on extensive annealing to exhibit the intrinsic properties of the respective copolymers. On the other hand, the extent of crystallization of nylon 6 in the blends is expected to be influenced only when the samples are not sufficiently annealed. When the blend is given sufficient time to crystallize, the presence of small amounts of copolymers should have negligible effect on the crystallinity of nylon 6 in the blend. Observation of the crystallization exotherms during cooling and the heats of melting during heating at constant rates were utilized to compare the thermal properties of nylon 6 in nylon 6/PEAA blends of different thermal history. Thermal analysis of fractions considered to be copolymers were also studied to observe the effect of subsequent heat treatments on these fractions.

## EXPERIMENTAL

### Materials

Nylon 6 with a number average molecular weight of 20,000 was supplied by Allied Corporation. PEAA, containing approximately 20 wt% acrylic acid was obtained from Dow Chemical Company. Its commercial trade name is Primacor™ 590 and the melt flow index is reported to be 55-75 g/(10 min) according to the ASTM test method D1238.

### Melt Blending

Melt blends of nylon 6 and PEAA were prepared by Allied Corporation. Nylon 6 and PEAA chips were tumble blended in a PK twin shell blender for 1 hour, processed through a 1 inch single screw extruder (RXT-2), quenched and pelletized. These blends are designated as NEAA x%, where the x refers to the percentage of PEAA in the blend by weight. Nylon 6 and NEAA 10% and 20% were processed at 255°C, NEAA 40% at 235°C and NEAA 50% at 200°C.

### Preparation of Samples of Different Thermal History

Melt cast films were obtained by heating the dried chips between the two heating plates of a Carver Laboratory Press with a 0.13 mm spacer at 550°K for 30 seconds without applying pressure, then pressing at 40 psi for 30 seconds and finally quenching in water (water-quenched samples). Water quenched films were also heated at 320K/min to 530K in a differential scanning calorimeter, kept at 530K for 1 minute, then removed and dropped into liquid nitrogen to obtain essentially noncrystalline samples (liquid nitrogen quenched samples). The chips were also annealed at 373K for 13 days to obtain a sample in which the nylon 6 component has had sufficient time to crystallize. The annealing temperature was selected to minimize degradation while maintaining a sufficient crystallization rate. Water-quenched films of NEAA 50% were also kept at 530K for varying amounts of time in the DSC to observe the effect of exposure to a higher temperature. The crystallinity of the samples was expected to be in the following order: liquid nitrogen quenched samples, water quenched samples, chips, and chips annealed at 373C for 13 days.

To obtain samples for x-ray analysis NEAA chips were also melt cast into tablets using an aluminum spacer ~1.5 mm thick with a 1.3 cm diameter circular hole as the mold and quenched with water. The tablets were annealed at 60°C for 24 hours before x-ray analysis. Films for dynamic mechanical analysis were prepared by melt casting and quenching in liquid nitrogen immediately before characterization. The liquid nitrogen quenched films were also annealed at 65°C for 50 hr to obtain dynamic mechanical analysis samples in which nylon 6 has had sufficient time to crystallize.

### Fractionation of NEAA Blends

Fractionation of the NEAA blends into the pure components, nylon 6 and PEAA, and a third component was also attempted in an effort to obtain evidence of actual chemical or physico-chemical reaction occurring between nylon 6 and PEAA. The nylon 6 was extracted with 90% formic acid and the PEAA extracted with o-xylene. Details of the fractionation process are described elsewhere.[28]

### Differential Scanning Calorimetry [DSC]

Thermal transitions were investigated using a Perkin Elmer DSC-2C differential scanning calorimeter interfaced with a 3600 data station and equipped with Intracooler II, which enabled cooling of the DSC heads down to ~210K. Calibration was carried out with standards available from Perkin Elmer: indium (mp = 429.7K, $\Delta H_m$ = 6.80 cal/g), potassium chromate (mp = 943.7K, $\Delta H_m$ = 8.50 cal/g) and tin (mp = 505.06K, $\Delta H_m$ = 14.45 cal/g) and a Fisher Scientific thermetric standard p-nitrotoluene (m.p.= 324.69°K). Sample weights of 3-5 mg were used in all cases to minimize heat transfer problems. Films were thin enough to be used as is, but the chips were cut into thin slices to ensure proper heat transfer. Samples obtained from fractionation experiments were in the form of powders and were analyzed after drying in vacuum at 40°C. Heating, cooling and reheating thermograms were obtained by heating a sample at 20C/min, then cooling the same sample at 20°C/min, and finally reheating the heated and cooled sample at 20K/min. Temperature ranges of either 230-530K or 215-515K were used.

The crystallinity of the nylon 6 component in nylon 6 and the NEAA blends were compared by a crystallinity index, $\chi_{N6}$, defined as the fraction of the heat of melting of the nylon 6 peak at ~490K, normalized with respect to the weight of nylon 6 in the blend, over the highest heat of fusion of bulk samples, 22.5 cal/g, reported in the literature.[29]

$$\chi_{N6} \; = \; \frac{\Delta H_{N6}}{\Delta H_{ref}}$$

where

$\Delta H_{N6}$ is the heat of fusion per unit weight of nylon 6

$$\Delta H_{N6} \; = \; \frac{\Delta H_{exp}}{1 - x/100}$$

$\Delta H_{exp}$ is the heat of fusion of the peak at ~490K based on the total weight of sample and x is the weight percent of the second component in blend

**X-ray Scattering and Dynamic Mechanical Analysis**

Diffractometer scans were carried out at halh-degrees $2\theta$ per minute on a Siemens x-ray spectrophotometer using Ni filtered Cu $K\alpha$ radiation at 30 KV and 20 mA. Dynamic mechanical measurements were made on an Autovibron (a Rheovibron equipped with automatic data processing and sample tensioning systems, IMASS) at 11 Hz with a 2°C/min heating rate. Higher heating rates could not be used on the Autovibron because of the difficulty in controlling the sample tension. Films of nylon 6 (NEAA 10, 20, 40, 50 %) were scanned over the temperature range of ~90 - 180°C. The liquid nitrogen quenched films (noncrystalline samples) were run immediately after preparation such that crystallization during storage did not influence the properties measured.

## RESULTS AND DISCUSSION

In our study, the Molau tests of nylon 6/PEAA blends were positive, indicating the possible presence of graft copolymers. Wide angle x-ray diffractometer scans of the blends annealed at 60°C for 24 hours exhibited a single peak, while the pure nylon 6 exhibited two peaks characteristic of the $\alpha$-structure (Figure 2). Melt blending with PEAA appears to restrict the transition of nylon 6 crystals in the blends to the $\alpha$-structure. Dynamic mechanical analysis of liquid nitrogen quenched nylon 6 and nylon 6/PEAA blend films exhibited an increase in the $\alpha$-transition temperatures of the nylon 6 component with higher PEAA content (Figure 3). On the other hand, the same films annealed for 50 hours at 65°C exhibited two transitions (Figure 4). These results suggest that this system is essentially immiscible, and that the thermal transitions of nylon 6 in the blends are dependent on the thermal history. DSC analysis was carried out on NEAA blends of various thermal histories and samples from fractionation experiments yielded additional information on the effect of interaction and/or reaction on the thermal properties.

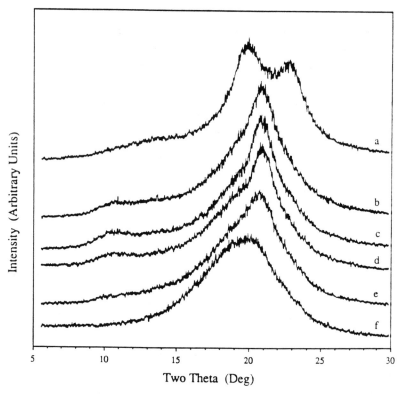

Figure 2.      Diffractometer scans of melt cast tablets of nylon 6,
poly(ethylene-co-acrylic acid) [PEAA] and their blends [NEAA]
annealed at 333K for 1 day: (a) nylon 6, (b) NEAA 10%,
(c) NEAA 20%, (d) NEAA 40%, (e) NEAA 50% and (f) PEAA.

## Liquid Nitrogen Quenched Samples

Thermograms of the $LN_2$ samples are shown in Figure 5.
Crysatllization exotherms are clearly shown as the temperature surpasses the
$T_g$. The melting points of the nylon 6 component in the blends are within
experimental error, and the heats of fusion of the nylon 6 component are low
and essentially in proportion to the amount of nylon 6 in the blend. A
gradual widening of the endothermic peak is reflected by the variation with
composition of the peak widths at half-intensity, as shown in Figure 6. Two
explanations appear to be possible. One is that PEAA interferes with the
crystallization of the nylon 6 in these samples, resulting in a wider
distribution of crystal sizes. A second explanation may be that most nylon 6
has crystallized in the same manner as in the pure state but that the dilution of
the nylon 6 in the blend interferes with the two competing processes,
crystallization and melting of nylon 6, during the heating scan itself. In either
case, the change in the melting behavior appears to result from the interference
in the crystallization of nylon 6 due to the interaction between the two
components in the noncrystalline state, an effect that may have resulted from
the fabrication of the original sample or from the heating in the DSC.

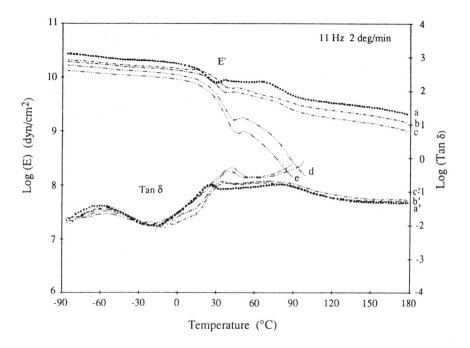

Figure 3.  Dynamic mechanical analysis of nylon 6 and nylon 6/poly (ethylene-co-acrylic acid) blend [NEAA] films, melt cast and quenched in liquid nitrogen:        (a') nylon 6(⋯⋯⋯),
(b')  NEAA  10%(–·–·–),        (c')  NEAA 20% (–·····–),
(d') NEAA 40% (–····–·)    and    ( e') NEAA 50% (–·······–).

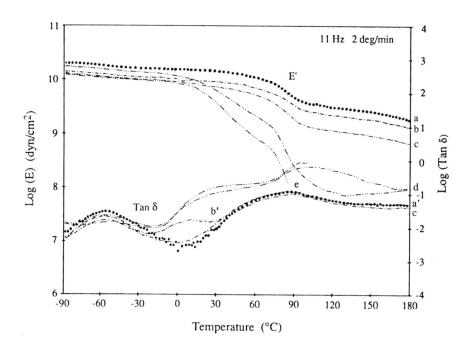

Figure 4.  Dynamic mechanical analysis of nylon 6 and nylon 6/poly (ethylene-co-acrylic acid) blend [NEAA] films, melt cast and quenched in liquid nitrogen then annealed at 65°C for 50 hours: (a') nylon 6(········), (b') NEAA 10%(-·--·-), (c') NEAA 20% (-·····-), (d') NEAA 40% (--------) and (e') NEAA 50% (----------).

## Water Quenched Samples

Thermograms of films melt cast on a Carver laboratory press and quenched in water are shown in Figure 7 and the analyzed results are in Table 1. The crystallinity index of the nylon 6 component in the first heating cycle is generally less than that of pure nylon 6. Changes in the melting temperatures are negligible, but significant decreases in the onset temperatures are present. It appears possible that, during the exposure to high temperatures, block and/or graft copolymers may be formed, especially since the Molau test also suggests the presence of graft copolymers. The presence of such species would enhance the degree of mixing of the component polymers and retard the rate of crystallization.

Retardation of the crystallization also results in lower crystallization temperatures of nylon 6 in the blend films compared to the nylon 6 (Figure 8). The heats of melting of the reheating endotherms are also lower compared to the predicted values based on their composition. Another interesting feature is that the lower end shoulders at ~454K (Figure 9), arising from the melting of the less perfect crystals of nylon 6,[29] shift towards lower temperatures and become broader with greater amounts of PEAA suggesting that the average size of these less perfect crystals decreases as a result of the retardation of crystallization.

## Chips

Results of the thermal analysis of the heating, cooling and reheating of NEAA chips at 20K/min are tabulated in Table 2. The heating thermograms in Figure 10 exhibit the respective melting endotherms of the two component polymers. Variations in the melting point and crystallinity index of the nylon 6 component in the blends are negligible (Table 2). This suggests that most nylon 6 crystals in the blend exist as a separate phase and that the interaction and/or reaction between the two polymers occur in the melt state and only in the noncrystalline regions in the solid state. The crystallization of the nylon 6 component in the blend occurs at lower temperatures with increasing amounts of PEAA in the blend and the exothermic areas are in proportion to the amount of nylon 6 in the blend. The crystallinity index of the reheating endotherm was lower than that of the pure nylon 6. This appears to be the result of the interaction and/or reaction between nylon 6 and PEAA which may reduce the rate at which nylon 6 can crystallize during the 20K/min cooling cycle and the reheating cycle. Shift of the peak temperatures of the lower melting, imperfect nylon 6 crystals[29] (~460K) toward lower temperatures was observed.

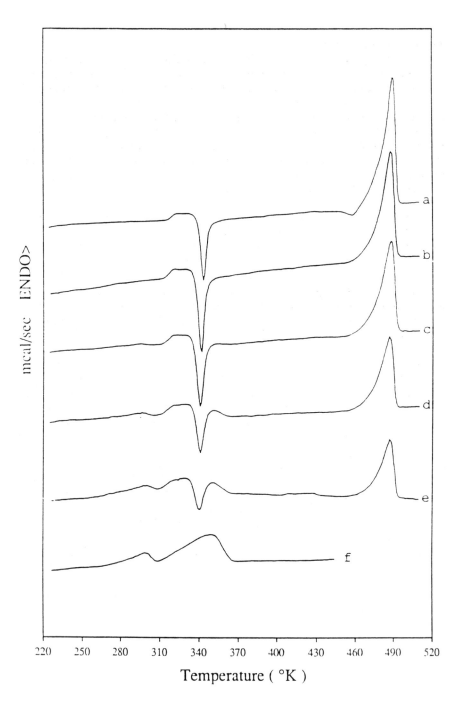

Figure 5.    Heating thermograms of nylon 6, poly(ethylene-co-acrylic acid) [PEAA] and their blends [NEAA], quenched in liquid nitrogen: (a) nylon 6,  (b) NEAA 10%,  (c) NEAA 20%, (d) NEAA 40%, (e) NEAA 50% and (f) PEAA.

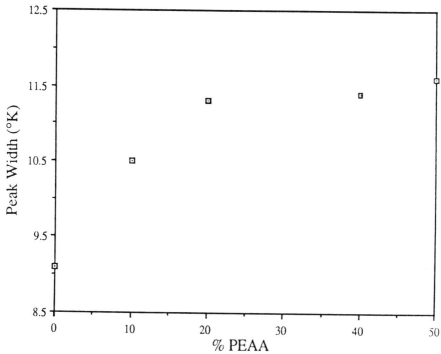

Figure 6.    Effect of blend composition on the peak widths at half-intensity of the melting endotherms of the nylon 6 component in the nylon 6/PEAA blends quenched in $LN_2$.

Table 1.   Thermal Properties of Nylon 6 in Water Quenched Films

| SAMPLE | HEAT | | | | | COOL | | REHEAT | | | |
|---|---|---|---|---|---|---|---|---|---|---|---|
| | Onset | $T_m$ | $\Delta H_m$ | $T_g$ | $\chi_{N6}$ [a] | $T_c$ | $\Delta H_c$ | $T_m$ | $\Delta H_m$ | $T_g$ | $\chi_{N6}$ |
| | (°K) | (°K) | cal/g | (°K) | | (°K) | cal/g | (°K) | cal/g | (°K) | |
| Nylon 6 | 479 | 490 | 16.8 | 307 | 0.75 | 451 | 17.1 | 490 | 20.3 | 324 | 0.90 |
| NEAA x=10% | 476 | 489 | 12.5 | 326 | 0.62 | 449 | 14.8 | 488 | 15.4 | 327 | 0.76 |
| NEAA x=20% | 475 | 489 | 11.6 | 311 | 0.65 | 450 | 13.6 | 488 | 14.7 | 327 | 0.81 |
| NEAA x=40% | 474 | 488 | 8.3 | * | 0.62 | 444 | 10.4 | 487 | 10.4 | * | 0.77 |
| NEAA x=50% | 474 | 488 | 7.8 | * | 0.69 | 444 | 8.3 | 487 | 8.8 | * | 0.78 |

a)  $\chi_{N6}$ (Crystallinity Index) = $\Delta H_{N6}/22.5$
    $\Delta H_{N6}$ = $\Delta H_m/(1-x/100)$, x: % PEAA
*  not distinguishable

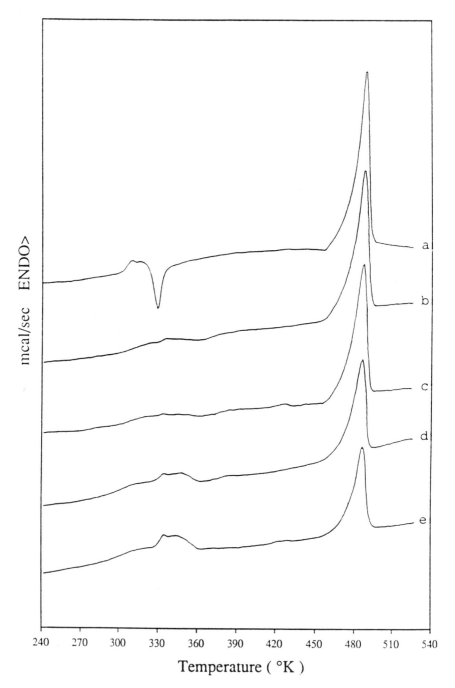

Figure 7.    Heating thermograms of nylon 6 and nylon 6/poly(ethylene-co-acrylic acid) blend [NEAA] films melt cast and quenched in water: (a) nylon 6,    (b) NEAA 10%,    (c) NEAA 20%, (d) NEAA 40% and (e) NEAA 50%

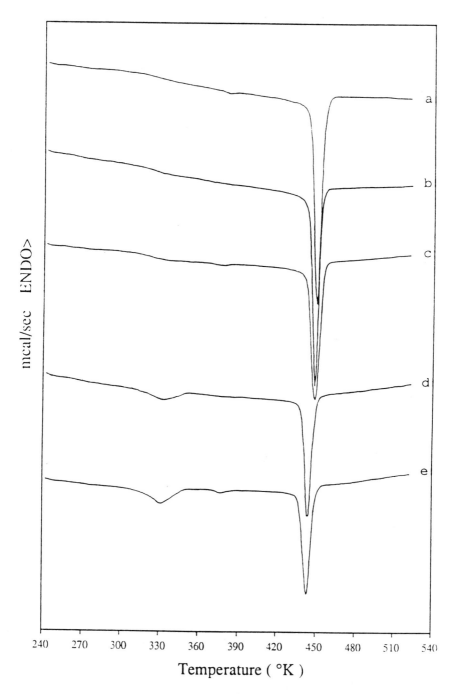

Figure 8.    Crystallization exotherms of nylon 6 and nylon 6/poly(ethylene-co-acrylic acid) blend [NEAA] films melt cast and quenched in water, after heating to 530°K at 20K/min: (a) nylon 6, (b) NEAA 10%, (c) NEAA 20%,  (d) NEAA 40% and (e) NEAA 50%.

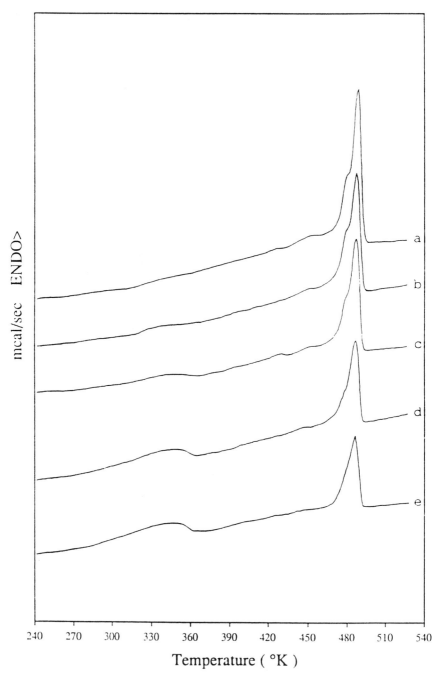

Figure 9.    Reheating thermograms of nylon 6 and nylon 6/poly(ethylene-co-acrylic acid) blend [NEAA] films melt cast and quenched in water, after heating to 530K and cooling to 230K at 20°C/min: (a) nylon 6, (b) NEAA 10%, (c) NEAA 20%, (d) NEAA 40% and (e) NEAA 50%.

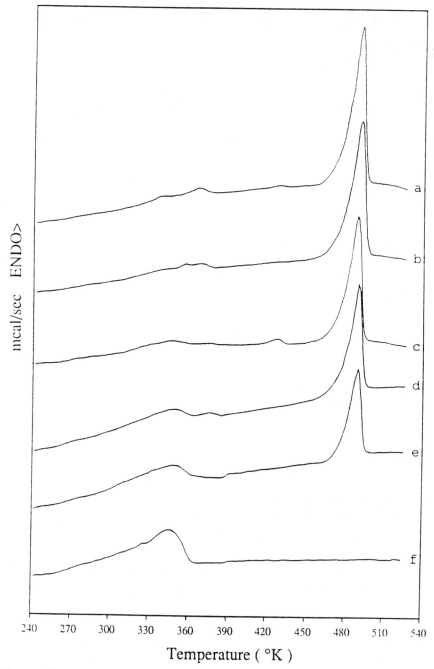

Figure 10.    Heating thermograms of nylon 6, poly(ethylene-co-acrylic acid) [PEAA] and their blend [NEAA] chips:    (a) nylon 6, (b) NEAA 10%,  (c) NEAA 20%, (d) NEAA 40%, (e) NEAA 50% and (f) PEAA.

Table 2. Thermal Properties of Nylon 6 in NEAA Chips

| SAM-PLE | HEAT | | | | COOL | | REHEAT | | | |
|---|---|---|---|---|---|---|---|---|---|---|
| | $T_m$ | $\Delta H_m$ | $T_g$ | $\chi_{N6}$ [a] | $T_c$ | $\Delta H_c$ | $T_m$ | $\Delta H_m$ | $T_g$ | $\chi_{N6}$ |
| | (°K) | (cal/g) | (°K) | | (°K) | (cal/g) | (°K) | cal/g | (°K) | |
| Nylon6 | 492 | 16.3 | 330 | 0.72 | 450 | 15.1 | 490 | 20.8 | 328 | 0.92 |
| NEAA x=10% | 493 | 14.9 | 326 | 0.74 | 449 | 13.9 | 489 | 14.9 | 326 | 0.74 |
| NEAA x=20% | 490 | 13.5 | 327 | 0.75 | 446 | 11.7 | 487 | 14.3 | 323 | 0.79 |
| NEAA x=40% | 491 | 10.4 | * | 0.77 | 443 | 9.2 | 488 | 10.4 | * | 0.77 |
| NEAA x=50% | 491 | 7.7 | * | 0.68 | 443 | 7.2 | 488 | 8.6 | * | 0.76 |
| PEAA | 347 | 10.2 | 264 | 0.00 | 319 | 11.2 | 348 | 10.1 | 263 | 0.00 |

a)  $\chi_{N6}$ (crystallinity index) = $\Delta H_{N6}/22.5$
    $\Delta H_{N6}$ = $\Delta H_m/(1-x/100)$, x: % PEAA
* not distinguishable

## Chips Annealed at 373K for 13 Days

The melting behaviour of nylon 6 in the samples discussed thus far have involved the kinetic factors associated with the formation of nylon 6 crystals rather than the equilibrium properties. The blend chips were thus annealed at a temperature (373K) where the crystallization rate of nylon 6 is relatively high and degradation is not significant. Results of the thermal analysis of the heating, cooling, reheating and subsequent cooling cycles of NEAA chips annealed at 373K for 13 days are tabulated in Table 3. The blend chips annealed at 373K for 13 days exhibit significant decreases in the melting temperatures of the nylon 6 component compared to annealed nylon 6 with greater amounts of PEAA (40 and 50% PEAA blends). The change in the melting behaviour is more prominent in the onset temperatures, with the largest decrease being ~12°C for the 40% NEAA blend. The larger decreases in the onset temperatures reflect the widening of the melting peaks in the blends (Figure 11), which may be a result of further interactions occurring between the two component polymers during the annealing process.

Table 3.  Thermal Properties of Nylon 6 in Annealed NEAA
Chips (373K, 13 days)

| SAM-PLE | HEAT | | | | | COOL | | REHEAT | | | |
|---|---|---|---|---|---|---|---|---|---|---|---|
| | On-set | $T_m$ | $\Delta H_m$ | $T_g$ | $\chi_{N6}{}^a$ | $T_c$ | $\Delta H_c$ | $T_m$ | $\Delta H_m$ | $T_g$ | $\chi_{N6}$ |
| | (K) | (K) | cal/g | (K) | | (K) | cal/g | (°K) | cal/g | (K) | |
| Nylon6 | 479 | 492 | 16.7 | 307 | 0.74 | 451 | 18.5 | 491 | 19.8 | 324 | 0.88 |
| NEAA x=10% | 476 | 491 | 15.2 | 312 | 0.75 | 451 | 16.6 | 489 | 17.7 | 326 | 0.87 |
| NEAA x=20% | 476 | 491 | 14.6 | 312 | 0.81 | 449 | 16.3 | 489 | 16.2 | 325 | 0.90 |
| NEAA x=40% | 468 | 486 | 11.1 | * | 0.82 | 438 | 10.2 | 484 | 11.5 | * | 0.85 |
| NEAA x=50% | 472 | 488 | 8.7 | * | 0.77 | 445 | 8.3 | 488 | 10.0 | * | 0.89 |

a) $\chi_{N6}$ (crystallinity index) $= \Delta H_{N6}/22.5$
   $\Delta H_{N6} = \Delta H_m/(1-x/100)$, x: % PEAA
* not distinguishable

The crystallization exotherms of the annealed chips heated to 530K at 20°C/min in the DSC ( Figure 12) display a significant change in their peak crystallization temperatures as well as the peak widths at half-intensity compared to the pure nylon 6. The crystallization peaks of the NEAA 40% and NEAA 50% blends annealed at 373K for 13 days are much broader than the nylon 6, NEAA 10% or NEAA 20%. There appears to be a gradual increase in the crystallization temperature distribution toward lower temperatures with increasing amounts of PEAA in the blend. In the case of the NEAA 40% blend a new peak at 438K is observed with the original peak occurring as a shoulder on the higher side. The peak crystallization temperature is about 13°C lower than the original temperature.

The crystallinity index values of NEAA blends summarized in Table 4 demonstrate that the kinetic factor is significant in the crystallization of nylon 6 in the NEAA blends. With essentially noncrystalline samples (liquid nitrogen quenched NEAA), even pure nylon 6 does not crystallize significantly during the initial heating, resulting in a low crystallinity index value similar to those of the blends. In the case of samples that did not have sufficient time to crystallize, melt blending of PEAA retarded

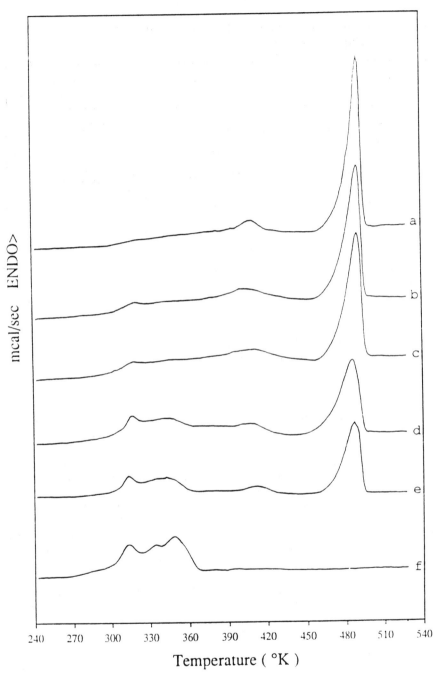

Figure 11.    Heating thermograms of nylon 6, poly(ethylene-co-acrylic acid) [PEAA] and their blend [NEAA] chips, annealed at 100°C for 13 days: (a) nylon 6, (b) NEAA 10%, (c) NEAA 20%, (d) NEAA 40%, (e) NEAA 50% and (f) PEAA.

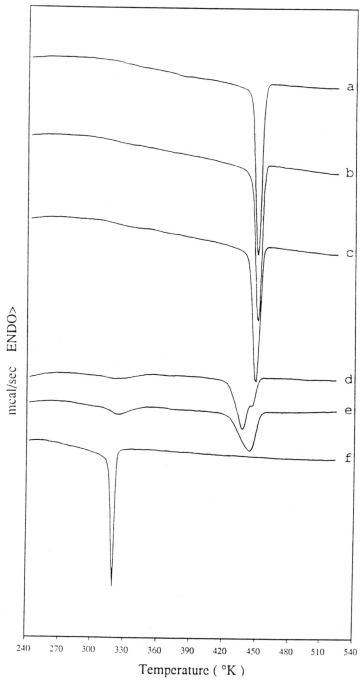

Figure 12.    Crystallization exotherms of nylon 6, poly(ethylene-co-acrylic acid) [PEAA] and their blend [NEAA] chips, annealed at 100°C for 13 days, after heating to 530°K at 20°C/min: (a) nylon 6, (b) NEAA 10%, (c) NEAA 20%, (d) NEAA 40%, (e) NEAA 50% and (f) PEAA.

the crystallization of nylon 6 in the blends, resulting in a lower crystallinity index for the blends compared with nylon 6. The crystallinity index of pure nylon 6 was similar to those of the blends when the samples were annealed at 100°C for 13 days. When the crystallization of nylon 6 in NEAA blends has been allowed to proceed to equilibrium, the interaction and/or reaction of nylon 6 with PEAA does not affect the ultimate crystallinity.

The development of a lower crystallization peak in NEAA 40% (Figure 12) suggests that the annealing of the blends may bring about further reaction among the two components. In an attempt to confirm that reaction occurs between the two components at elevated temperatures, the annealed NEAA 40% sample was subjected to further cycles of heating and cooling (Figure 13). Further decrease in the nylon 6 crystallization exotherm occurs with each cycle (438, 437 and 435K for the first, second and third cooling, respectively). The original nylon 6 crystallization shoulder at 446K is resolved as a separate peak at 448 and 44K in the second and third cooling cycles, respectively.

### NEAA 50% Film Annealed at 530K

When the water quenched NEAA 50% films annealed at 530K for varying amounts of time are cooled at 20°K/min, the sample annealed for 35 minutes displayed an exotherm at 438K with a shoulder at 445K upon cooling (Figure 14). This behavior was similar to that observed in the NEAA 40% chips annealed at 373°K for 13 days. The difference in the crystallization behavior of these blends may possibly be due to reaction of PEAA with nylon 6 that retards the nucleation of nylon 6 and thus lowers the crystallization temperature. A reaction that results in the formation of small amounts of graft or block copolymers appears to be a possible explanation for the observed change in the crystallization behavior. The presence of small amounts of such copolymers would not have a significant effect on the heats of melting but would decrease the average size of the nylon 6 crystals in as much as they would restrict the accessibility of the chain to the growing crystal.

When the samples, annealed at 530K for varying amounts of time and subsequently cooled to 230K at 20°C/min, are heated at 20°C/min, the melting temperature of nylon 6 decreases to 481K in the sample crystallized after annealing for 35 minutes. The onset temperatures varies from 476 to 461K, for the NEAA 50% annealed at 530K for 1 to 35 minutes reflecting the widening of the melting peak. The nylon 6 and PEAA in the blends appear to undergo further reaction to result in small amounts of block and/or graft copolymers, especially after prolonged exposure to high temperatures.

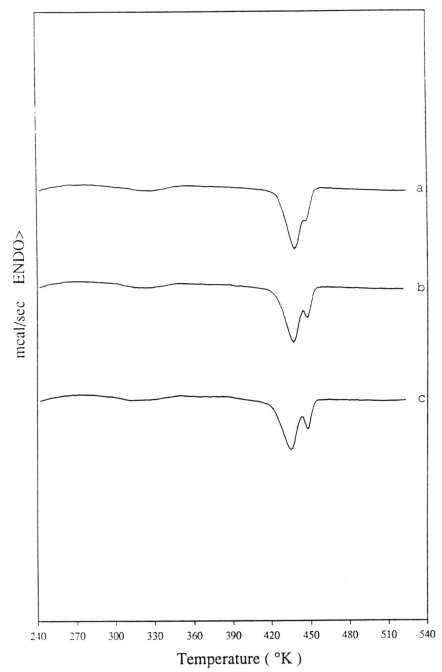

Figure 13.    Crystallization exotherms of nylon 6/poly(ethylene-co-acrylic acid), 40% blend chips, annealed at 373K for 13 days: (a) first cooling after heating to 530K at 20°C/min, (b) second cooling after heating, cooling and reheating at 20°C/min and (c) third cooling after heating sample.

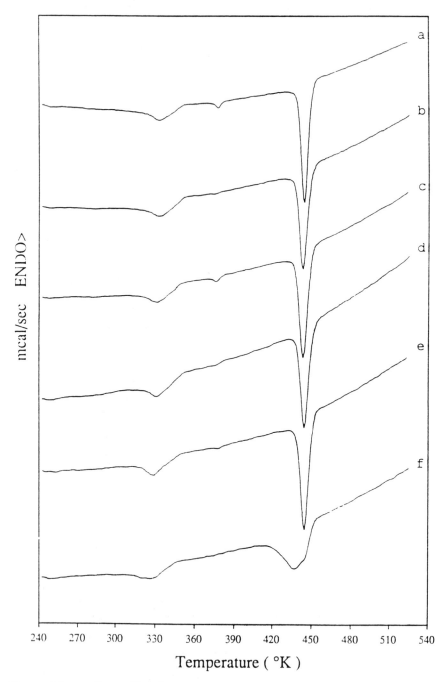

Figure 14.    Crystallization exotherms of melt cast and water quenched nylon 6/poly(ethylene-co-acrylic acid) 50% blend film annealed at 530K for: (a) 1 minute,   (b) recool of sample (a), (c) 3 minutes, (d) 5 minutes, (e) 10 minutes and (f) 35 minutes.

Table 4. Crystallinity Indices ($\chi_{N6}$) of Nylon 6 in NEAA Blends

| Sample | | $\chi_{N6}$ [a] | |
|---|---|---|---|
| | | Heat | Reheat |
| Liquid Nitrogen Quenched Films | Nylon 6 | 0.66 | |
| | NEAA 10% | 0.68 | |
| | NEAA 20% | 0.69 | |
| | NEAA 40% | 0.70 | |
| | NEAA 50% | 0.68 | |
| Water Quenched Films | Nylon 6 | 0.75 | 0.90 |
| | NEAA 10% | 0.62 | 0.76 |
| | NEAA 20% | 0.65 | 0.81 |
| | NEAA 40% | 0.62 | 0.77 |
| | NEAA 50% | 0.69 | 0.78 |
| Chips | Nylon 6 | 0.72 | 0.92 |
| | NEAA 10% | 0.74 | 0.74 |
| | NEAA 20% | 0.75 | 0.79 |
| | NEAA 40% | 0.77 | 0.77 |
| | NEAA 50% | 0.68 | 0.76 |
| Annealed Chips | Nylon 6 | 0.74 | 0.88 |
| | NEAA 10% | 0.75 | 0.87 |
| | NEAA 20% | 0.81 | 0.90 |
| | NEAA 40% | 0.82 | 0.85 |
| | NEAA 50% | 0.77 | 0.89 |

a   $\chi_{N6}$ (crystallinity index) $= \Delta H_{N6}/22.5$;
$\Delta H_{N6} = \Delta H_m/(1-x/100)$, x: % PEAA,


## Samples from Fractionation of NEAA

PEAA has approximately 1 acrylic acid monomer per 11 ethylene monomer units, so there are enough carboxylic acid groups for reaction with nylon 6.  Although the reaction may not be extensive, the mobility of nylon 6 in the blends will be restricted, resulting in retardation of crystallization.  Thermal analysis of fractions obtained by extracting nylon 6 with formic acid and PEAA with o-xylene suggested that reaction between nylon 6 and PEAA occurs and continues on subsequent heat treatments.

In the case of NEAA 10%, the residue of extraction exhibits multiple endotherms at 337 and 349K corresponding to PEAA and a small peak at 427°K in the initial heating (Figure 15). The endotherm at 427K is very broad and its area is 2.29 cal/g.

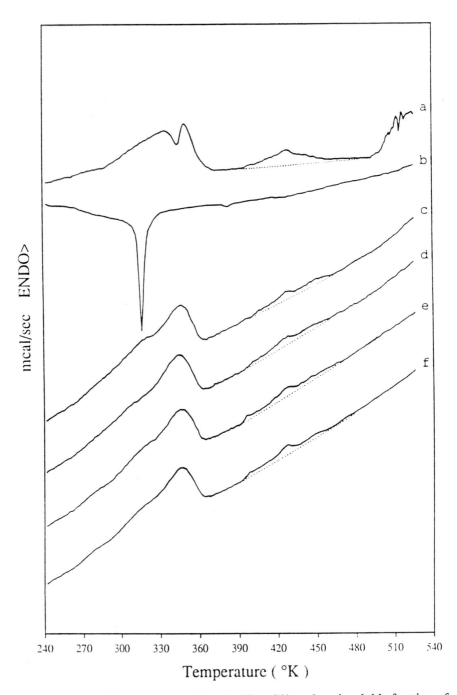

Figure 15.    Thermograms of the formic acid/o-xylene insoluble fraction of
NEAA 10% chips: (a) first heating, (b) cooling, (c) second heating,
(d) third heating, (e) heating after annealing at 385°K for 15 minutes
and (f) heating after annealing at 385K for 2 hours.

There appears to be a reaction occurring at ~497K as can be seen from the onset of the jagged peak, which disappears on remelting. On cooling, the crystallization exotherm corresponding to the endothermic peak at 427K is very weak and appears at ~383K. Upon reheating, the endothermic peak at 427K is broadened even more and the area is decreased to 1.13 cal/g with a second peak appearing at 451K. It appears possible that this endotherm may represent the presence of nylon 6/PEAA copolymers, which further react with PEAA to bring about a decrease in the peak area along with a decrease in the area of the PEAA endothermic peak on subsequent annealing and rerun. No further changes are observed after additional heating and cooling cycles.

The formic acid and o-xylene insoluble fraction of NEAA 20% also consists of nylon 6 and PEAA (Figure 16). In the first scan a broad nylon 6 peak at 488K is seen with a separate peak at 439K, which shifts down to 433K on the second heating with a new peak occurring at ~461K. The maximum of the nylon 6 melting peak shifts to 491K but the fractional area below 488K has increased from 0.58 to 0.65. There is also a significant drop in the area of the PEAA peak from 3.54 cal/g in the first run down to 0.38 cal/g in the second run. The PEAA appears to react with nylon 6 to produce fewer PEAA crystals and more lower melting nylon 6 crystals.

The formic acid and o-xylene insoluble fraction of NEAA 40% shows the respective melting endotherms of PEAA and nylon 6 at 345 and 492K (Figure 17). The broad crystallization peak of the nylon 6 component occurs at 442K and that of the PEAA component at approximately 310K. Upon reheating, the area of the lower melting endotherm decreases significantly from 5.71 cal/g for the initial run to 1.16 cal/g. The higher endotherm corresponding to nylon 6 occurs at~7°C lower than that of the first scan and a notable extension of the endothermic peak to lower temperatures can be seen. It appears that PEAA further reacts with nylon 6 to form copolymers or crosslinked polymers to restrict the mobility of the polymer chains.

The extraction residue of NEAA 50% exhibits an endotherm at 485K, about 5 degrees below the usual melting temperature of nylon 6 and the crystallization exotherm is at 429K, approximately 23°C lower than the crystallization temperature of nylon 6 (Figure 18). The position of the higher endotherm further decreases to 482, 480 and 479°K on the second, third and fourth heating cycles. When this fraction was annealed at 430 and 330K (the original peak crystallization temperatures) for 25 minutes each, the endotherm at ~480K corresponding to nylon 6 crystals disappeared completely and a small broad endotherm around 443K was observed. Further annealing at 430K for 12 hours did not result in a significant change in the peak location but the peak area was further reduced. It appears that reaction between the two components resulted in drastic changes in the size and consequently the melting temperature of the nylon 6 crystals.

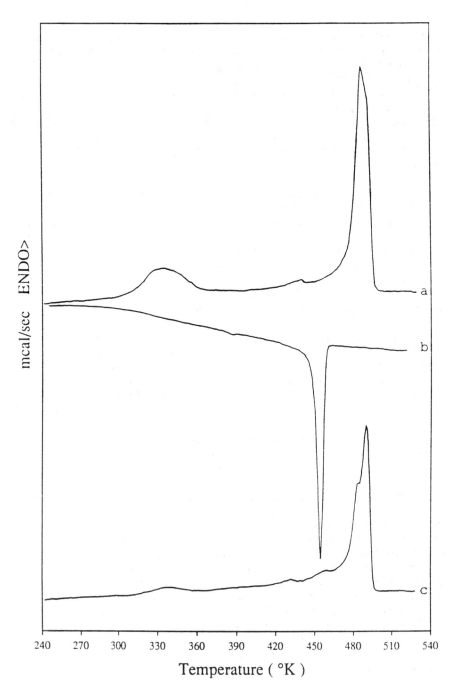

Figure 16.    Thermograms of the formic acid/o-xylene insoluble fraction of
NEAA 20% chips: (a) first heating, (b) cooling after heating and
(c) reheating after heating and cooling.

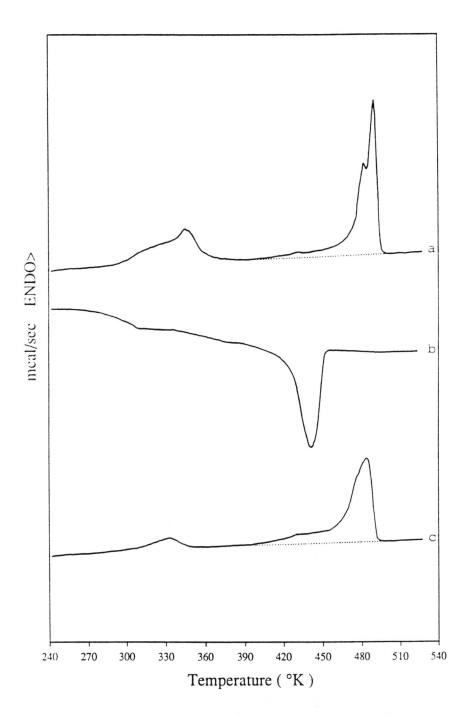

Figure 17.    Thermograms of the formic acid/o-xylene insoluble fraction of NEAA 40% chips: (a) first heating, (b) cooling after heating and (c) reheating after heating and cooling.

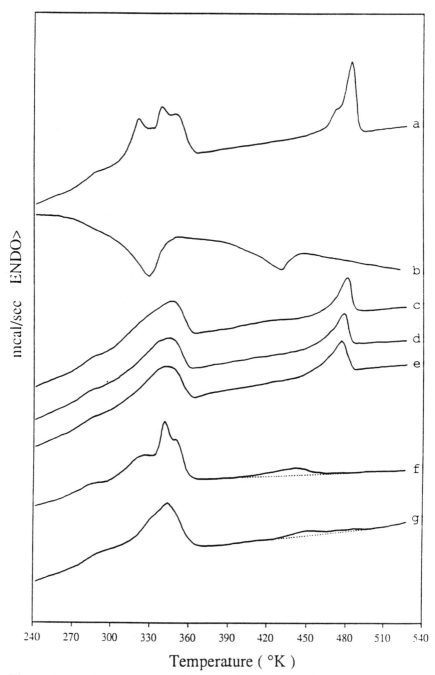

Figure 18.    Thermograms of the formic acid/o-xylene insoluble fraction of NEAA 50% chips: (a) first heating, (b) cooling, (c) second heating, (d) third heating, (e) fourth heating after rapid quenching of (d) in the DSC, (f) heating after annealing at 430 and 330K for 25 minutes each and (g) heating of (f) after annealing at 430K for 12 hrs.

## SUMMARY

Nylon 6 and poly(ethylene-co-acrylic acid) crystallize separately to form the respective crystals of nylon 6 and poly(ethylene-co-acrylic acid). But the noncrystalline polymer chains appear to be intimately mixed, as evidenced by retarded crystallization. The crystallization behavior of nylon 6 in the blend was dependent on the thermal history of the sample. When crystallization was carried out at elevated temperatures ($< T_m$), no difference in ultimate degree of crystallization of the nylon 6 component was present between the nylon 6 in the blends and pure nylon 6. Thermal analysis of the fractionation samples suggest that chemical reaction between nylon 6 and poly(ethylene-co-acrylic acid) results in small amounts of graft or block copolymers, and the reaction appears to continue on subsequent heat treatment.

## REFERENCES

1.    O. Olabisi, L.M. Robeson, and M.T. Shaw, "Polymer-Polymer Miscibility", Academic Press, New York, 1979.
2.    D.R. Paul, in "Polymer Blends", Vol. 1, D.R. Paul and S. Newman eds., Academic Press, New York, 1978, p. 2.
3.    G. M. Venkatesh, R. E. Fornes, and R. D. Gilbert, *J. Appl. Polym. Sci., 28,* 2247 (1983).
4.    G. M. Venkatesh, R. D. Gilbert and R. E. Fornes, *Polymer, 26,* 45 (1985).
5.    S. W. Shalaby, S. Sifniades, and D. Sheehan, *J. Polym. Sci., Polym. Chem. Ed., 14,* 2675 (1976).
6.    S. W. Shalaby, S. Sifniades, and D. Sheehan, *Amer. Chem. Soc., Div. Polym. Chem., Polym. Prepr., 16(1),* 688 (1975).
7.    K. Dimov and J. Georgiev, *Faserforsch. Textiltech., 24(3),* 120 (1973).
8.    K. Dimov and J. Georgiev, Angew. *Makromol. Chem., 39(1),* 21 (1974).
9.    K. Dimov and J. Georgiev, Faserforsch. *Textiltech., 26(10),* 479 (1975).
10.   K. Dimov, Ya. Georgiev, R. Garvanska, and M. Savov, *Fibre Chem., 8(3),* 261 (1976).
11.   F. Ide and A. Hasegawa, *J. Appl. Polym. Sci., 18,* 963 (1974).
12.   F. Ide, T. Kodama, and A. Hasegawa, *Kobunshi Kagaku, 29(4),* 259 (1972).
13.   F. Ide, T. Kodama, and A. Hasegawa, *Kobunshi Kagaku, 29(4),* 265 (1972).
14.   F. Ide. I. Sasaki, *Kobunshi Kagaku, 30(10),* 641 (1973).
15.   M. Matzner, D. L. Schober, R. N. Johnson, L. M. Robeson, and J. E. McGrath, in "Permeability of Plastic Films and Coatings", H. B. Hopfenberg ed., Plenum Press, New York, 1975,p. 125.
16.   G. Illing, in "Polymer Blends", E. Martuscelli, R. Palumbo, and M. Kryszewski eds., Plenum Press, New York, 1980.
17.   G. Illing, Angew. Makromol. Chem., 95, 83 (1981), p. 167.

18. E. C. Szamborski, *Org. Coat. Plast. Chem., 37(2)*, 17 (1977).
19. C. D. Han and H.-K. Chuang, *J. Appl. Polym. Sci., 30*, 2431 (1985).
20. H.-K. Chuang and C. D. Han, *J. Appl. Polym. Sci., 30*, 2457 (1985).
21. S. W. Seo, W. H. Jo, and W. S. Ha, *J. Appl. Polym. Sci., 29*, 567 (1984).
22. W. J. MacKnight, R. W. Lenz, P. V. Musto and R. J. Somani, *Polym. Eng. Sci., 25(18)*, 1124 (1985).
23. S. Cimmino, F. Coppola, L. D'Orazio, R. Greco, G. Maglio, M. Malinconico, C. Mancarella, E. Martuscelli, and G. Ragosta, *Polymer, 27*, 1874 (1986).
24. J. J. Maurer, D. J. Eustace, and C. T. Ratcliffe, *Macromolecules, 20*, 196 (1987).
25. V. P. Kabanov, V. A. Dubnitskaya, and S. N. Khar'kov, *Polym. Sci. U.S.S.R., 17*, 1848 (1975).
26. R. Greco, N. Lanzetta, G. Maglio, M. Malinconico, E. Martuscelli, R. Palumbo, G. Ragosta, and G. Scarinzi, *Polymer, 27*, 299 (1986).
27. D.R. Paul, in "Polymer Blends", Vol. 2, D.R. Paul and S. Newman eds., Academic Press, New York, 1978, p. 135.
28. K. J. Yoon, "Blends of Nylon 6 with Poly(ethylene-co-acrylic acid)", Ph.D. thesis, North Carolina State University, Raleigh, 1989.
29. N. Avramova and S. Fakirov, *Acta Polymerica, 32(6)*, 318 (1981).

# Chapter 9

# A STUDY OF THE MOLECULAR MOTIONS IN POLY(ETHYLENE TEREPHTHALATE) FILMS BY THE ESR SPIN PROBE TECHNIQUE

JOSEPH G. DOOLAN

*New Product Development Department*
*RHEOX, Inc.*
*Hightstown, New Jersey 08520*

## INTRODUCTION

In the study of macromolecular systems, the electron spin resonance (ESR) spin probe technique consists of incorporating a stable, paramagnetic molecule into a polymer matrix. The rotational motions of this molecule are then measured by ESR to produce an ESR spectrum that is extremely sensitive to the nature and rate of the motions that the molecule is undergoing. Any changes in the environment of the molecule that affect its rotational mobility will be reflected in its ESR spectra. Consequently, the molecule acts as a "probe" of the physical structure of the polymer and of the intensity of the molecular motions and relaxations occurring in it.

This chapter reports the use of this relatively new experimental technique to study the molecular motions occurring in poly(ethylene terephthalate), commonly known as PET or polyester. The results of a study involving six nitroxide free radical spin probe molecules, differing in size and structure, introduced into semi-crystalline PET films will be presented. ESR spectra for each probe were recorded over a wide temperature range, and the various regions of probe mobility and transition points have been correlated with the relaxation mechanisms detected by other experimental methods.

## EXPERIMENTAL

### Spin Probes

The six spin probes, along with their molecular weights and molar volumes calculated according to Slonimskii et al,.[1] are presented in Figure 1. The four smaller probes, SP-H, SP-O, SP-OH, and SP-NH$_2$, were commercially available in sufficiently high purity levels to be used as received. SP-I was synthesized from a procedure outlined in Japanese Patent 7,100,104; details of this procedure are described elsewhere.[2] SP-II was supplied by Dr. Ralph McGregor, North Carolina State University; it was synthesized according to Komzolova et al.[3]

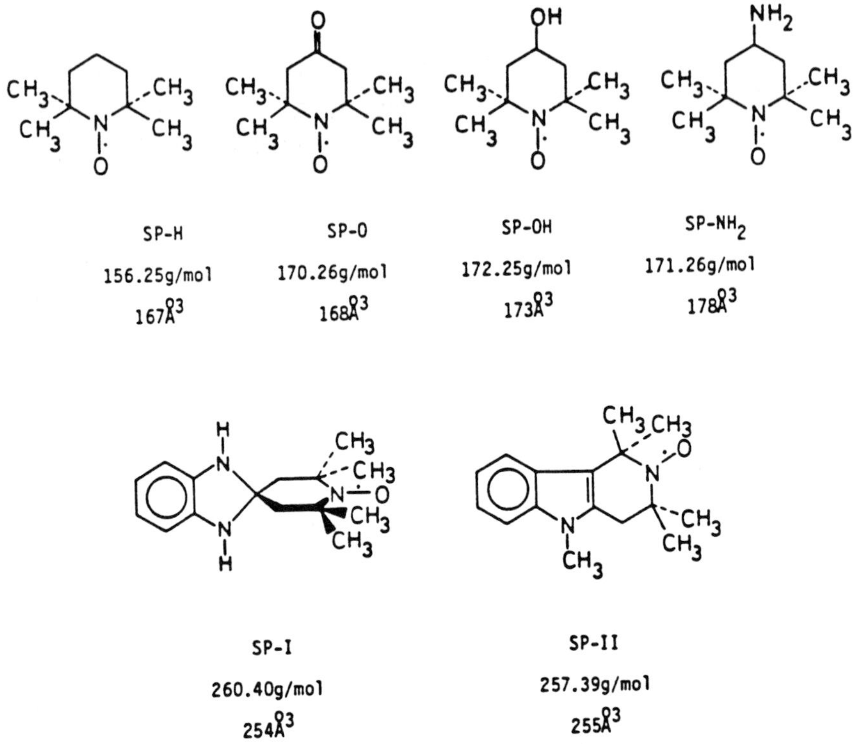

Figure 1. The spin probes.

## PET Films

The poly(ethylene terephthalate) films used in this study were kindly provided by the Toray Chemical Co., Ltd., Japan. These films contained no fillers or other foreign materials. The Tg of these films, as measured by DSC, was determined to be 82°C; the Tg measured by dynamic mechanical analysis at a constant frequency of 11 Hz was determined to be 111°C. All the spin probes were introduced into the PET films by sorption from aqueous solution at 75°C for 24 hours.

## ESR Measurements

The ESR spectrometer system used in this study was a JEOL JES-FE3X Spectrometer equipped with a JES-VT-3A2 variable temperature controller. Measurements were made in the X-band (9.1-9.5 GHz) at a modulation frequency of 100 kHz. Spectra were recorded in temperature increments of 10-20°C from -140 to +190°C; samples were allowed to equilibrate for a minimum of 7 minutes at each temperature before measurement. Spectra were recorded against a $Mn^{2+}/MgO$ standard for which an inner extremum separation of $86.9 \pm 0.1$ G had been established.

## RESULTS AND DISCUSSION

### ESR Spectra

The line shape of an ESR spectrum and the various changes that occur in a series of spectra with increasing temperature provide important information about the probe's rotational motion. The spectra of SP-I recorded at various temperatures are given in Figure 2. SP-NH$_2$ and SP-II have spectra very similar to SP-I, although the actual temperatures at which changes in the spectra occur differ for the three probes.

The spectra recorded at the highest temperatures (i.e., 170 and 190°C) are comparatively sharp and symmetrical and show the probe to be rotating in the rapid tumbling region. As the temperature is lowered, the rotational motion of the probe begins to slow down and the ESR spectrum becomes increasingly asymmetric, with the positions of the outer extrema increasing toward their rigid limit values.

In the temperature range fof 110-170°C, the spectra of SP-I appear to be superposition spectra where a rapid tumbling component is superimposed upon a slow tumbling spectrum. This is also true for SP-NH$_2$ and SP-II. The decision that the spectra of the SP-I group, in particular temperature ranges, are superposition spectra was made on the basis of the changes that occur in the low-field peak with increasing temperature. For SP-I at 110°C and above, a peak, at the position of the rapid tumbling low-field peak, begins to appear. As the temperature is increased, the area under this peak increases as the area under the slow tumbling peak decreases. At 140°C, a majority of the probes are in the rapid tumbling region until, at 170°C and above, all the probes appear to be rotating rapidly.

In contrast, the spectra of the SP-H, SP-O and SP-OH do not appear to be superimposition spectra at any temperature range. Also, the change from the slow to the rapid tumbling region occurs at considerably lower temperatures for these three smaller probes.

### Extremum Separation

$2A'_z$, the extremum separation as shown in Figure 2, gives a qualitative measure of the rotational mobility of a spin probe. With increasing temperature, the value of $2A'_z$ decreases, reflecting an increase in the rotational motion of the probe.

A plot of S (= $2A'_z/2A^R_z$) vs. T(K)/Tg(K) is shown in Figure 3; $2A^R_z$ is the rigid limit value of the extremum separation. Plotting S rather than $2A'_z$ essentially normalizes the data and allows a better comparison of the various probes. A Tg of 384K (111°C) was taken from the dynamic mechanical measurements. $T_{50G}$, the temperature at which the extremum separation is equal to 50 G, is also given for each probe in Figure 3.

Probe size clearly has an effect on the value of $2A'_z$ and the rotational mobility of the probes. The two larger probes, SP-I and SP-II, have much

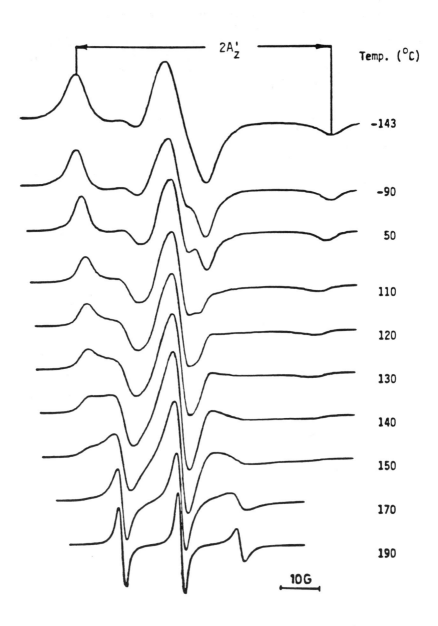

Figure 2. ESR spectra of SP-I in undrawn PET. showing a higher value of $T_{50G}$ than PET using other methods.

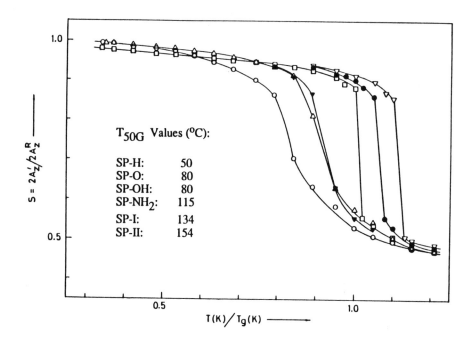

Figure 3.  $S = 2A'_z/2A^R_z$ vs. $T/T_g$ for all probes in the semi-crystalline PET films. $2A^R_z$ is the rigid limit value of the extremum separation; $T_g = 384$ K as determined by dynamic mechanical analysis. $T_{50G}$ is the temperature at which the extremum separation has a value of 50 G. o: SP-H; $\Delta$: SP-OH; ▼: SP-O; $\square$: SP-NH$_2$; ● : SP-I; and $\nabla$: SP-II.

higher $T_{50G}$ values than the smaller probes. Despite its size, SP-NH$_2$ also has a higher value of $T_{50G}$. Although the molar volumes of SP-H, SP-O and SP-OH are approximately equal at 167, 168 and 173nm$^3$, respectively, there is a difference of 30°C in the $T_{50G}$ values of SP-H compared to SP-O and SP-OH. These two latter probes essentially show the same behavior with increasing temperature and have equal $T_{50G}$ values.

SP-NH$_2$, SP-I and SP-II show a very sharp drop in the value of S over a very small temperature range. The other three probes, however, show a more gradual decrease in S with temperature. This is due to the fact that SP-NH2, SP-I, and SP-II exhibit superposition spectra. In the temperature range in which the superposition spectra appear, the value of $2A'_z$ is determined from the slow tumbling component until it is overtaken by the rapid tumbling component. It is interesting to note that, in the case of SP-NH2, the last temperature at which a slow tumbling component can be measured (i.e., 110°C) is essentially equal to the $T_g$ of 111°C as measured by the dynamic mechanical analysis of the undrawn PET film.

### Determination of the Model of Rotational Diffusion

The approach of Kuznetsov and Ebert[4] was used to determine the appropriate rotational diffusion model for each of the probes in the slow tumbling region.  In this approach, the change in position of the low field peak and high field minimum relative to their rigid limit positions is determined.  The change in position of the high field minimum is model-dependent allowing a determination of the model of rotational diffusion from a plot of $R = \Delta(-)/\Delta(+)$ vs. $\Delta(+)$, where $\Delta(-)$ and $\Delta(+)$ are the changes in the high and low field peaks in gauss, respectively.  These results are given in Figure 4.

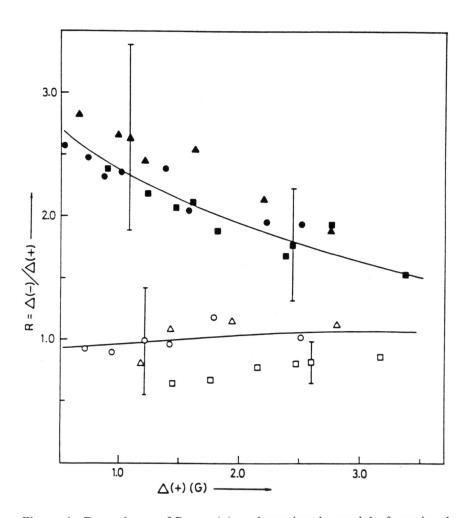

Figure 4.  Dependence of R on $\Delta(+)$ to determine the model of rotational diffusion.  The limits shown represent 95% confidence limits.  ▲: SP-II; ●: SP-I; ■: SP-NH$_2$; ○: SP-H; △: SP-O; and ▫: SP-OH.

The two curves in Figure 4 were taken from earlier work[4]; the upper curve was calculated for Brownian rotational diffusion, while the lower curve represents the strong jump model, also known as arbitrary jump tumbling. For probes SP-NH$_2$, SP-I and SP-II, all of the experimental R values are between 1.5 and 2.8 and closely correspond to the theoretical values for the Brownian diffusion model. For SP-H, SP-O, and SP-OH, R lies in the range between 0.6 and 1.2, approximately corresponding to the strong jump model. Similar deviations from the strong jump curve have been reported by Wasserman et al.[5]

Results reported in the literature have shown that probe size has a strong effect on the rotational model.[4-6] Larger probes tend to undergo Brownian rotational diffusion, while smaller probes are best described by the arbitrary or strong jump model. The present results are consistent with these conclusions. It should be noted, however, that SP-NH$_2$, whose size is relatively smaller than both SP-I and SP-II, is also undergoing Brownian diffusion.

The error limits shown in Figure 4 represent 95% confidence limits. Although the error reported in using this method can be quite large[7], these confidence limits also reflect the small sample size used in computing them.

**Frequency of Rotational Motion**

The rotational correlation time $\tau_R$ can be defined as the average time required for a reorientation of the probe molecule through an angle of one radian.[7] There is, however, no single encompassing theoretical treatment which allows for the calculation of $\tau_R$ over the entire range of motion of nitroxide probes in polymeric media. Consequently, different approaches are used if the probe is in the rapid tumbling region ($5 \times 10^{-11}$ s $< \tau_R < 10^{-9}$ s) or the slow tumbling region ($10^{-9}$ s $< \tau_R < 10^{-6}$ s). The approach used in the slow tumbling region is sensitive to the rotational diffusion mechanism of the probe molecule with three possible models described as Brownian rotational diffusion, free diffusion, and strong jump diffusion.[8] For a complete derivation of the equations used to calculate the rotational correlation times in these two different regions, the reader is referred to other work.[2, 7, 8]

A plot, in the typical Arrhenius fashion, of log $\tau_R$ against the reciprocal temperature for SP-II and SP-H is shown in Figure 5. The lines drawn in these plots represent a least-squares fit of the data points with a correlation coefficient of 0.97 or higher. For both probes, the three different models converge to a value of $\tau_R = 10^{-8}$ s; this behavior is typical of all six probes. In the case of SP-II, all three models predict a transition near 80°C; a second transition at 150°C is also predicted upon extrapolation of the rapid tumbling values up to the slow tumbling region. SP-NH$_2$ and SP-I give results similar to SP-II, i.e., a transition in the slow tumbling region followed

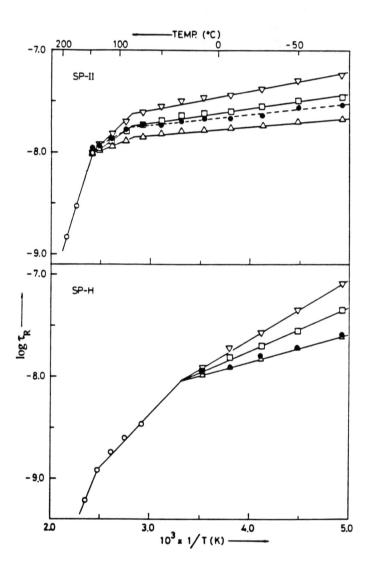

Figure 5.  Log $\tau_R$ vs. 1/T for SP-II and SP-H in the semicrystalline PET films.  From the method of Freed[8]:  ∇: Brownian; □: free; and △: strong jump diffusion;  ●: from the method of Kuznetsov[15]; ○: rapid tumbling region.[2, 7]

by a second transition at a higher temperature as the probe enters the rapid tumbling region. Transitions occurring at 30 and 130°C are predicted for SP-H. SP-O and SP-OH show similar behavior where, in this case, the low temperature transition represents the onset of rapid tumbling, followed by a second transition at 130-150°C.

The frequency of rotational motion f may also be calculated (f = $(2\pi\tau_R)^{-1}$) allowing a comparison of the ESR results with PET relaxation data determined from NMR, dielectric, dynamic mechanical and DSC measurements. Figure 6 shows an Arrhenius plot of the frequency of rotation for the six probes. As in the $2A'_z$ results, size appears to have a strong effect on the probe mobility; the two larger probes have comparatively lower frequency values than the smaller probes. The SP-O and SP-OH results are virtually superimposed on each another, and SP-NH$_2$ gives results very similar to the larger spin probes.

All the probes show two transition points: $T_L$ and $T_H$ transitions occurring at a low and high temperature, respectively. A summary of the transition points predicted by these plots along with the rotation diffusion models used to determine $\tau_R$ in the slow tumbling region and the $T_{50G}$ values are given in Table 1.

Table 1. Summary of transition points. $T_{50G}$ is the temperature at which the extremum separation is equal to 50G. $T_L$ and $T_H$ are the low and high temperature transitions, respectively, determined from the log f vs. reciprocol temperature plots.

| Probe | Diffusion Model | T50G (°C) | $T_L$ (°C) | $T_H$ (°C) |
|---|---|---|---|---|
| SP-H | Strong Jump | 50 | 30 | 130 |
| SP-O | Strong Jump | 80 | 61 | 150 |
| SP-OH | Strong Jump | 80 | 54 | 150 |
| SP-NH$_2$ | Brownian | 115 | 73 | 110 |
| SP-I | Brownian | 134 | 82 | 130 |
| SP-II | Brownian | 154 | 80 | 150 |

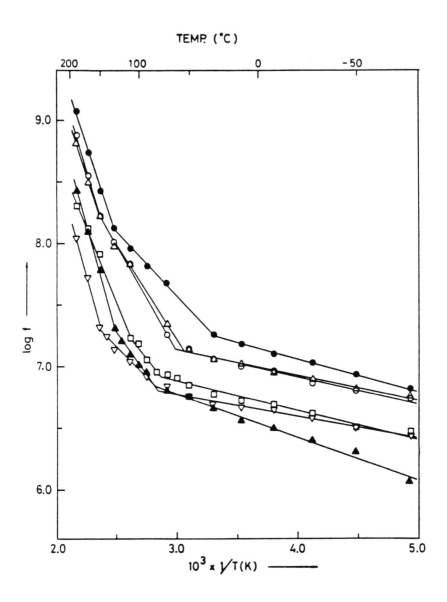

Figure 6. Log f vs. 1/T for all probes in the semicrystalline PET films.
●: SP-H; ○: SP-O; △ SP-OH; □: SP-NH2; ▲: SP-I; ▽: SP-II.

### Relaxation Map of PET

In the preceding section, each probe was shown to exhibit two transitions in the log f vs. reciprocal temperature plots. In view of the fact that the motion of a probe molecule will be affected by the motions of the polymer segments, a significant aid in the interpretation of these ESR results can be realized by constructing a relaxation map of PET. This map is a culmination of data from different experimental techniques performed at various temperatures and frequencies. For the case of undrawn, semicrystalline PET, this relaxation map is shown in Figure 7.

For all probes, the low temperature transition lies between 30-80°C with the high temperature transition occurring from 110-150°C. These break points in the ESR data lie very close to the point where the dielectric and dynamic mechanical data intersect the ESR results. It seems probable, therefore, that $T_L$ represents the probes' response to the $\gamma$ relaxation, while $T_H$ represents a response to the glass transition. In view of this, the data presented in the previous sections can now be explained.

Probe size clearly affects the rotational mobility of the six probes. The two larger probes, SP-I and SP-II, follow the Brownian diffusion model and have the highest $T_{50G}$ values. The superposition spectra of these two probes apparently arise from a combination of their size and the crystalline nature of the PET films. For SP-I and SP-II, the $T_{50G}$ and $T_H$ values of each probe are approximately equal. It is above these temperatures that the probes are in the rapid tumbling region. The fact that these probes respond to the predicted Tg of the polymer by undergoing rapid rotation is consistent with the results of Kumler et al.[9]. This group has shown that larger probe molecules have a tendency to respond to the glass transition, while smaller probes respond to relaxations in the polymer occurring at lower temperatures.

Although the molar volumes and molecular weights of the probes are almost equal, the value of $T_{50G}$ for SP-I is equal to 134°C, while that of SP-II is 154°C, i.e., the motion of SP-II remains effectively restricted until higher temperatures. A possible explanation for this difference is the structural configurations of the two probe molecules. SP-II, being a relatively planar molecule, can diffuse into those areas of the noncrystalline regions where the polymer chains have a more restricted motion. Being a nonplanar molecule, SP-I is prevented from doing this, which is reflected in its lower $T_{50G}$ and $T_H$ values. The structural configuration of SP-I also appears to prevent this probe from interacting with the carbonyl groups of the polymer through hydrogen bonds.

The rotational motions of the smaller probes, SP-H, SP-O and SP-OH, are quite different from the results of the larger probes. All three probes follow the strong jump diffusion model, and their spectra do not appear to be superposition spectra. This is consistent with the work reported in the literature with these three probes.[6, 10-12] Due to the size of the probes, their $T_{50G}$ values are much lower than those of SP-I and SP-II.

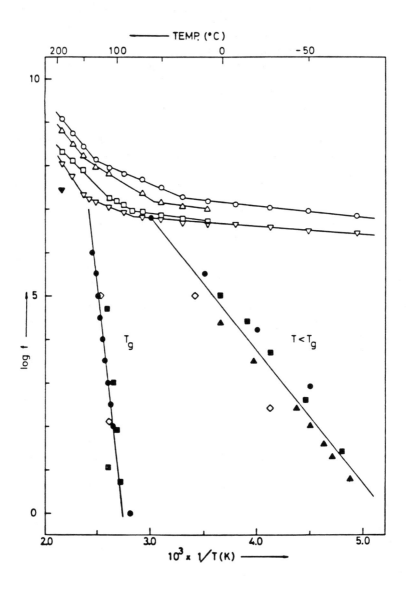

Figure 7. Relaxation map of semi-crystalline PET films. ESR: ○, SP-H; △, SP-OH; ☐, SP-NH2; ▽, SP-II. Dielectric: ●, Ref. 16; ■, Ref. 17; ▲, Ref. 18. Dynamic mechanical: ◇, Ref. 19. NMR: ▼: Ref. 20.

The main difference between these probes and the larger probes is their response to the $\gamma$ relaxation. Contrary to SP-I and SP-II, the $T_{50G}$ values do not represent the probes' response to the Tg of the polymer but instead reflect a response to the $\gamma$ relaxation. The molar volumes of the probe molecules are small enough that the local motions along the polymer chains provide the probes with enough excess volume to bring about the onset of rotation in the rapid tumbling region. Kovarskii et al.[13] have shown that, for the case of SP-H in polyethylene, the excess volume required to provide the freedom for probe reorientation is only about 8-10% of the probe volume. The current results are also consistent with the prediction of Kumler[9] that the $T_{50G}$ of smaller probes reflect their response to subglass transitions.

SP-NH$_2$ presents a very different case from the other three smaller probes. It undergoes Brownian rotational diffusion, it has a relatively higher $T_{50G}$ value (115C compared to 50-80C for the other probes of comparable size), and its spectra are superposition spectra. It effectively acts like a larger probe molecule. The reason for this is that SP-NH$_2$ can hydrogen-bond with the carbonyl groups of the PET chain segments. This same result has been reported in the literature.[14] The fact that the amino group has two hydrogens allows the probe to interact with the polymer in such a way that the rest of the probe molecule does not disrupt the interaction.

## CONCLUSIONS

The ESR spin probe technique has been applied to the study of molecular motions in semicrystalline poly(ethylene terephthalate) films. Probe size and ability to interact with the polymer through hydrogen bonds were clearly shown to have an effect on the spectral line shapes, rotational diffusion mechanism and the various transition points predicted by the ESR data. A relaxation map of PET was constructed from literature data of various experimental techniques conducted at different frequencies. Comparing the ESR results to this relaxation map showed that the six probes used in this study all responded to the glass transition and a subglass transition at lower temperatures.

## ACKNOWLEDGMENTS

The author would like to express his deep appreciation to Drs. Richard Gilbert and Ralph McGregor, the cochairmen of his Ph.D thesis advisory committee at North Carolina State University. Appreciation is also expressed to the other members of that committee: Drs. Suzanne Purrington, Vivian Stannett, and Raymond Fornes.

The author would also like to express his sincere gratitude to Dr. Toshiro Iijima of the Department of Polymer Science at the Tokyo Institute of Technology, Tokyo, Japan. A large amount of the experimental work of this chapter was completed in his laboratories with the help of Drs. K. Hamada and T. Seo. Also acknowledged is the financial support of the Fulbright Program of the Japan-United States Educational Commission.

## REFERENCES

1. G. L. Slonimskii, A. A. Askadskii, and A. I. Kitaigorodskii, *Vysokomol. Soyed, Vol. A12*, No. 3, pp. 494-512, 1970.
2. J. G. Doolan, A Study of the Molecular Motions in Poly(ethylene terephthalate) Films by the Electron Spin Resonance Spin Probe Technique, Ph.D. thesis, North Carolina State University, Raleigh, North Carolina, 1987.
3. N. N. Komzolova, N. F. Kucherova, and V. A. Zagorevskii, *Zhurnal Obshchei Khimii, Vol. 34, No. 7*, pp. 2383-2387, 1964. See also: E. G. Rozantsev, A. B. Shapiro, and N. N. Komzolova, *Izvestiya Akademii Nauk SSSR, Seriya Khimicheskaya, No. 6*, pp. 1100-1102, 1965.
4. A. N. Kuznetsov and B. Ebert, *Chem. Phys. Letters, Vol. 25, No. 3*, pp. 342-345, 1974.
5. A. M. Wasserman, T. A. Alexandrova, A. L. Buchachenko, *Eur. Polym. J., Vol. 12*, pp. 691-695, 1976.
6. P. Meurisse, C. Friedrich, M. Dvolaitzky, F. Lauprete, C. Noel, and L. Monnerie, *Macromolecules, Vol. 17*, pp. 72-83, 1984.
7. R. McGregor, T. Iijima, T. Sakai, R. D. Gilbert, and K. Hamada, *J. of Membrane Sci., Vol. 18*, pp. 129-160, 1984.
8. J. H. Freed, Theory of Slow Tumbling ESR Spectra for Nitroxides, Chap. 3 in *Spin Labeling: Theory and Applications*, L. J. Berliner, ed., Academic Press, New York, 1976.
9. P. L. Kumler, ESR Spin-Probe Studies of Polymer Transitions, in *Molecular Motion in Polymers by ESR*, R. F. Boyer and S. E. Keinath, eds., Harwood Academic Publishers, London, 1980.
10. C. Noel, F. Laupretre, C. Friedrich, C. Leonard, J. L. Halary, and L. Monnerie, *Macromolecules, Vol. 19*, pp. 201-210, 1986.
11. I. I. Barashkova, A. L. Kovarskii, and A. M. Vasserman, *Vysokomol. soyed., Vol. A24, No. 1*, pp. 91-95, 1982.
12. A. L. Kovarskii, A. M. Vasserman, and A. L. Buchachenko, *Vysokomol. soyed., Vol. A13, No. 7*, pp. 1647-1653, 1971.
13. A. L. Kovarskii, A. M. Wasserman, and A. L. Buchachenko, Spin Probe Studies in Polymer Solids, in *Molecular Motion in Polymers by ESR*, R. F. Boyer and S. E. Keinath, ed., Harwood Academic Publishers, London, 1980.
14. Z. Veksli and W. G. Miller, *Macromolecules, Vol. 10, No. 6*, pp. 1245-1250, 1977.
15. A. N. Kuznetsov, A. M. Wasserman, A. U. Volkov, and N. N. Korst, *Chem. Phys. Letters, Vol. 12, No. 1*, pp. 103-106, 1971.
16. W. Reddish, *Trans. Faraday Soc., Vol. 46*, pp. 459-475, 1950.
17. Y. Ishida, K. Yamafuji, H. Ito, and M. Takayanagi, *Kolloid-Z, u. Z. Polymere, Vol. 184*, pp. 97-108, 1962.
18. E. Ito and Y. Kobayashi, *J. Appl. Polym. Sci., Vol. 25*, pp. 2145-2157, 1980.
19. J. Bateman, R. E. Richards, G. Farrow, and I. M. Ward, *Polymer, Vol. 1*, pp. 63-71, 1960.
20. G. P. Mikhalov and V. A. Shevelen, *Vysokomol. soyed., Vol. 8*, pp. 763-768, 1966.

# Chapter 10

# PARAMETRIC DEPENDENCE OF $^{13}$C SOLID STATE MAS NMR SPLITTING DUE TO DIPOLAR INTERACTION WITH A $^{14}$N NUCLEUS

J. D. MEMORY

*University of North Carolina*
*General Administration,*
*Chapel Hill, North Carolina 27515*

## INTRODUCTION

Splitting of $^{13}$C NMR lines from solid samples undergoing magic angle spinning (MAS) due to a dipolar interaction with a $^{14}$N nucleus has been observed in many samples.[1-6] The effect is well understood: the quadrupole interaction of $^{14}$N is strong enough so that the effect on the $^{13}$C due to its dipolar interaction with $^{14}$N is not averaged by the magic angle spinning. The theory is developed thoroughly in a paper by Hexem, Frey and Opella.[6]

It is the purpose of this note to generalize a comment in an earlier work.[6] It was shown in section IIIF of that paper that: if the $^{14}$N quadrupole interaction may be appropriately treated as a perturbation on the $^{14}$N Zeeman interaction, and (2) if the electric field gradient is axially symmetric ($\eta=0$), then the $^{13}$C splitting is proportional to the $^{14}$N quadrupole coupling constant, inversely proportional to $RCN^3$, and inversely proportional to the external magnetic field. It will be shown here that only hypothesis (1) is needed for the conclusion.

## THEORETICAL TREATMENT

In the perturbation treatment, the perturbed $^{14}$N spin states have coefficients

$$C_{nm} = HQ_{nm}/(E_m^{(0)}-E_n^{(0)}), n \neq m \tag{1}$$

and $C_{nn} = 1$, in the notation of that paper. This leads, for the axially symmetric case, to the following expressions for the $^{13}C$ carbon shifts for the three nitrogen spin states:

$$\omega^C_{1,3} = -(9/8)\,(\omega_D\omega_Q/\omega_N)\sin2\beta^D\sin2\beta^Q \tag{2}$$

$$\omega^C_2 = (9/4)\,(\omega_D\omega_Q/\omega_N)\sin2\beta^D\sin2\beta^Q \tag{3}$$

where

$$\omega_D = \gamma_C\gamma_N\hbar/R_{CN}^3,\ \ \omega_Q = e^2Qq/2S(2S-1)\hbar,$$

$$\omega_N = \gamma_N B \tag{4}$$

$\beta^D$ is the angle between the rotor axis and the internuclear vector, and $\beta^Q$ is the angle between the rotor axis and the principal axis of the electric field gradient.

If one allows $\eta$ to be different from zero in equations (2) and (4) of the work of Hexem and coworkers[6], substitutes in equation (14) of that paper for the $^{13}C$ shifts (using the appropriate expressions for the Wigner rotation matrices), and uses the Hexem treatment (neglecting second order products and averaging the first term in equation (14) to zero), one obtains

$$\omega^C_{1,3} = -(9/8)(\omega_D\omega_Q/\omega_N)\sin2\beta^D\sin2\beta^Q(1-6^{-1/2}\eta\cos\gamma^Q) \tag{5}$$

and

$$\omega^C_2 = (9/4)\,(\omega_D\omega_Q/\omega_N)\sin2\beta^D\sin2\beta^Q(1-6^{-1/2}\eta\cos\gamma^Q) \tag{6}$$

where $\gamma^Q$ is the third Euler angle orienting the transverse axis of the electric field gradient system. Clearly, when $\eta = 0$, equations (5) and (6) reduce to equations (2) and (3).

Of principal interest is the fact that the splitting of the asymmetric doublet is proportional to the quadrupole coupling constant, inversely proportional to $R_{CN}^3$, and inversely proportional to the magnetic field, even if the field gradient is not axially symmetric. The dependence on the magnetic field is of particular importance, since the inverse proportionality of the splitting (in Hz, not ppm), is a signature for this type of interaction. This is also consistent with the accurate, but perhaps misleading, Figure 5 of reference 6. In that figure it is shown that the logarithm of the splitting plotted versus the logarithm of $B^{-2}$ is a straight line for a wide range of numerically calculated values and experimental results. Clearly, if $\delta = k/B$ so that $\log \delta = \log k - \log B$, then $\log \delta$ plotted versus $\log (B^{-2}) = -2 \log B$ will be a straight line.

One final observation: in the case of axial symmetry and with the internuclear vector collinear with the principal axis of the electric field gradient, one can carry out in closed form the average of $\sin 2\beta^D \sin 2\beta^Q$ over all orientations in a powder sample, and over azimuthal angle to account for magic angle spinning. Starting with equation (18) of reference 6, a certain amount of trigonometric manipulation leads in a straightforward fashion to

$$<\sin 2\beta^D \sin 2\beta^Q> = 8/15 = 0.5333. \tag{7}$$

This checks with the numbers given in Figure 2 by Hexem and coworkers.[6]

These methods have been used, for example, for the studies of protein structure, as extensively reviewed by Opella, Stewart and Valentine.[7]

## REFERENCES

1. C.G. Moreland, E.O. Stejskal, S.C.J. Sumner, J.D. Memory, F.I. Carroll, G.A. Brine, and P.S. Portoghese, *J. Magn. Reson. 83*, 173 (1989)

2. M. Alla, E. Kundla, and E. Lipmaa, *JEPT Lett. 27*, 208 (1978).

3. S.J. Opella, M.H. Frey, and T.A. Cross, J. Am.Chem Soc 101, 5856 (1980).

4. J.G. Hexem, M.H. Frey, and S.J. Opella, *J. Am.Chem. Soc. 103*, 224 (1981)

5. S.J. Opella, J.G. Hexem, M.H. Frey, and T.A.Cross, *Philos. Trans. Roy. Soc. London Ser. A.299*, 665 1981).

6. J.G. Hexem, M.H. Frey, and S.J. Opella, *J. Chem. Phys. 77*, 3847 (1982)

7. S.J. Opella, P.L. Stewart, and K.J.Valentine, *Quarterly Reviews of Biophysics, 19*: 7-49 (1987)

# Chapter 11

# SOLID STATE $^{13}$C CPMAS-NMR STUDY OF AN A/B BLOCK COPOLYMER OF POLYSTRYENE (PS) AND POLY(METHYL METHACRYLATE) (PMMA)

S. S. SANKAR, E. O. STEJSKAL AND R. E. FORNES

*Departments of Chemistry and Physics*
*North Carolina State University,*
*Raleigh, NC   27695*

and

W. W. FLEMING, T. P. RUSSELL AND C. G. WADE

*IBM Almaden Research Center,*
*San Jose, CA   95120*

## INTRODUCTION

In this study we have used $^{13}$C-detected $T_{1\rho}(^1H)$ to examine symmetric diblock phase separation in copolymers of polystyrene (PS) and poly(methyl methacrylate) (PMMA). In a homogeneously dispersed system, spin diffusion between protons causes all the protons to relax at about the same rate. However, in a phase separated system, if the domains are large enough, it may be possible to observe different proton relaxation rates in the different regions of the sample. This is most easily done by measuring the proton relaxation via the carbon system, if carbon resonances can be found that may be associated with one block, and not the other. The particular PS/PMMA block copolymers studied have sufficiently high molecular weight that phase separation occurs during the initial preparation of the specimen. However, annealing markedly enhances the phase separation as well as changes the rate of chain motion. Similar effects can be induced by dispersing high surface area glass fibers in the polymer. These effects are detected by changes in both $T_{1\rho}(^1H)$ and $T_{1\rho}(^{13}C)$, which are not subject to averaging by spin diffusion and are thus more directly influenced by molecular motion.

### CP/MAS applications to polymers

The combination of cross-polarization (CP), high power proton decoupling and magic-angle sample spinning (MAS) makes it possible to obtain liquid like, high resolution, rare spin (usually $^{13}$C, but sometimes $^{15}$N,

$^{29}$Si, $^{31}$P, et al.) NMR spectra in organic solids.[1] This technique is particularly well suited to the study of polymers and other complex solids. It can help determine primary chemical structures, but can also yield information about secondary structures and has been used to characterize molecular motion.[2] Since NMR is nondestructive, it lends itself to the study of a sequence of changes, chemical or physical, all on the same specimen. CP/MAS NMR methods have been most often applied to systems that cannot be put into solution or melted without unacceptable changes. This includes systems without a convenient solvent that would have to be changed chemically to be made liquid. Crosslinked systems and biopolymers are two examples. Other systems, such as engineering polymers, may be soluble but are of interest only as solid materials. In these systems, it is often molecular motion that is most interesting. Phase separation in the solid state can also be studied[3-5] and is the principal concern of the work reported here.

### Characterization of Molecular Motion

The CP/MAS experiment requires the manipulation of both the abundant spin (usually $^1$H, but sometimes $^{19}$F) and the rare spin along the radiofrequency fields appropriate to each. These manipulations depend upon several rotating-frame relaxation times, most often $T_{1\rho}(^1H)$ and $T_{1\rho}(^{13}C)$, that are sensitive to the frequency and amplitude of the motion of the spin system being studied. Frequencies in the range from several to a few hundred kilo Hertz are the most important in determining these relaxation times; near room temperature, this frequency range reflects the motion of short segments of the polymer main chain, as well as side chain motion. Anything that alters these motions may also alter one of the rotating frame relaxation times. Suitable experiments exist for the independent determination of the several relaxation times.[2]

$T_{1\rho}(^1H)$ is different from $T_{1\rho}(^{13}C)$ in that $T_{1\rho}(^1H)$ is sensitive to the motion of the $^1$H system averaged over a short distance while $T_{1\rho}(^{13}C)$ relaxation reflects the motion of individual $^{13}$C spins.[3] Thus the $^1$H relaxation is sensitive to macroscopic variations in motion while the $^{13}$C relaxation is site specific. Together, these two relaxation times detect not only changes in molecular motion but also motional inhomogeneity. Even in fundamentally homogeneous systems, these two relaxation times still behave differently in that $T_{1\rho}(^1H)$ tends to be dominated by the most efficient site of relaxation even if it is a minority site. On the other hand, $T_{1\rho}(^{13}C)$ reports the most common relaxation rate and de-emphasizes minor sources of relaxation. Since each $T_{1\rho}(^{13}C)$ measured is associated with a particular carbon, it is totally insensitive to motion of any other chemically distinct species even if that species may dominate $T_{1\rho}(^1H)$.

When relaxation times are altered by changes in a system it may be due to either or both of two fundamentally different changes: either the rate of motion may change in relation to the characteristic frequency of the applied radiofrequency field, or the character of the motion (amplitude or type) may change. These different sources of change can be sorted out by studying relaxation as a function of the amplitude of the radiofrequency field and fitting the resultant dependence to a theoretical curve. There are also a number of line shape measurements that are affected by motion.

## Polystyrene/Poly(methyl methacrylate) A/B Block Copolymers

Studies of films of PS/PMMA block copolymers cast on glass or gold surfaces, by dynamic secondary ion mass spectroscopy[6,7] and neutron reflectivity[8-10] show considerable organization into alternating layers of polystyrene and poly(methyl methacrylate) parallel to the solid surface. When the films are annealed in situ, the layers become even more clearly defined, with a half layer of PMMA near the solid surface, a half-layer of PS at the air interface and full layers alternating in between. The thickness of the full layers are consistent values found in the bulk[11] and are essentially equal for the two components.

The initial objective of this research project was to study the effect of annealing on the physical state of a symmetric diblock copolymer of polystyrene and poly (methyl methacrylate ). The two fundamental tools we have to apply to this problem are: (1) [13]C-CP/MAS NMR, with its variety of relaxation times for characterizing molecular and spin dynamics, and (2) selective perdeuteration of either or both of the blocks, to aid in the identification and separation of the sources of relaxation. Further work has begun to address the influence of solid surfaces during the solidification and annealing of these systems.

## EXPERIMENTAL

The A/B block copolymer samples of polystyrene/poly(methyl methacrylate) or PS/PMMA (prepared by Polymer Laboratories, Ltd, U. K.) were of relatively narrow molecular weight distribution ($M_w$ = ~1.1 $M_n$) with each block MW ~ $5 \times 10^4$. Four deuteration patterns were prepared : both blocks protonated (PS/PMMA), PS only perdeuterated (PSD/PMMA), PMMA only perdeuterated (PS/PMMAD) and both blocks perdeuterated (PSD/PMMAD). PS and PMMA homopolymers of MW ~ $10^5$ were also studied. Originally, the polymers were studied in a quenched (most disordered) state obtained by freeze drying from benzene solution. To see the effects of annealing, the quenched specimens were annealed in an inert atmosphere for 24 hours at 150°C. At these molecular weights, the block copolymers are partially phase separated even in the quenched state. To see the effects of solidification near a solid surface, some samples were prepared from a solution mixed with short lengths of 10μm diameter glass fibers.

The CP/MAS [13]C-NMR spectrum of polystyrene/poly(methyl methacrylate) block copolymer (both blocks protonated) is shown in Figure 1, and the spectrum of the polymer with deuterated polystyrene in Figure 2. In these spectra, polystyrene lines are marked **s** and poly(methyl methacrylate) lines are marked **m**. The highest field (17 ppm, methyl carbon) and lowest field (177 ppm, carbonyl carbon) were due to the poly(methyl methacrylate)

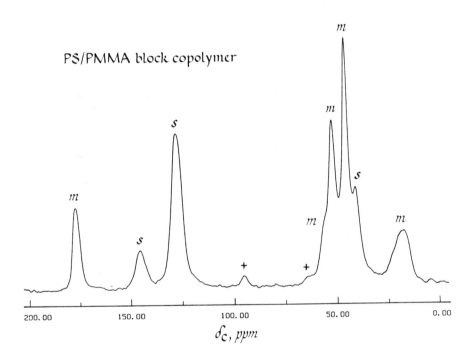

Figure 1.  50 MHz CP/MAS $^{13}$C NMR spectrum of polystyrene/poly(methyl methacrylate) block copolymer (PS/PMMA).  m, PMMA; s, PS; and +, spinning sideband.

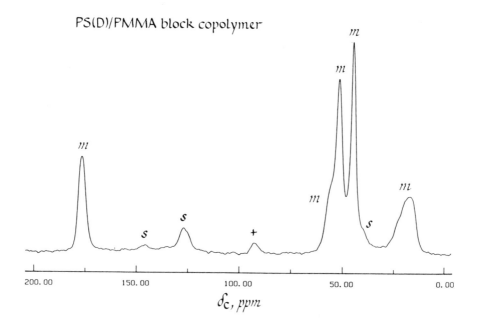

Figure 2. 50 MHz CP/MAS $^{13}$C NMR spectrum of perdeuteropolystyrene/ poly(methyl methacrylate) block copolymer (PSD/PMMA).  m, PMMA; s, PS; and +, spinning sideband.

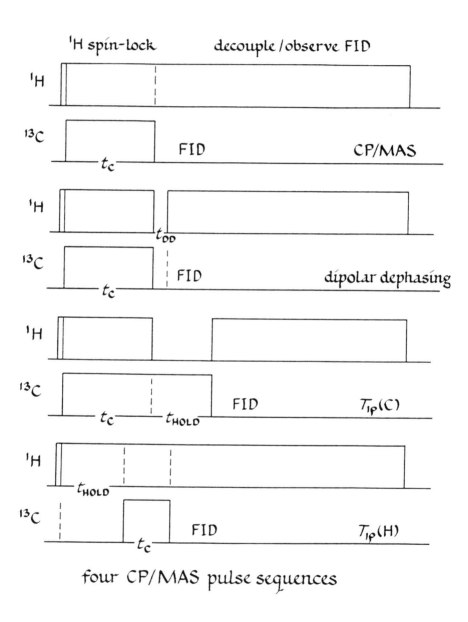

four CP/MAS pulse sequences

Figure 3.  Four cross-polarization/magic angle spinning pulse sequences: CP/MAS to obtain high resolution NMR spectra;  dipolar dephasing to suppress protonated carbons;  sequences to measure two different rotating frame relaxation times $T_{1\rho}(^{13}C)$ and $T_{1\rho}(^{1}H)$.

(PMMA) block. The protonated and nonprotonated aromatic polystyrene (PS) block signals resonate around 127 ppm and 145 ppm, respectively. The signals in the 40-60 ppm region were due to the remaining aliphatic carbons of both PMMA and PS blocks.

The solid state $^{13}C$ NMR spectra were obtained under matched Hartmann-Hahn conditions from 32 to 60 kHz (65 kHz in one instance) on a Chemagnetics CMC 200S NMR spectrometer equipped with a CP/MAS probe. The pulse sequences employed in this study are shown in Figure 3. Not shown is a pulse in either the proton or carbon channel, located clear of the data collection, that maintains equal rf heating when the length of the time variable is changed. To choose the appropriate contact time for the $T_{1\rho}(^1H)$ and $T_{1\rho}(^{13}C)$ measurements, a number of standard CP experiments were performed by varying the matched spin-lock contact times. The carbon signal intensity was plotted as a function of spin-lock contact times. The contact time, for which the signal intensities of the protonated blocks had either reached maximum or started to decay, was chosen for the $T_{1\rho}$ measurements. Two thousand (2000) transients (block-averaged in groups of 200) were collected for each $\tau$ value over a spectral width of 15 kHz using 2K data points, and with a spinning rate of approximately 4.3 kHz. The $T_{1\rho}$'s were initially evaluated from the slope of the semilog plots of carbon intensity as a function of $\tau$ value. Later they were fitted to several model distributions of first order relaxation behavior by means of a PC-XT program, RELAX(ation), written by Robert Skarjune, 3M Corporation.

Table 1.  $T_{1\rho}(^{13}C)$, ms at 50 kHz

| | Line (ppm) | |
|---|---|---|
| Sample/Line | 17 ppm | 127 ppm |
| PS | - - - | 17.1 |
| PS* | - - - | 30.9 |
| PMMA | 22.0 | - - - |
| PMMA* | 16.4 | - - - |
| PS/PMMA | 17.4 | 25.4 |
| PS/PMMA* | 17.4 | 22.3 |
| PS/PMMA/Glass | 14.9 | 16.7 |
| PS/PMMA/Glass* | 16.1 | 16.8 |
| PSD/PMMA | 24.1 | (75) |
| PSD/PMMA* | 17.9 | (108) |

( )   Data in parentheses inadequate for accurate determination.
  *    Annealed at 150° C for 24 hours.

The results of $T_{1\rho}(^{13}C)$ and $T_{1\rho}(^{1}H)$ measurements made with a radiofrequency (rf) field intensity of 50 kHz are presented in Tables 1 and 2. To estimate the approximate rate of motion producing these relaxation times, and to make it possible to interpret changes in relaxation times, $T_{1\rho}(^{13}C)$ was also studied as a function of rf field from 32 to 60 kHz. These data are presented in Table 3. Each relaxation time measurement is made by observing the intensity of a single line in the spectrum as a function of hold time; these lines are identified in the tables. The accuracy of these measurements is limited to approximately ±0.5 msec, as can be deduced by comparing $T_{1\rho}(^{1}H)$ measurements made on the same phase as seen by different carbons.

Table 2. $^{13}$C-Detected $T_{1\rho}(^{1}H)$, m at 50 kHz

| Sample/Line | Line (ppm) | | | |
|---|---|---|---|---|
| | 17 | 127 | 145 | 177 |
| PS | - - - | 6.0 | 5.6 | - - - |
| PS* | - - - | 5.8 | 5.7 | - - - |
| PMMA | 18.9 | - - | -- | 18.4 |
| PMMA* | 17.3 | - - | - - | 17.6 |
| PS/PMMA | 14.5 | 7.4 | 7.5 | 15.1 |
| PS/PMMA* | 16.7 | 6.7 | 5.8 | 17.1 |
| PS/PMMA/Glass | 15.7 | 6.5 | 6.6 | 14.8 |
| PS/PMMA/Glass* | 16.2 | 6.2 | 5.7 | 14.4 |
| PSD/PMMA | 17.3 | 21.0 | ( ) | 17.0 |
| PSD/PMMA* | 17.9 | 18.2 | ( ) | 17.9 |
| PS/PMMAD | ( ) | 6.5 | 5.7 | ( ) |

( )   Data in parentheses inadequate for accurate determination.

 *    Annealed at 150° C for 24 hours.

Table 3. $T_{1\rho}(^{13}C)$, m in PS/PMMA as a Function of RF Field Intensity.

| | RF Field (Hz) | | | |
|---|---|---|---|---|
| Line/RF Field | 32 kHz | 40 kHz | 50 kHz | 60 kHz |
| 17 ppm | 14.1 | 18.4 | 23.8 | 26.6 |
| 127 ppm | 11.3 | 22.8 | 33.3 | 45.6 |
| 145 ppm | (35) | (106) | (113) | (160) |
| 177 ppm | (50) | (100) | (115) | (177) |

( )  Data in parentheses inadequate for accurate determination.

## DISCUSSION

$T_{1\rho}(^{13}C)$ senses local motion when measured at high enough radiofrequency fields. Before analyzing changes in $T_{1\rho}(^{13}C)$ on the basis of molecular motion, it is desirable to know the relationship between $\omega_c$ the correlation frequency of the prevalent motion and $\omega_1$ the rotating frame frequency. If $\omega_c \gg \omega_1$, then the relaxation time will be independent of $\omega_1$; and, if $\omega_c \ll \omega_1$, then the relaxation time will vary with the square of $\omega_1$. The dependence of $T_{1\rho}(^{13}C)$ in PS/PMMA on the amplitude of the rotating frame radiofrequency is shown in Table 3, which shows that the relaxation times in PMMA (17 ppm line) vary at less than the square of the rotating frame frequency, while the relaxation times in PS (127 ppm line) vary as the square of the rotating frame frequency. This suggests that the correlation frequency for the motion determining the relaxation time is greater for PMMA than the rotating frame frequency ($\omega_c > \omega_1$), while the reverse is true for PS ($\omega_c < \omega_1$). Thus a similar change in the rate of motion will have the reverse effect on relaxation rates in the two polymers. This is explained by the fact that $T_{1\rho}(^{13}C)$ is minimum when $\omega_c \approx \omega_1$ so that increasing $\omega_c$, for instance, will push PS nearer the minimum and PMMA farther from it.

### $T_{1\rho}(^{13}C)$ at 50 kHz

The carbon $T_{1\rho}$ values obtained for the various samples are shown in Table 1. $T_{1\rho}(^{13}C)$ for the protonated phenyl carbons in PS (127 ppm) does show a striking effect of annealing on that group. This is interpreted as a slowing down, even though the relaxation time increases on annealing. This is because the correlation frequency is less than the rotating frame frequency ($\omega_c < \omega_1$). PMMA (17ppm) shows a decrease on annealing. This is also interpreted as a slowing down, since the correlation frequency is greater than the rotating frame frequency. The effect is not as great as for PS annealing, since the PMMA methyl reflects polymer backbone motion and is not as much affected as the side chain PS phenyl motion. Since cross-relaxation is not a factor among rare spins, the changes going from quenched homopolymer to copolymer, which indicate that both PS and PMMA slow down, may be ascribed to entanglement of two less-than-compatible chains. Annealing and increased phase separation cause motion in PS/PMMA to change little since enhanced packing due to annealing and disentanglement resulting from phase separation offset one another.

$T_{1\rho}(^{13}C)$ decreases for both blocks in the copolymer coated on glass fibers in comparison to the block copolymer, packed as powder. Glass surface attracts the PMMA, hence would have an influence on both molecular motion and phase separation. The decrease in $T_{1\rho}(^{13}C)$ for 17 ppm line (PMMA), in comparison to both homopolymer (powder ) and block copolymer (powder), is interpreted as arising from reduced motion of PMMA, due most likely to attachment to the glass surface which would cause increased entanglement. $T_{1\rho}(^{13}C)$ for 127 ppm line (PS) is similar to that of the homopolymer. This is interpreted as enhanced phase separation due to the glass surface. $T_{1\rho}(^{13}C)$ for PMMA (17 ppm) increases slightly on annealing. Disentanglement due to increased phase separation causes the motion to increase in PMMA [i.e.: to be more like PMMA* (see Table 2), hence the $T_{1\rho}(^{13}C)$ increases. $T_{1\rho}(^{13}C)$ of PS (127 ppm) does not change on annealing; perhaps the PMMA block prevents the extreme packing of PS*.

Deuteration of the PS returns the PMMA rate to near that of the homopolymer. Phase separation on annealing reconcentrates the protons and improves relaxation. These last observations are puzzling, since deuteration of the PS should have little effect on $T_{1\rho}(^{13}C)$ in the PMMA.

## $T_{1\rho}(^{1}H)$ at 50 kHz

The proton $T_{1\rho}$ values obtained for the various samples are shown in Table 2. The homopolymers PS and PMMA clearly have different relaxation rates, perhaps due to basic differences in molecular motion. Spin diffusion within each is good enough to yield a single proton relaxation rate. However, in the PS/PMMA copolymer there is a pulling together of the two rates since the difference between the relaxation rates measured at 127 and 17 ppm is less than in the homopolymers. This suggests that some communication occurs between spins in each component but not enough to give a single relaxation time. This likely results from phase separation. When the same polymer is annealed (PS/PMMA*), there is a partial return toward the homopolymer values. Upon phase separation, the relaxation rate seems to be uniform, even as seen by the nonprotonated carbons, which must sense the nearest available protons.

$T_{1\rho}(^{1}H)$ increases for the 17 ppm line (PMMA) and decreases for the PS lines (127 and 145 ppm) in PS/PMMA/glass, in comparison to PS/PMMA. This is due to increased phase separation in PS/PMMA/glass. Annealing PS/PMMA/glass further enhances phase separation, and reduces cross relaxation between PS and PMMA blocks. As a result, the $T_{1\rho}(^{1}H)$ increases for the 17 ppm line (PMMA) and decreases for the 127 and 145 ppm lines (PS). The reason for the behavior of the $T_{1\rho}(^{1}H)$ of the 177 ppm line in PS/PMMA/glass is not quite clear; it is similar to that in PS/PMMA. It decreases slightly on annealing, which is similar to the behavior of homopolymer on annealing.

Deuteration of the PS returns the PMMA rate to near that of the homopolymer. Furthermore, the $T_{1\rho}(^1H)$ observed through the PS at 127 ppm is approximately the same as the PMMA rate. This is expected since this signal originates through cross-polarization from nearby protons in the PMMA.

## CONCLUSIONS

$T_{1\rho}(^1H)$ measurements show that the carbons in each block of the copolymers see primarily their own protons, indicating a degree of phase separation. Spin diffusion tends to decrease the difference in the $T_{1\rho}(^1H)$ values for the two blocks in the copolymer compared with the homopolymer values. Annealing enhances phase separation between blocks, as seen by $T_{1\rho}(^1H)$ measurements returning toward the homopolymer values. Measurements of $T_{1\rho}(^{13}C)$ at various field strengths (32 to 65 kHz) suggest that the correlation frequency at room temperature is less than 50 kHz for the PS while slightly greater than 65 kHz for the PMMA component. $T_{1\rho}$ measurements on PS/PMMA/glass show that the glass surface has an influence on both molecular motion and phase separation.

## REFERENCES

1.  J. Schaefer and E. O. Stejskal, *J. Am. Chem. Soc.*, *98*, 1031 (1976).
2.  J. Schaefer, E. O. Stejskal, and R. Buchdahl, *Macromolecules, 10*, 384 (1977).
3.  E. O. Stejskal, J. Schaefer, M. D. Sefcik, and R. A. McKay, *Macromolecules, 14*, 275 (1981).
4.  J. Schaefer, M. D. Sefcik, E. O. Stejskal, and R. A. McKay, Macromolecules, 14, 188 (1981).
5.  G. C. Gobbi, R. Silvestri, T. P. Russell, J. R. Lyerla, and W. W. Fleming, J. Polymer Sci., *Part C, Polym. Lett., 25*, 60 (1987).
6.  G. Coulon, T. P. Russell, V. R. Deline and P. F. Green, *Macromolecules, 22*, 2581 (1989).
7.  T. P. Russell, G. Coulon, V. R. Deline and D. C. Miller, *Macromolecules, 22*, 4600 (1989).
8.  S. H. Anastasiadis, T. P. Russell, S. K. Satija and C. F. Majkrzak, *Phys. Rev. Lett., 62*, 1852 (1989).
9.  S. H. Anastasiadis, T. P. Russell, S. K. Satija and C. F. Majkrzak, J. *Chem. Phys., 92*, 5677 (1990).
10. T. P. Russell, A. Menelle, W. A. Hamilton, G. C. Smith, S. K. Satija and C. F. Majkrzak, *Macromolecules*, (in press).
11. P. F. Green, T. P. Russell, R. Jerome and M. Granville, *Macromolecules, 22*, 908 (1989).

# Chapter 12

# EFFECT OF TEMPERATURE, SOLVENT AND SUBSTITUTION ON THE CHOLESTERIC PITCH OF (ETHYL)CELLULOSE MESOPHASES

DEREK R. BUDGELL AND DEREK G. GRAY

*Pulp and Paper Research Institute of Canada
and Department of Chemistry
McGill University
3420 University Street
Montreal Canada, H3A 2A7.*

## INTRODUCTION

Many cellulose derivatives form lyotropic and thermotropic liquid crystals.[1-6] Cellulosic mesophases are normally cholesteric; measurement of their optical properties provides a means to probe their structures. For a planar cholesteric liquid crystal the pitch P of the supramolecular helicoidal assembly of chain molecules is related to the reflection band wavelength by the de Vries equation.[7]

$$\lambda_0 = nP \tag{1}$$

where n is the mean refractive index and $\lambda_0$ is the wavelength of maximum reflection.

Two of the primary optical methods for studying cholesteric liquid crystals are optical rotatory dispersion (ORD) and circular dichroism (CD). ORD and CD measure, as a function of wavelength, the rotation of plane-polarized light and the differential absorption of left- and right-handed circularly polarized light, respectively. These techniques provide information on both the pitch and the twist sense of the cholesteric structure. A negative CD signal and an ORD curve that goes from positive to negative with increasing wavelength indicate the existence of a righted-handed cholesteric structure while a positive CD signal and a negative to positive change in optical rotation with increasing wavelength are characteristic of a left-handed cholesteric structure. The majority of cellulosic mesophases that have been studied to date are right-handed cholesterics, but a few left-handed systems also have been reported.[8-11] The observation of a left-handed (ethyl) cellulose (EC) mesophase by Zugenmaier[8,9] was perhaps unexpected, because all cellulosic polymers share the same chiral backbone. Other studies on EC mesophases have focused on phase separation with a second polymer,[12] dielectric behavior,[13] birefringence,[14] rheological behavior,[9,15] morphology[16] and pitch.[8]   In this work, the pitch and twist sense of EC mesophases in several solvents is examined as a function of temperature and degree of substitution.

## EXPERIMENTAL

The EC was obtained from Aldrich Ltd and had an ethoxyl content of 48% by weight (corresponding to a DS of 2.29), and a mass average molar mass of 65,000 as determined by low angle laser light scattering in dioxane with a Chromatix KMX-6 LALLS photometer.   Samples with increased ethoxyl substitution were prepared from this material, as follows.  Dried EC (DS=2.29), 5 g, was dissolved in 60 mL of tetrahydrofuran under nitrogen. Sodium hydride, 0.7 g, was added and the solution stirred for one hour before the dropwise addition of iodoethane as alkylating agent.   The temperature (in the range 40-60°C), reaction time and quantity of iodoethane were varied to give samples with a range of DS (Table 1).  After cooling, 50 mL of methanol was slowly added to the reaction mixture to remove unreacted sodium hydride, and the polymer was precipitated by pouring into 200 mL of distilled water containing 1g of sodium thiosulfate. After washing with water and drying under vacuum at 60°C, the precipitation step was repeated at least twice from tetrahydrofuran solution, the final time without thiosulfate.  Yields ranged between 70% and 90%.  The DS values were determined from [1]H NMR spectra run on Varian XL200 or XL300 spectrometers, in CDCl3 with tetramethyl silane as internal reference.  The DS values were calculated from the ratio of the methyl component of the ethyl group peak area at 1.18 ppm to the area of the peaks of all other protons from 2.7 to 5.0 ppm.  The molar mass of the sample with a DS of 3, measured as above in benzene, was 62,000, so degradation during the alkylation was not severe.

Table 1. Preparation of EC Samples

| EC (g) | Iodoethane (mL) | Temperature (°C) | Reaction Time (h) | DS |
|---|---|---|---|---|
| 5.0 | 1 | 40 | 3 | 2.55 |
| 5.0 | 2 | 40 | 3 | 2.65 |
| 5.0 | 5 | 40 | 3 | 2.70 |
| 5.0 | 6.5 | 40 | 3 | 2.8 |
| 5.0 | 8 | 40 | 3 | 2.88 |
| 5.0 | 8 | 55 | 5 | 3.00 |
| 20 | 28 | 60 | 12 | 3.00 |

Samples of EC in m-cresol, aqueous phenol (Anachemia, 8% water) and dichloroacetic acid (DCA) were prepared by weighing measured amounts of polymer and solvent into a small vial. Mixing of the sample was aided by centrifuging back and forth in the vial. After mesophase formation the EC/m-cresol and EC/aqueous phenol liquid crystals were stored in the dark and allowed to stand for a minimum of two months. Thin samples of EC and m-cresol, aqueous phenol and DCA were prepared for optical measurements by sandwiching the mesophase between a glass slide and coverplate. The coverplate was sealed to the slide with Devcon epoxy and allowed to stand for two days. Samples of EC in spectral grade chloroform were prepared by flame-sealing one end of a 0.4 mm thick glass microslide (Vitro Dynamics Inc., Rockaway, NJ, 07866), packing the slide with a known mass of polymer and injecting a known mass of solvent with a microsyringe. The sample was then cooled in dry ice and the other end of the microslide was flame-sealed. Mixing in the microslides was also aided by centrifugation. Samples of EC in chloroform were measured by placing the microslides in the beam path. A single sample of EC in chloroacetic acid was prepared by mixing the polymer and solid chloroacetic acid in a small vial and heating to $\sim$60 $^{\circ}$C. The polymer dissolved in the molten acid, but because some solvent evaporated and crystallized on the sides of the vial, the concentration was not determined, and only the cholesteric twist sense and temperature dependence of the reflection wavelength were identified.

The handedness and wavelength of the cholesteric reflection bands were measured with Jasco J-500C or Jasco ORD/UV-5 spectropolarimeters. Photoacoustic Fourier transform infrared (PAS-FTIR) spectra were run with a Mattson Instruments Cygnus 25 spectrometer and an EG&G Princeton Applied Research Model 6003 photoacoustic cell. The spectra were recorded at 8 cm$^{-1}$ resolution with 100 scans and were ratioed to carbon black (Norit A, BDH).

## RESULTS AND DISCUSSION

### Handedness in Different Solvents

Most cellulosic mesophases reported to date are right-handed and all right-handed cellulosic liquid crystals have been found to increase in pitch with increase in temperature. EC however, is reported to form left-handed mesophases in acetic acid and a right-handed one in dichloroacetic acid[9]. Using a circular reflectance (CR) technique to identify the twist sense, we have found (Table 2) that the left-handed structure appears to be the norm for EC, as it forms left-handed mesophases in chloroform, dichloromethane, phenol, aqueous phenol, m-cresol and chloroacetic acid, but it was confirmed that EC in DCA forms a right-handed mesophase. The solvent dependence of the twist sense is well known for the polypeptides.[19] Samulski and Samulski[20] have developed a theory explaining the cholesteric sense in terms of solvent dielectric constants. However, the solvent dielectric constant of the right-handed system, EC in dichloroacetic acid, is bracketed by the dielectric constants of solvents forming left-handed mesophases, so this is not the key factor for this cellulosic system.   Another possibility is that there is a  reaction

Table 2.  Dielectric Constant, Ethyl Cellulose Mesophase Twist Sense and
          Temperature Dependence of the Pitch

| | Dielectric constant[17] | Twist sense | $d\lambda_0/dT$ |
|---|---|---|---|
| (Ethyl)cellulose | 3.09 (25ºC,1 kHz) | ---- | ------- |
| **Solvent** | | | |
| Chloroacetic acid | 12.3 (60 ºC) | Left | Increases |
| m-Cresol | 11.8 | Left | Increases |
| Phenol | 9.78 (60 ºC) | Left | Increases |
| Dichloromethane | 9.08 | Left | No Report |
| Dichloroacetic acid | 8.2 | Right[9] | Increases |
| Acetic acid | 6.15 | Left[9] | Decreases[8] |
| Chloroform | 4.81 | Left | Decreases |

between the solvent and the polymer.  Guo and Gray[18] found that minor substitution of acetate groups onto EC molecules can change the mesophase from left- to right-handed in the same solvent.  Given the acidic strength of DCA, substitution of dichloroacetyl groups at unreacted hydroxyls to give (dichloroacetyl)(ethyl)cellulose could take place.  However, after diluting an EC/DCA mesophase to the isotropic state by the addition of dioxane, precipitating the polymer out of solution by the addition of diethyl glycol, a nonsolvent, and then washing the polymer with absolute ethanol, the PAS-FTIR spectra of the starting material and the precipitated polymer were found to be nearly identical, suggesting that substitution of dichloroacetate groups onto the polymer did not occur.  However, hydrolysis to regenerate the hydroxyl substituents during the work-up cannot be completely discounted.

   The temperature dependence of the cholesteric pitch of cellulosic mesophases has apparently been consistent, with all right-handed mesophases increasing in pitch with an increase in temperature, independent of solvent. However, the left-handed EC mesophases show a more complex solvent-dependent behavior (Table 2).  As illustrated in Figure 1, in chloroform and bromoform the wavelength of maximum reflection decreases with an increase in temperature, but in m-cresol and aqueous phenol, which also form left-handed systems, the reflection wavelength increases with increasing temperature.  These temperature effects are reversible, with little hysteresis.

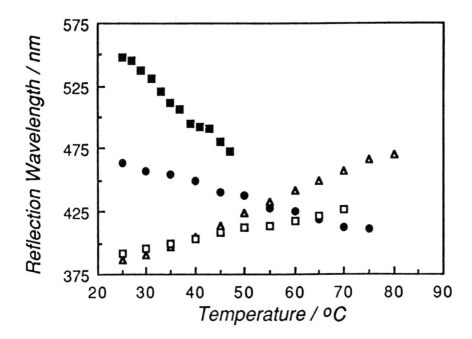

Figure 1. Variation of reflection wavelength with temperature for left-handed EC (DS = 2.29) mesophases in various solvents. EC in chloroform (43.7 % w/w, filled square), in bromoform (23.0 % w/w, filled circle), in *m*-cresol (54.4 % w/w, open square) and in aqueous phenol (50.9 % w/w, open triangle).

### Effect of Degree of Substitution on Handedness

The role of the degree of substitution in determining the chiral nematic properties of cellulose derivatives is largely unexplored. Mesophases were prepared from EC samples with DS varying from 2.29 to 3.00 in several solvents, and the CR technique was used to monitor the reflection wavelength and cholesteric twist sense. Unfortunately, the solubility of EC samples depends on the DS, so highly substituted EC samples cannot be examined in the same solvents as the commercial product. Table 3 lists how the DS affects the cholesteric twist sense for several solvents.

The impact of the DS on the cholesteric twist sense is most explicitly illustrated using chloroform or dichloromethane as the solvent. The sample with DS of 3.00 in chloroform and dichloromethane forms mesophases whose cholesteric twist senses are right-handed, in contrast to the samples of lower DS that form left-handed mesophases in these same solvents. The critical DS at which the switch from a left-handed to a right-handed cholesteric twist sense occurs for these solvents is approximately 2.88. Samples with a

Table 3.    Cholesteric Twist Sense of EC in Several Solvents
as a Function of DS

| | Degree of Substitution | | | | | | |
|---|---|---|---|---|---|---|---|
| Solvent | 2.29 | 2.55 | 2.65 | 2.70 | 2.80 | 2.88 | 3.00 |
| Aq. Phenol | Left | Left | Left | Left | Left | - | - |
| Bromoform | Left | Left | Left | Left | Left | - | - |
| Chloroform | Left | Left | Left | Left | Left | - | Right |
| m-Cresol | Left | Left | Left | Left | Left | - | - |
| Dichloromethane | Left | Left | Left | Left | Left | - | Right |
| DCA | Right | Right | Right | Right | Right | Right | Right |

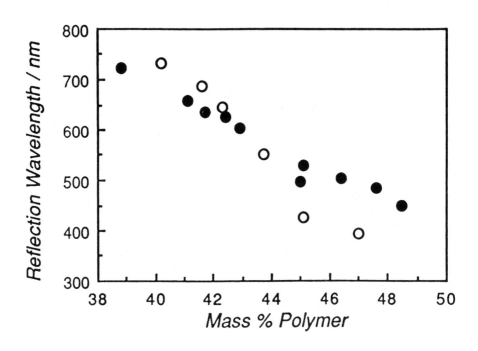

Figure 2. Variation of reflection wavelength with concentration for right-handed EC (DS = 3.0, filled circles) and left-handed (DS = 2.29, open circles) mesophases of EC in chloroform.

DS of less than 2.88 form left-handed mesophases with reflection colours while the sample with DS of 3.00 forms right-handed mesophases with reflection colors.   As the polymer concentration of the sample with DS of 2.88 is increased from that of an isotropic solution to the pure film, there is no appearance of visible reflection colors or even of fingerprint texture when viewed under a polarizing microscope, suggesting that  the mesophase is approaching the untwisted nematic state.   Both the left-handed (DS = 2.29) and right-handed (DS = 3) mesophases in chloroform show a similar dependence of reflection wavelength on concentration (Figure 2).   In fact, for the left-handed mesophases examined here, at fixed polymer concentration and temperature, an increase in the DS produces an increase in the reflection wavelength.  The greatest sensitivity of the reflection wavelength to DS is observed in m-cresol.   The degree to which changes in the DS of the EC / m-cresol mesophases can affect the reflection wavelength is illustrated by Figure 3, which shows that an increase in the DS of 0.5 at constant concentration can have as have much influence on the reflection wavelength as significant changes in concentration at a single DS.   There has been no significant decrease in the molar mass during the preparation of the EC derivatives, so the increase is due exclusively to the change in DS.  For the right-handed mesophases of EC in DCA, a decrease is observed in the reflection wavelength of the mesophase as the DS is increased .   This behavior is opposite to that  observed for the left-handed EC mesophases.

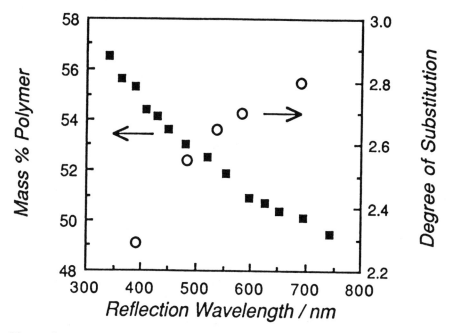

Figure 3.      Relative effect of changes in concentration at DS = 2.29 and changes in DS at 54.5% (w/w) on the reflection wavelengths of EC / m-cresol mesophases.

The DS also influences the the temperature dependence of the reflection wavelength, $d\lambda_0/dT$. The effect of temperature on the reflection wavelengths of EC samples with DS of 2.29 and 3.00 in chloroform can be seen from Figure 4. The reflection wavelengths of the left-handed mesophases decrease with increasing temperature, while the reflection wavelengths for the right-handed mesophases increase. The reflection wavelength of the right-handed trisubstituted sample is more sensitive to a change in temperature than the left-handed sample of lower DS.

A change in sign of $d\lambda_0/dT$ does not necessarily involve a change in handedness. Figure 5 shows that for the exclusively left-handed EC mesophases in m-cresol, the reflection wavelength increases slightly with increasing temperature for samples with DS of 2.29, while the reflection wavelength decreases with increasing temperature for samples with DS of 2.70 and 2.80. At an intermediate DS of 2.55 the reflection wavelength is independent of temperature. The negative $d\lambda_0 / dT$ for EC at high DS in m-cresol now matches that for samples with DS of 2.29 in other solvents.

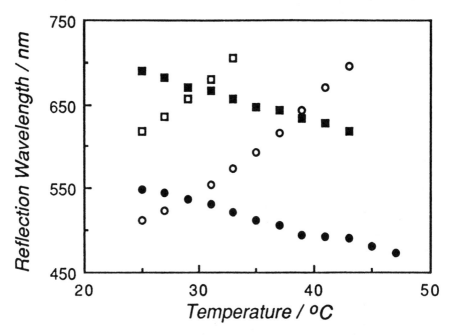

Figure 4. Variation of reflection wavelength with temperature for mesophases of EC with DS of 2.29 and 3.00 in chloroform. EC with DS of 2.29 (43.7% w/w, filled circle), EC with DS of 2.29 (41.6% w/w, filled square), EC with DS of 3.00 (47.6% w/w, open circle) and EC with DS of 3.00 (42.9% w/w, open square). Mesophases of EC with DS of 2.29 and 3.00 in chloroform are left-handed and right-handed, respectively.

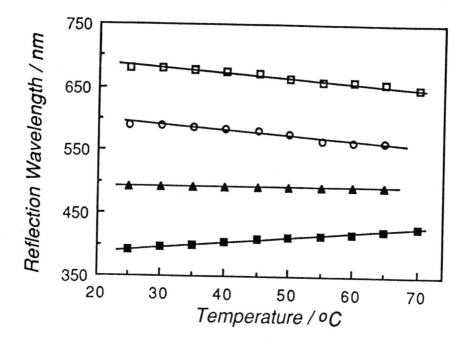

Figure 5. Effect of temperature on the reflection wavelength of EC / m-cresol mesophases.   DS = 2.29 (54.4% w/w, squares), DS = 2.55 (54.4% w/w, triangles), DS = 2.65 (54.4% w/w, open circles) and DS = 2.80 (54.4% w/w, open squares).  The lines indicate the trend of the data.

## Other Alkyl Derivatives of Ethyl Cellulose

Very high levels of substitution of commercial EC by ethyl groups are required to change the cholesteric twist sense from left-handed to right-handed.  The twist sense reversal is also seen if acetyl groups are introduced onto the EC chain, but complete substitution is not required for the change to occur.[18]  The acidic solvents DCA and TFA also reverse the usual twist sense of EC and CA, so it is possible that the reversal is due to strong interactions with polar substituents or solvents.  However, it is also possible that the reversal in twist sense is related to the size of the substituent group or the total DS of the polymer.  By introducing other alkyl groups onto the EC sample of DS = 2.29, it is possible to assess whether the variation in cholesteric twist sense with increasing levels of substitution is related merely to the level of substitution of the EC chain or whether the size of the alkyl substituent plays a role.

Introducing methyl groups onto the EC chain produces O-(ethyl)(methyl)cellulose (EMC) while the introduction of n-butyl groups gives O-(n-butyl)(ethyl)cellulose (BEC).[21]  Table 4 lists the total DS of the samples examined and the cholesteric twist sense of their mesophases in three solvents.

Table 4.  Cholesteric Twist Sense of EC, EMC and BEC Mesophases as a Function of DS

| Solvent | Polymer (Total Degree of Substitution) | | | | | |
|---|---|---|---|---|---|---|
| | EC 2.29 | EC 3.0 | EMC 3.0 | BEC 2.6 | BEC 2.9 | BEC 3.0 |
| Chloroform | Left | Right | Left | Left | Left | Right |
| Dichloromethane | Left | Right | Left | Left | Left | Right |
| DCA | Right | Right | Right | - | - | Right |

Unlike the other trisubstituted derivatives, EMC in chloroform and dichloromethane forms a left-handed mesophase. This suggests that the complete substitution of the hydroxyl groups is not necessarily sufficient to induce a reversal in twist change. For the BEC samples with total DS of 2.6, 2.9 and 3.0, only the trisubstituted derivative possesses a right-handed twist sense in chloroform and dichloromethane. It would appear that the size of the substituents introduced onto the EC chain does play a role in determining the twist sense.

Given that DCA is at present the only solvent in which EC with a DS of 2.29 is observed to form a right-handed mesophase and the possibility exists that this is due to the interaction of the solvent with the hydroxyl groups, it is instructive to examine the effect of DCA on the trisubstituted derivatives where the possibility of hydroxyl group interaction is eliminated. Table 4 shows that like EC with DS of 2.29, EMC, tri-O-ethyl cellulose and BEC with total DS of 3.0 in DCA all produce right-handed cholesteric structures, so specific hydroxyl interactions are not necessary. Also, the cholesteric twist sense for EMC is truly solvent dependent; it forms a left-handed cholesteric mesophase in chloroform and dichloromethane and a right-handed one in DCA.

## CONCLUSION

These qualitative results show that the handedness and temperature dependence of the pitch of chiral nematic mesophases of ethyl cellulose display a very marked dependence on solvent and DS.  A striking reversal of handedness is observed at high DS values in some solvents. This reversal does not appear to be related to DS-induced changes in handedness of the helicity of individual chains; x-ray diffraction studies of crystalline EC fibres[9] have established that a change in DS from 2.5 to 3.0 has no effect on the molecular conformations (both are left-handed threefold (3/2) helices). Results for some (alkyl)(ethyl)cellulose samples suggests that substituent size plays a role in determining the handedness of these lyotropic mesophases.

## ACKNOWLEDGEMENTS

We thank the Natural Sciences and Engineering Research Council of Canada for support. These observations form part of the Ph.D. thesis of DRB.[21]

## REFERENCES

1.  R.S. Werbowyj and D.G. Gray, *Mol. Cryst. Liq. Cryst., 34,* 97 (1976).
2.  M. Panar and O.B. Willcox, Offenlegungsschrift, German Federal Republic Patent 2,705,381 (1977).
3.  J. Maeno ,United States Patent 4 132 464 (1979)
4.  J. Bheda, J.F. Fellers and J.L. White, *Coll. and Polym Sci., 258,* 1335 (1980).
5.  R.D. Gilbert and P.A. Patton, *Progr. Polym. Sci., 37,* 179 (1983).
6.  D.G. Gray, *J. Appl. Polym. Sci., 37,* 179 (1983).
7.  Hl. de Vries, *Acta Cryst., 4,* 219 (1951).
8.  U. Vogt and P. Zugenmaier, *Ber. Bunsenges. Phys. Chem., 89,* 1217 (1985).
9.  P. Zugenmaier and P. Haurand, *Carbohydr. Res., 160,* 369 (1987).
10. A. M. Ritcey, K. R. Holme and D. G. Gray, *Macromolecules, 21,* 2914 (1988).
11. W.A. Pawlowski, R.D. Gilbert, R.E. Fornes and S.T. Purrington, *J. Polym. Sci.: Part B, 25,* 2293 (1987).
12. S. Ambrosino, T. Khallala, M. J. Seurin, A. Ten Bosch, F. Fried, P. Maissa and P. Sixou, *J. Polym. Sci.: Part C, 25,* 351 (1987).
13. K. Araki, Y. Iida and Y. Imamura, Makromol. Chem., Rapid Commun., 5, 99 (1984).
14. S. Suto, H. Ise and M. Karasawa, J. Polym. Sci.: Part B, 24, 1515 (1986).
15. S. Suto, J.L. White and J. F. Fellers, *Rheol. Acta, 21,* 62 (1982).
16. Y. Nishio, S. Susuki and T. Takahashi, *Polym. J., 6,* 753(1985).
17. *CRC Handbook of Chemistry and Physics,* R.C. Weast, Ed., CRC Press, Cleveland, Ohio, 56th Ed, 1976, p E-55.
18. J.-X. Guo and D.G. Gray, *Macromolecules, 22,* 2086 (1989).
19. I. Uematsu and Y. Uematsu, *Adv. Polym. Sci., 59,* 37 (1984).
20. T.V. Samulski and E.T. Samulski, *J. Chem. Phys., 67,* 824 (1977).
21. D.R. Budgell, Ph.D. thesis, McGill University, Montreal, 1989.

# Chapter 13

# MECHANISTIC CONSIDERATIONS FOR HOMOGENEOUS REACTIONS OF CELLULOSE IN LiCl/N,N-DIMETHYLACETAMIDE

TIMOTHY R. DAWSEY

*Tennessee Eastman Company*
*P.O. Box 511*
*Kingsport, TN 37662*

## INTRODUCTION

Cellulose, or poly(1-4)-α-D-glucose is the most abundant of our renewable organic raw materials.[1] The molecular structure (Figure 1) consists of cellobiose repeating units, which allow chain-packing by intermolecular[2] and intramolecular[3] hydrogen bonding. Such strong interactions are responsible for its excellent inherent mechanical properties, yet at the same time, they interfere with efforts to process or modify the material. Only in a few instances have cellulose derivatives been exploited commercially and certainly not to the extent predicted from raw material availability and cost. Controllable, uniform derivatization has been hampered by the lack of suitable, nondegrading solvents or by a limited range of synthetic reactions within these solvents.

In 1979 the dissolution of cellulose in N,N-dimethylacetamide (DMAc) and lithium chloride (LiCl) was first reported.[4-6] A number of synthetic reactions that could be carried out in a facile manner under homogeneous reaction conditions were described as well. Since this initial discovery, numerous studies have reported the utility of the DMAc/LiCl solvent for cellulose reactions, characterization and processing.[7]

The polar aprotic nature of the lithium chloride/N,N-dimethylacetamide (LiCl/DMAc) solvent allows a range of organic reactions including typical alcohol modification reactions such as esterification and carbamate formation. The preparation of cellulose esters and carbamates is facile, utilizing acid chlorides and isocyanates, respectively.[8] Cyclic anhydrides have also been used to prepare cellulose monoesters. Cellulose has been reacted as well with toluenesulfonyl chloride in LiCl/DMAc to produce the sulfonate ester and chlorodeoxycellulose.[9]

The reaction of cellulose with epoxide reagents can be used to prepare cellulose ethers in LiCl/DMAc.[8,10] This typically requires high temperatures, long reaction times and strong alkaline conditions.  As a result, considerable polymer degradation results.

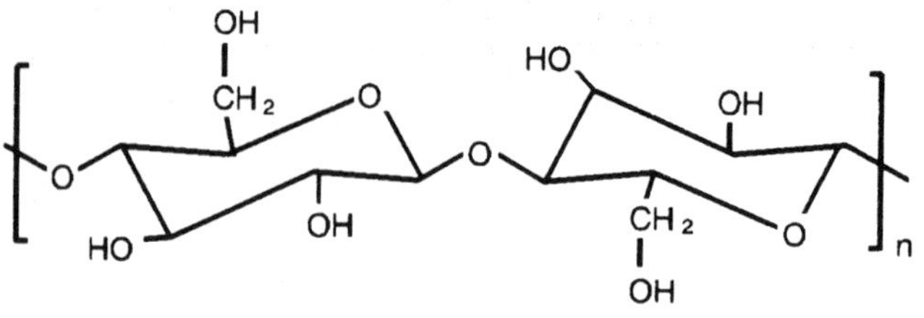

Figure 1. Cellulose Structure

In this work we discuss a range of reagents that have been evaluated as candidates for cellulose derivatization in LiCl/DMAc.  The successes and failures of these efforts provide clues to the mechanisms of reaction within this medium.

## THE SOLVENT

In 1979 it was reported that mixtures of LiCl in DMAc produced homogeneous solutions of cellulose under moderate conditions with little or no degradation.[4,5,11]  The LiCl/DMAc solvent was used to dissolve proteins, synthetic polyamides[12] and chitin[13] prior to its application to cellulose.  A laboratory dissolution procedure in LiCl/DMAc resulted in solutions containing from 15 to 17% cellulose.[14,15]  LiCl content was varied from 3 to 11% (w/w) with optimum concentration in the 5 to 9% range.

Dissolution of cellulose was promoted by prior activation of the cellulose with a solvent exchange series:  water, methanol, DMAc.  This procedure considerably reduced the time required for dissolution and allowed formation of clear solutions in less than an hour for concentrations of 6-15%.  The rate of dissolution decreased as the degree of polymerization (DP) of the cellulose sample increased.[10]

The absence of additional signals in the [13]C nuclear magnetic resonance (NMR) spectra from cellulose solutions in LiCl/DMAc has led to the conclusion that this is a true solvent[14,16] - one in which the polymer is not derivatized.[17]  Evidence suggests that LiCl/DMAc might best be described as a cellulose complexing solvent.  More than a decade after its discovery, however, all facets of the mode of action of this binary solvent in cellulose dissolution have not been fully clarified.

It has been suggested that the solution mechanism involves the complexation of Li$^+$ with the carbonyl oxygen atoms of up to four DMAc molecules to produce a macrocation, leaving the chloride anion (Cl$^-$) free to hydrogen-bond with the N-H protons of the polyamide. A number of different LiCl/DMAc/cellulose complex schemes have been proposed (Figure 2). Common to all schemes are the complexation of $Li^+$ with the carbonyl of DMAc and hydrogen bonding of $Cl^-$ with the cellulosic hydroxyl groups.

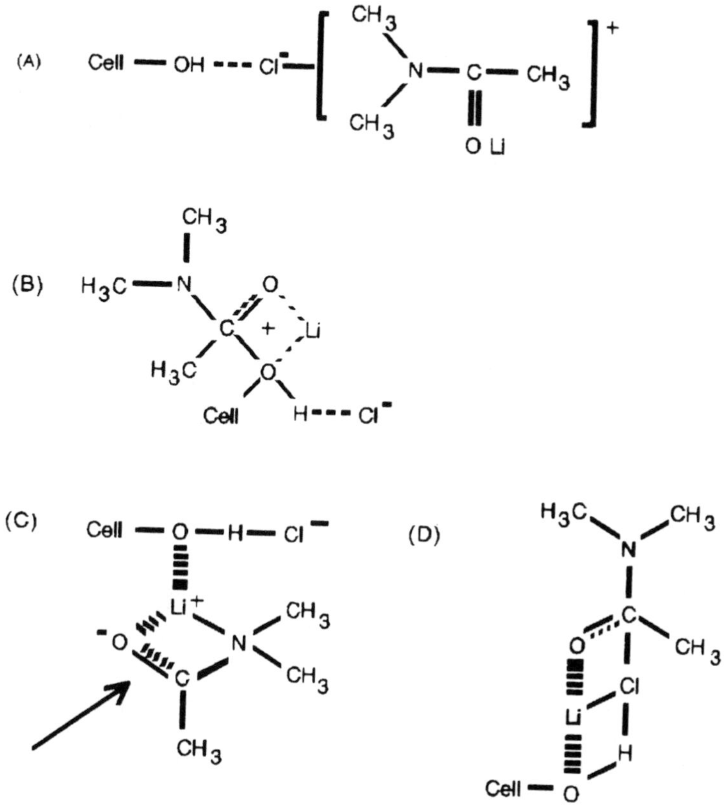

Figure 2. Mechanistic schemes.

The existence of the complex is consistent with reported increases in viscosity upon addition of LiCl to DMAc.[14,18,19] In $^{13}C$-NMR studies, this viscosity increase has been associated with reduced $T_1$'s for all carbons of the DMAc solvent molecule and the cellulose solute molecule.[20]

Various solvation schemes proposed in the literature are shown in Figure 2. Interpretation of available evidence has led to differing descriptions of the interactions. In McCormick's proposed version (Scheme A),[14] the $Cl^-$ is jointly associated with the cellulose hydroxyl proton and the macrocation. The LiCl/DMAc/cellulose complex as seen by El-Kafrawy (Scheme B)[16] depicts the $Li^+$ associated with both the DMAc carbonyl oxygen and the cellulose hydroxyl oxygen, leaving the Cl⁻ free. According to Turbak (Scheme C),[21] the Li⁺ is extensively complexed with the oxygen and nitrogen atoms of the DMAc molecule, as well as with the oxygen atom of the cellulose hydroxyl. This scheme, like that of El-Kafrawy's, shows an untethered Cl⁻. Vincendon[22] pictures a "sandwich" structure for the complex (Scheme D).

Despite the uncertainties in describing the precise complex structure, one fact remains: the only alkali metal salt that clearly leads to cellulose dissolution to date is LiCl. Other lithium salts[11,14,23] including bromide, iodide, nitrate, and sulfate and chloride salts[11,14] of sodium, potassium, barium, calcium and zinc have been ineffective. Additionally, DMAc[4,5] and N-methyl-2-pyrrollidinone (NMP)[24] are the primary solvents that act with LiCl to dissolve cellulose.

## DERIVATIZATION

The current applications for cellulose derivatives are diverse, with cellulose esters and ethers in highest demand. The esters are typically used as fibers, films and coatings whereas cellulose ethers are most commonly employed as water-soluble viscosifiers for aqueous media. Other derivatives such as carbamates, silyl ethers, sulfonate esters and halodeoxy compounds are produced on a limited scale and are typically used as aids in structure elucidation; some are purely of research interest. More specialized applications in the future, however, will increase the demand for more diverse cellulose derivatives. Applications such as selective membranes, controlled release of bioactive agents and fire retardancy all require new synthetic capabilities and increased control and predictability of the polymer modification process.

Commercial production of cellulose derivatives normally involves use of a heterogeneous slurry of cellulose in a solvent that swells but does not dissolve the polymer. The physical nature of the reaction must be considered at two levels: (1) the solid cellulose is suspended in a liquid reaction medium, and (2) the cellulose is heterogeneous in nature, with different areas of its suprastructure displaying different morphologies and, consequently, different accessibilities to the same reagent.[25] Heterogeneous derivatization can be pictured as occurring in increments. The surface of each cellulose particle must first become sufficiently substituted to produce a soluble product. This layer is then dissolved and the next layer becomes accessible to the reaction medium (Figure 3).

Each subsequent layer is similarly derivatized until the reaction is complete. Such a model illustrates that control of the reaction at any point short of completion is difficult. A polymer modified in such a manner to provide a degree of substitution (DS) of 2, for instance, would have segments with complete substitution (DS = 3), some with intermediate levels of attachment (DS = 1 or 2), and others that are unsubstituted (DS = 0). It can be safely postulated that unless the extent of reaction is 100%, the product is not uniform under such heterogeneous conditions.[25]

The commercial significance of nonuniform substitution can be illustrated in the production of acetone-soluble cellulose acetate, an important fiber source. Cellulose acetate is soluble in acetone only within a narrow range, about DS = 2.4.[26] The degree of uniformity of such a partially substituted derivative has a profound effect on the resulting properties. Two separately prepared cellulose acetates with average DS values of 2.5 may have vastly different solubilities. One may be soluble and the other insoluble in acetone.[25]

A direct production of cellulose acetate with a DS = 2.4 is not possible from a heterogeneous reaction medium. Instead, to acquire an acetone-soluble cellulose acetate, cellulose is completely derivatized to produce cellulose triacetate (DS = 2.76-3.00).[27] This product is then saponified to a DS of 2.4. Such a process is not only inefficient but also results in decreased molecular weight.

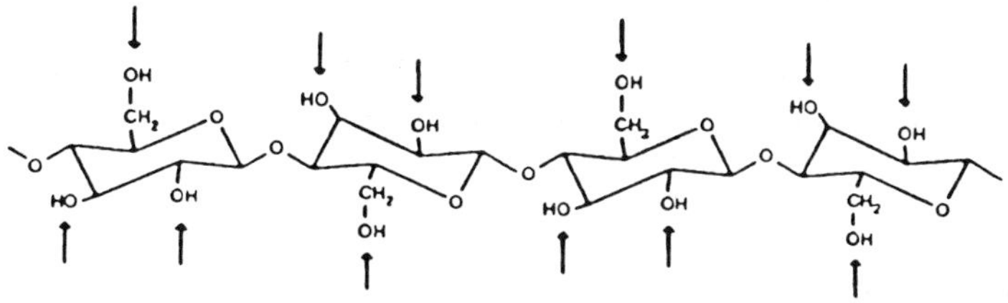

Homogeneous System: reaction sites ( ⟶ ) are uniform and fully accessible.

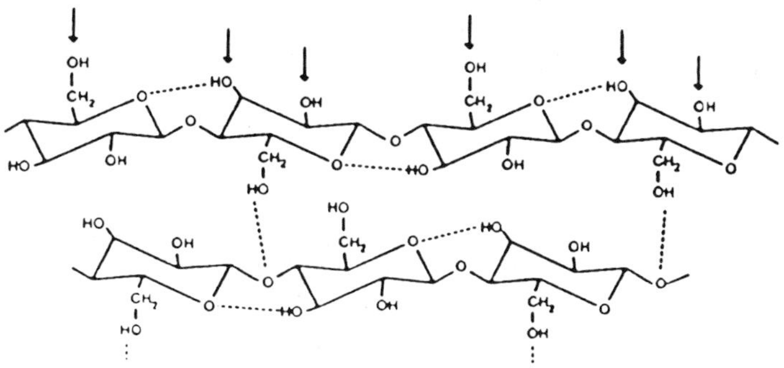

Heterogeneous System: reaction sites ( ⟶ ) are non-uniform and only on the surface.

Figure 3. Homogeneous vs heterogeneous reaction media.

The homogeneous reaction conditions in LiCl/DMAc provide uniform substituent distribution[28] and controlled degrees of substitution.[8] This becomes an especially important asset when a low DS is desirable. Uniformly substituted cellulosics should exhibit predictable product properties and enhanced resistance to enzymatic degradation. This last characteristic is attributable to the absence of regions of underivatized polymer, which provide an identifiable substrate for the enzyme.

The cellulose molecule contains three hydroxyl groups per anhydroglucose residue (AGR) in the polymer chain. Each of these alcohols [one primary, OH(6) and two secondary, OH(2) and OH(3)] can react in the same manner as low molecular weight substances of similar composition;[25] however, their relative reactivities vary considerably.

In most instances, the 1° hydroxyl group, OH(6), is found to react much more readily than the two 2° hydroxyl groups. This difference in reactivities has been utilized to tag the OH(6) by tosylation with negligible reaction at OH(2) and OH(3). The usual order of reaction is considered to be OH(6) >> OH(2) > OH(3). Considering the primary nature of OH(6) and the intramolecular hydrogen bonding known to be characteristic of OH(3), this order appears to be logical.

The polar aprotic character of LiCl/DMAc allows a range of organic reactions, including most of the typical alcohol modification reactions. As we will see, however, there are limitations.

## CELLULOSE ESTERS

Cellulose undergoes esterification with acids in the presence of a dehydrating agent or by reaction with acid chlorides. The production of cellulose esters in LiCl/DMAc has been extensively investigated. A typical reaction pathway is shown in Scheme 1.

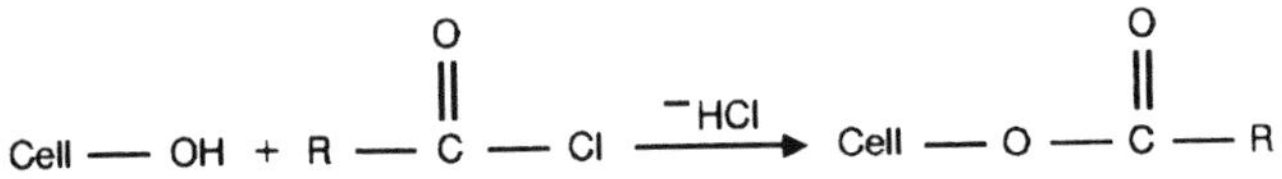

$$\text{Cell} - \text{OH} + \text{R} - \overset{\overset{\displaystyle O}{\|}}{\text{C}} - \text{Cl} \xrightarrow{\ -HCl\ } \text{Cell} - \text{O} - \overset{\overset{\displaystyle O}{\|}}{\text{C}} - \text{R}$$

Scheme 1

Among the first esters produced in the LiCL/DMAc solvent were potential controlled-release systems for herbicides.[4] The acid chlorides of 2,4-dichlorophenoxyacetic acid (2,4-D) and dichloropropionic acid (dalapon) were reacted with cellulose to produce polymers bearing 28 to 61% (w/w) herbicide (DS = 0.5-1.2).[29] This reaction medium was shown to be similarly appropriate in derivatization of other biodegradable polysaccharides.[30]

Further studies have concentrated on the synthesis of a range of cellulose esters for a broad spectrum of applications.[5,8,11,26,30-37] The reaction of acid chlorides with cellulose in LiCl/DMAc is facile, especially in the presence of tertiary amines (DS = 0.5 - 2.8).

## CELLULOSE CARBAMATES

In reactions complementary to the esterifications, cellulose carbamates have been produced (Scheme 2).[8,11,29-31,38,39] The reactions of cellulose with various isocyanates in the presence of a tertiary amine yield the corresponding carbamates with high degrees of substitution (DS = 1.0 - 2.7).[8] As with the esters, the first polysaccharide carbamates synthesized in the LiCl/DMAc solvent were herbicide derivatives being investigated for potential controlled-release applications.[4]

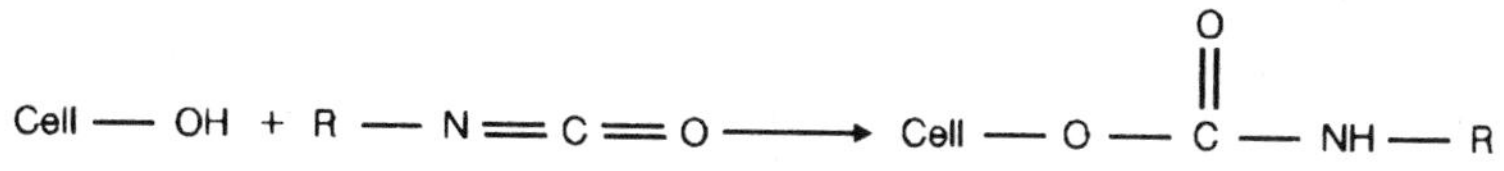

$$\text{Cell} - \text{OH} + \text{R} - \text{N} = \text{C} = \text{O} \longrightarrow \text{Cell} - \text{O} - \overset{\displaystyle O}{\overset{\displaystyle \|}{\text{C}}} - \text{NH} - \text{R}$$

Scheme 2

## CELLULOSE ETHERS

The preparation of cellulose ethers is difficult relative to the esters and carbamates in LiCl/DMAc. Cellulose ethers are typically prepared by the reaction of alkali cellulose with alkyl halides or epoxides (Schemes 3 and 4). Since most typical bases (NaOH and KOH) are insoluble in LiCl/DMAc, it is difficult to obtain the alkali cellulose needed to initiate the reaction. Efforts to synthesize cellulose ethers in the absence of base have been unsuccessful.[31] It is apparent that the hydroxyl functionality of cellulose is not a strong enough nucleophile for these reactions to proceed at reasonable temperatures.

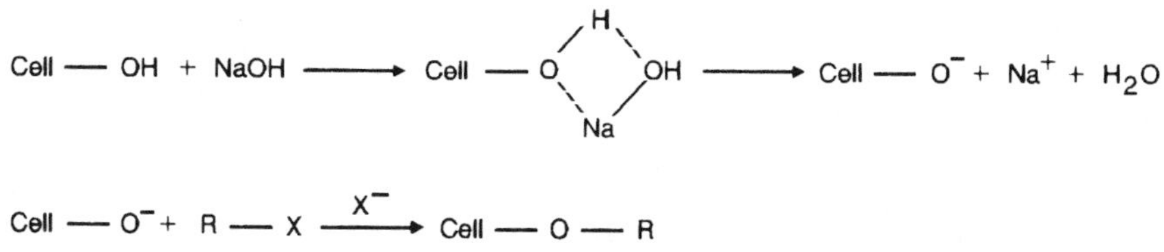

$$\text{Cell} - \text{OH} + \text{NaOH} \longrightarrow \text{Cell} - \text{O} \overset{H}{\underset{Na}{\diamond}} \text{OH} \longrightarrow \text{Cell} - \text{O}^- + \text{Na}^+ + \text{H}_2\text{O}$$

$$\text{Cell} - \text{O}^- + \text{R} - \text{X} \xrightarrow{\text{X}^-} \text{Cell} - \text{O} - \text{R}$$

Scheme 3

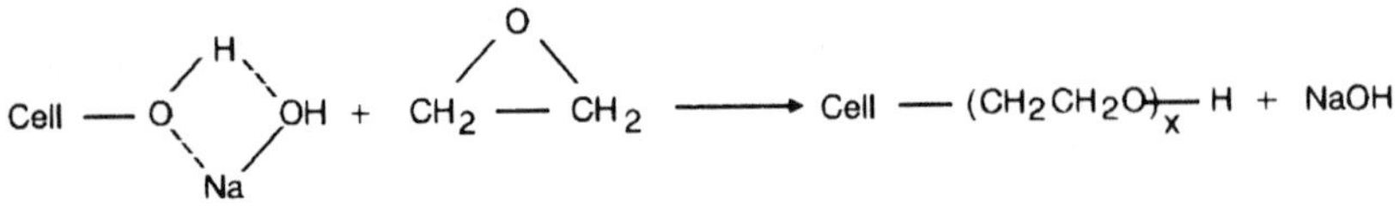

Cell — O $\overset{\text{H}}{\diagdown}$ OH + CH$_2$ — CH$_2$ $\longrightarrow$ Cell — (CH$_2$CH$_2$O)$_x$— H + NaOH

Scheme 4

Despite these difficulties, a number of cellulose ether derivatives have been prepared under heterogeneous conditions.[8,11,28,40] These reactions are typically conducted by reacting cellulose with a 3 to 5 molar excess of reagent. A base, usually KOH or NaOH powder, is slurried in the reaction mixture at elevated temperatures (60-80°C) for 48-72 hours.[8] Degrees of substitution are usually lower (DS = 1.1-1.7) than those attained in the previously discussed reactions.

The use of strong bases, high reaction temperatures (relative to the room temperature reactions noted earlier) and long reaction times unfortunately results in chain degradation.[8,38,40,41] Considering the lack of availability of soluble, strong bases for use in LiCl/DMAc, there appears to be no particular advantage of this system over conventional heterogeneous processes for cellulose ether production.[8]

## TOSYLATION

Sulfonic acid esters of various carbohydrates, including cellulose, have been extensively reported in the literature. These represent a technologically important class of materials that have been utilized as intermediates for nucleophilic displacement reactions. The p-toluenesulfonate derivative (tosylate) has been the most extensively studied of these intermediates, followed by the methanesulfonate (mesylate) and benzenesulfonate derivatives.

Klein and Snowden, as early as 1958, reported the preparation of a number of sulfonic acid esters of cellulose (both mesylate and tosylate).[42] The sulfonates were reacted in displacement reactions with potassium halides, potassium phthalimide, p-toluenesulfonamide, propylamine, p-methylthiophenol, saccharin, potassium thiocyanate, pentabromophenol, bis(2-propyl)dithiophosphoric acid, bis(dibromopropyl)phosphoric acid, nitropropane and potassium cyanide.

In addition to serving as leaving groups in substitution reactions, tosylate substituents serve as valuable hydroxyl protecting groups for carbohydrates.[43,44] Tosylates are stable under a variety of reaction conditions, allowing a broad range of transformations, yet photolytic cleavage is facile. Tosylate-protected hydroxyl groups are reported to withstand treatment with acids and bases, alkylation, hydrogenation and nitrous acid deamination.[44]

Historically, the p-toluenesulfonylation (tosylation) of cellulose has been carried out heterogeneously in pyridine.[42,45,46] Large excesses of the sulfonyl chloride are required. Reaction times of 2 to 4 days typically result in DS values of 0.08 to 1.36.

As an extension of previous work on dissolution and derivatization of cellulose in LiCl/DMAc,[4,5,8,11,14,29,30,33] McCormick and Callais reported the *in situ* synthesis and subsequent characterization of chlorodeoxycellulose.[8] This reaction, conducted under homogeneous conditions, was presumed to have proceeded through a cellulose tosylate (Scheme 5) intermediate (DS = 2.3). Further research, including isolation of intermediates, indicated that the originally proposed sequence was overly simplified. Subsequent investigations examined in more detail the tosylation reaction of cellulose under homogeneous reaction conditions in LiCl/DMAc.[9,47] These results provide insight into the conditions necessary for the competitive formation of chlorodeoxycellulose.

$$\text{Cell} - \text{OH} + \text{ArSO}_2\text{Cl} \xrightarrow{\text{pyr.}} \text{Cell} - \text{OSO}_2\text{Ar} \xrightarrow[\substack{\text{LiCl}^- \\ \text{DMAc}}]{\text{heat}} \text{Cell} - \text{Cl}$$

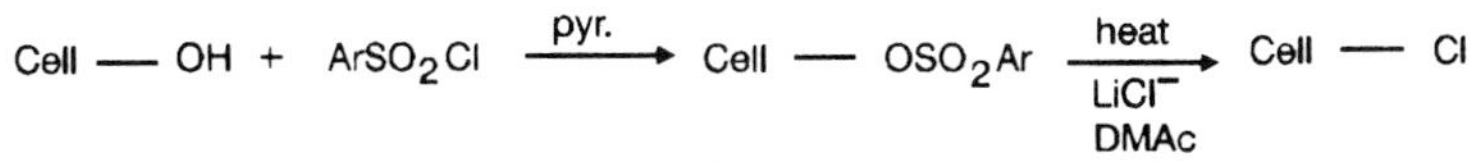

Scheme 5

Initial efforts to isolate a cellulose tosylate under previously reported conditions yielded an unexpected water-soluble product. Since no other solvent could be found, it was concluded that the polymer had been converted into an ionic derivative. [13]C-NMR analysis of the product in aqueous solution showed two unexpected peaks in the carbonyl region, indicating that side reactions were involved.

Attempts to obtain a resolvable FT-NMR spectrum in water were unsuccessful due to formation of an insoluble product before sufficient scans could be obtained. The polymer solution became visibly viscous after about one hour in the NMR tube, eventually phase-separating into a clear, swollen gel.

A literature search of ionic cellulose derivatives capable of hydrolysis reactions was important to the mechanistic considerations. Vigo, Daigle, and Welch[48] had reported the reactions of cellulose with chlorodimethylforminium chloride in DMF to yield either N,N-dimethylformimidate chloride or cellulose formate, depending on the isolation procedure. Two different products were rationalized by those authors as shown in Scheme 6. The intermediate ionic product could be isolated in an anhydrous solvent such as benzene; the hydrolysis product, on the other hand, would be the formate. Elevating the temperature led to displacement of DMF by the chloride ion and produced a chlorodeoxycellulose. Consideration of this reaction scheme and the reactions of formamides and acetamides with sulfonyl chlorides to yield iminium salts via the Vilsmeier-Haack reaction[49] led to consideration of the series of reactions shown in Scheme 7.

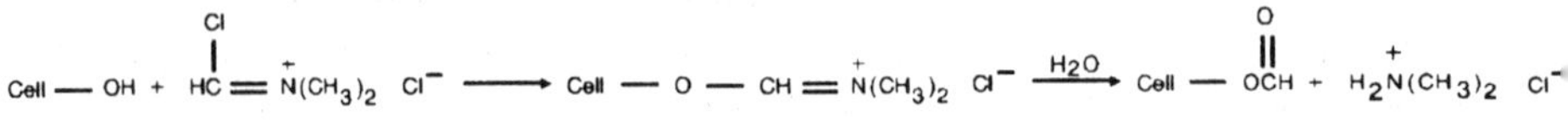

**Scheme 6**

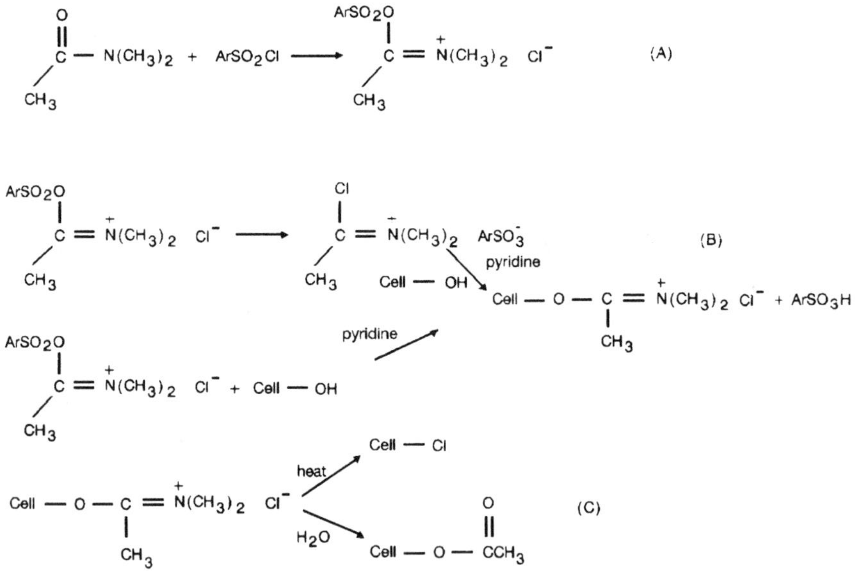

**Scheme 7**

Step one appears to be the formation of the O-(p-toluenesulfonyl)-N,N-dimethylacetiminium salt. The second step, nucleophilic displacement of the tosylate, has two possible routes [Scheme 7, Eq. (B)]. The chloride ion may attack the iminium carbon, displacing the sulfonate, with subsequent displacement of the chloride by cellulose. Alternatively, the cellulose hydroxyl group could attack the tosylated iminium species, directly displacing the tosylate. The chloroiminium cellulose intermediate [Scheme 7, equation (C)] could then be attacked by the chloride ion to produce chlorodeoxycellulose, or hydrolyzed (as in the case of Vigo et al.[48]) to produce acetate.

_Cellulose p-toluenesulfonate._ - Facile preparation of an easily isolated cellulose tosylate under homogeneous conditions continued to be a priority research effort. A key to the successful reaction seemed to be avoiding removal of the tosylate group from the reaction pathway. The approach was to investigate the effects of base strength and nucleophilicity on reaction products.

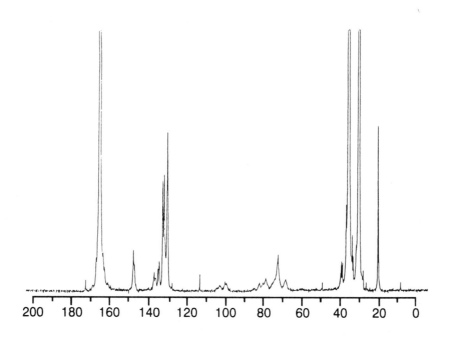

Figure 4.        [13]C-NMR spectrum of tosyl ester of cellulose.

Simple substitution of triethylamine (TEA) for pyridine in the reaction produced the desired results. The tosyl ester, a white solid, was prepared in good yield. The [13]C-NMR spectrum is shown in Figure 4. Resonances assigned to the tosylate methyl and aromatic ring carbons are visible at 21 ppm and 128-145 ppm, respectively. The carbons of the cellulose backbone are seen from 65 to 105 ppm. Peaks at 30, 34 and 162 ppm are DMF from the solvent. Elemental analysis shows minimal incorporation of chlorine. DS values for the tosylate from the initial experiment were 1.1, as determined by elemental analysis.

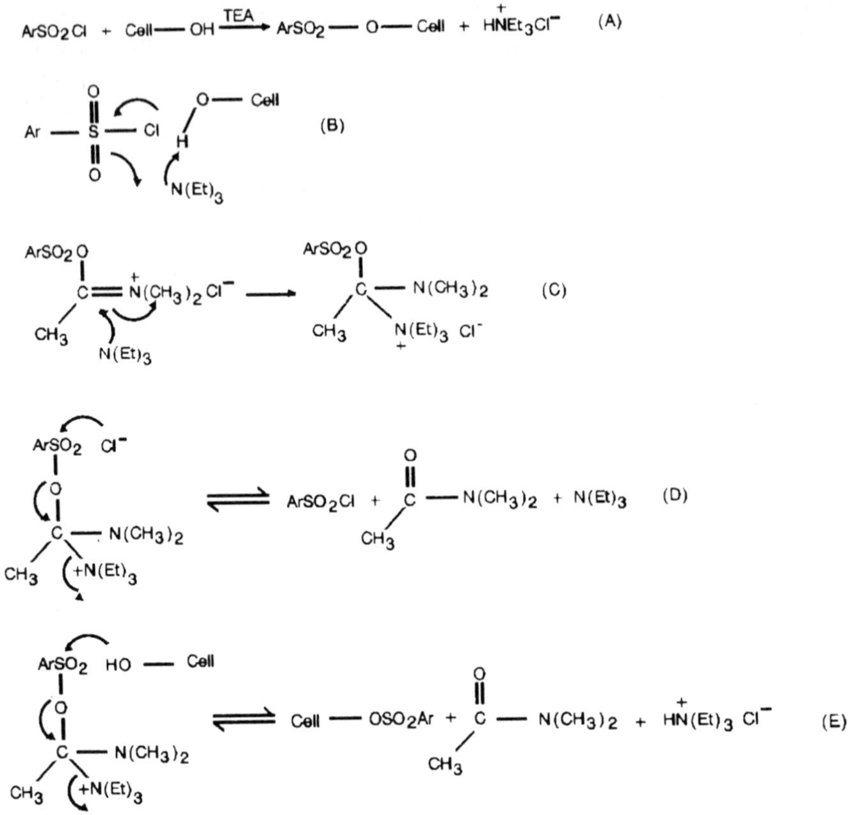

$$ArSO_2Cl + Cell\!-\!\!-OH \xrightarrow{TEA} ArSO_2\!-\!\!-O\!-\!\!-Cell + HN\overset{+}{E}t_3Cl^- \quad (A)$$

(B)

(C)

(D)

(E)

Scheme 8

Obviously TEA, unlike pyridine, favors formation of cellulose tosylate [Scheme 8, equation (A)] over the cellulose iminium species [Scheme 7, equation (B)]. To date the precise reasons for differential reactivity of the two bases has not been determined. However, at least two explanations seem reasonable. TEA may simply be more effective as a catalyst or acid scavenger [Scheme 8, equations (A) and (B)] than pyridine under these reaction conditions. Alternatively, the TEA may react efficiently with a tosylated iminium species [Scheme 8, equation (C)] to produce a tetragonal species that is less accessible to attack at the central carbon by cellulose. The tosylate could be reactivated by reaction with a chloride anion [Scheme 8, equation (D)], or a cellulose hydroxyl group could attack the activated tosylate complex as shown in Scheme 8, equation (E).

Electronic or solvation factors stabilizing similar pyridine complexes may inhibit reactions comparable to equations (D) or (E) in Scheme 8.

## RING-OPENING REACTIONS

**Anhydrides.**   One ring-opening reaction of interest is that of cellulose with cyclic anhydrides as illustrated in Scheme 9. Reactions of this type produce half-acid esters.  Classically, these reactions have been carried out heterogeneously on cotton fabrics or regenerated cellulose at 150-160°C and have resulted in DS values of 0.13-1.08.[50-52]  The products generally displayed increased moisture retention and in few cases (DS = 0.5) were water soluble.[52]

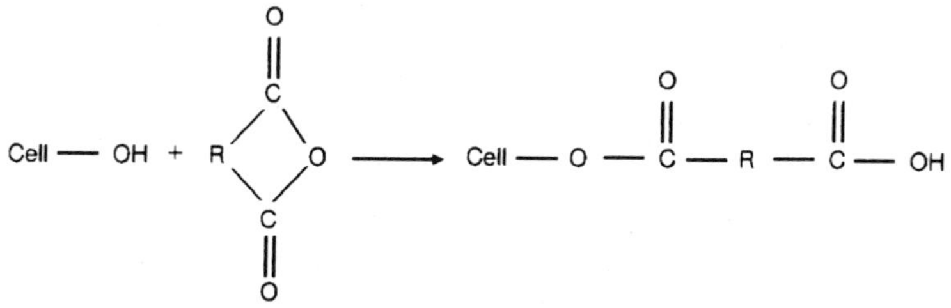

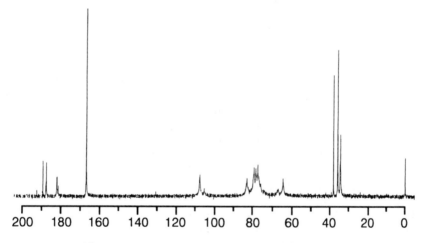

Scheme 9

In order to minimize polymer degradation, sodium carbonate can be slurried into the reaction mixture.[53]  Water-soluble derivatives are produced at 80°C after 20 hours.  Higher temperatures (85-120°C) result in reduced reaction times required to produce water-soluble cellulosics (12.5 to 0.25 hours, respectively).  The $^{13}$C-NMR spectra of these products show the expected two carbonyl resonances of the ester and acid functionalities at 163 and 177 ppm, respectively (Figure 5).

Figure 5.      $^{13}$C NMR spectra of degradation products of cellulose.

When pyridine was used, the reaction temperature was reduced to 80°C with a reaction time of 3 hours. The resulting product was a white, water-soluble powder with a DS = 1.4. Later, TEA was used as an activator/H+ scavenger. Water-soluble cellulosics were prepared in <10 minutes at room temperature (DS = 1). This reaction provides an exception to most reactions in the LiCl/DMAc solvent. As mentioned previously, most cellulose derivatives remain soluble in the LiCl/DMAC system. TEA salts, however, are insoluble in this solvent. As a result, this reaction proceeds until the polymer precipitates. The limiting DS is, in this case, determined by the solubility of the TEA salt of the cellulose half-acid ester.

Maleic anhydride was also reacted with cellulose in the presence of TEA. Precipitate formation was noted to be even more rapid (~2 minutes) than with succinic anhydride. This water-soluble product (DS = 1.2) represented the first successful effort to introduce a pendent site of unsaturation onto the cellulose molecule in the LiCl/DMAc solvent.

Although no other amines were studied, a tertiary amine with properties intermediate to those of TEA and pyridine may lead to better control of reactivity. (Dimethylamino)pyridine may be such a reagent.

**Sultones.** The reactivity of anhydride reagents encouraged the pursuit of similar derivatives with sulfonic acid groups through the ring opening of 1,3-propane sultone. An additional incentive for the use of this reagent was the possibility of preparing cellulose ether linkages (Scheme 10).

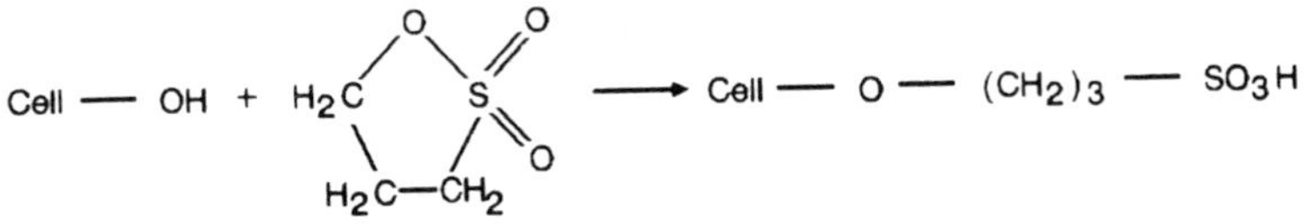

Scheme 10

Derivatization of cellulose with propane sultone was not achieved in LiCl/DMAc. Analysis of the reaction mixture by $^{13}$C-NMR indicated that the sultone ring had opened; however, no evidence of a cellulose substitution reaction was found. NMR observation of the 1,3-propane sultone indicated that the LiCl/DMAc solvent had induced ring opening at room temperature. Preliminary indications were that the chloride ion attacked the sultone ring in preference to the cellulose hydroxyl group. The enhanced nucleophilicity of dissociated anions in polar aprotic solvents is well documented.[54]

**Ring Stability Comparison (Sultone versus Anhydride).** To confirm that the chloride ion was the interfering nucleophile, the sultone was dissolved in neat DMAc and the reaction monitored by $^{13}$C-NMR with increasing temperature. The sultone ring remained intact even at 120°C.

Addition of LiCl to the sultone/DMAc solution, however, resulted in immediate opening of the ring. Substitution of LiBr for LiCl produced the same results with notable shifts in the product $^{13}C$ spectrum, confirming the involvement of the halogen in the reaction.

Attack of Cl⁻ on carbon 3 of the sultone ring produces the ring-opened 3-chloro-1-propanesulfonate (Scheme 11). This newly formed compound has an alkyl halide reactive site. In principle, $S_N2$ displacement of Cl⁻ by the cellulose hydroxyl could result in formation of the sulfonate derivative (Scheme 12). Such a reaction was never observed, consistent with other attempted alkylations and is likely due to the low nucleophilicity of the cellulose hydroxyl groups.

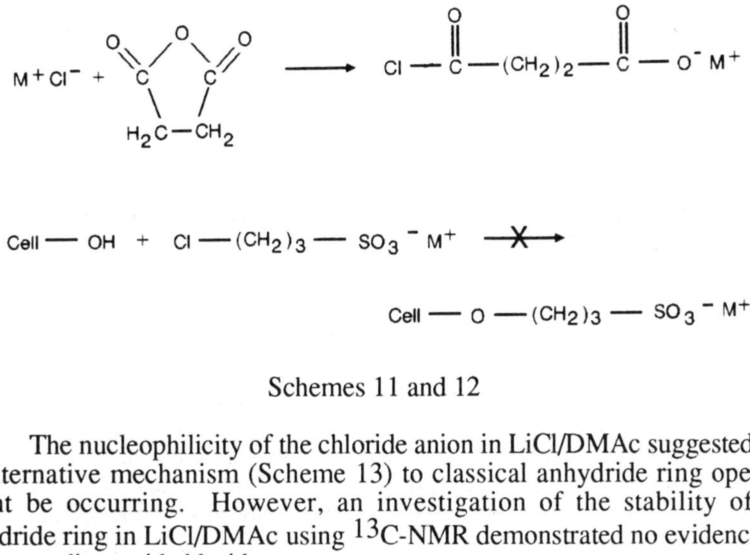

Schemes 11 and 12

The nucleophilicity of the chloride anion in LiCl/DMAc suggested that an alternative mechanism (Scheme 13) to classical anhydride ring opening might be occurring. However, an investigation of the stability of the anhydride ring in LiCl/DMAc using $^{13}$C-NMR demonstrated no evidence for an intermediate acid chloride.

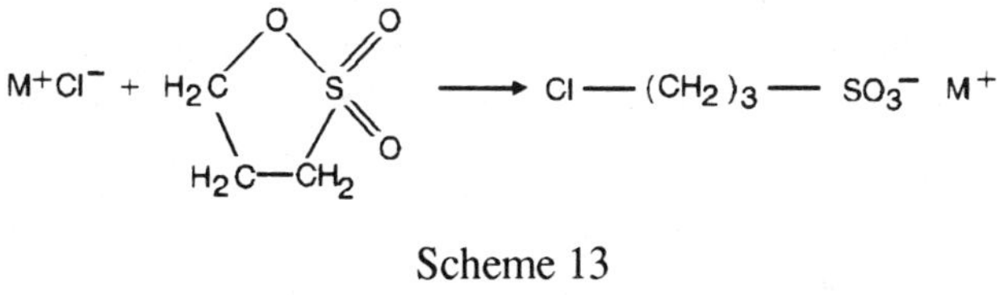

Scheme 13

**Lactones.** Reaction of an anhydride reagent involves attack by the hydroxyl group on the carbonyl carbon to form the ester (Scheme 9). The mechanism likely proceeds through a tetrahedral intermediate. Nucleophilic attack of Cl⁻ on the ring carbon, C-3, of propane sultone, on the other hand, occurs with nucleophilic displacement of the sulfonate group (Scheme 10). Of interest was the reaction pathway for a reagent providing sites for both mechanisms. Lactones provide such models. Nucleophilic attack of the cellulose hydroxyl group on the ε-caprolactone ring could, in principle, follow either of two routes: (1) attack at the carbonyl site A (Scheme 14) to produce the ester linkage and a terminal hydroxyl group or (2) attack on the ring carbon next to the heteroatom site B (Scheme 15) yielding an ether linkage and a terminal carboxylic acid.

Schemes 14 and 15

If the lactone were susceptible to attack at site B, reaction with Cl⁻ might result in simple ring opening with no cellulose substitution (as observed with propane sultone). However, attack at the carbonyl carbon, site B (whether by Cl⁻ or by the cellulose OH), would result in esterification. Interestingly, the reaction of cellulose with ε-caprolactone in LiCl/DMAc yielded the cellulose ester shown in Scheme 14. The reaction was conducted at 80°C in the presence of triethylamine. There is a marked decrease in reactivity as compared to the anhydrides, likely a reflection of the poor leaving group character of the alkoxide anion, in comparison to the carboxylate anion. The resulting derivative was found to be DMSO soluble and water insoluble. After an 18 hour reaction time, the level of substitution remained low (DS = 0.8).[53] This low DS probably accounts for the water insolubility of the product.

**Pyrrolidinone.**   We have already discussed the competitive formation of cellulose tosylate and cellulose deoxychloride proceeding via the Vilsmeier-Haack type of reaction. Further work has led to the discovery of a novel reaction of cellulose with a cyclic intermediate reagent of a Vilsmeier-Haack type of reaction of N-methylpyrrolidinone with toluenesulfonyl chloride. This reaction likely proceeds via an ionic intermediate that further reacts with cellulose, leading to a cyclic iminium chloride (Scheme 16). Subsequent hydrolysis of this derivative could follow two possible pathways shown in Scheme 17.  One route would regenerate N-methylpyrrolidinone and the underivatized cellulose. Alternatively, the ring could open, producing an ester linkage and the salt of a secondary amine. Evidence from [13]C-NMR observations suggests the latter case.[53] This hydrolysis occurs much more rapidly than does that of the similar N,N-dimethylacetamide derivative. The iminium carbon resonance disappears rapidly with the simultaneous appearance of the ester carbonyl. Although the upfield peaks are more subtle, the resulting peaks correlate well with calculated values for the opened ring. The hydrolysis product from this reaction remains water soluble.

Schemes 16 and 17

## CONCLUSION

The LiCl/DMAc solvent for cellulose is clearly a valuable tool for cellulose derivatization. It is not, however, without limitations. As we have seen, the chloride ion is a formidable nucleophile in this medium. Likewise, the carbonyl oxygens of N,N-dimethylacetamide and N-methylpyrrolidinone have been found to be susceptible to reaction. It is, therefore, essential to consider the possibility of side reactions when planning any reaction using the LiCl/DMAc solvent. It has also been shown that, in some instances at least, these side reactions can be circumvented or capitalized upon.

## ACKNOWLEDGMENT

This work is based on research conducted at the University of Southern Mississippi, Department of Polymer Science, and is in no way connected with Eastman Kodak or Tennessee Eastman companies.

## REFERENCES

1.  Nissan, A.H., Hunger, G.K., and Sternstein, S.S., in *Encyclopedia of Polymer Science and Technology*, Vol. III, Bikales, N.M. (Ed.), Wiley Interscience, New York, pp. 131-226 (1965).
2.  Hon, David N.-S., *Polymer News*, *13*, p. 134 (1988).
3.  Krassig, H. in *Cellulose and its Derivatives*, Kennedy, J.F., Phillips, G.O., Wedlock, D.J., and Williams, P.A. (Eds.), Ellis Norwood Ltd., New York. pp. 3-25 (1985).
4.  McCormick, C.L., Lichatowich, D.K., *J. Polym. Sci., Polym. Lett. Ed.*, *17*, pp. 479-484 (1979).
5.  McCormick, C.L., U.S. Patent, 4,278,790 (1981).
6.  McCormick, C.L., Lichatowich, D.K., Pelezo, J.A., and Anderson, K.W., *Polym. Prep., Am. Chem. Soc., Div. Polym. Chem.*, *21*, p. 109 (1980).
7.  Dawsey, T.R. and McCormick, C.L., *J. Mol. Sci-Rev. Macromol. Chem. Phys.*, *C30(3&4)*, p. 405-440 (1990).
8.  McCormick, C. and Callais, P., *Polymer*, *28*, pp. 2317-2323 (1987).
9.  Dawsey, T., PhD. Dissertation, University of Southern Mississippi, Hattiesburg, MS, (1989).
10. Callais, P., Ph.D. Dissertation, University of Southern Mississippi, Hattiesburg, MS (1986).
11. McCormick, C.L., and Shen, T.S. in *Macromolecular Solutions*, (Eds. R. B. Seymour and G. S. Stahl), Pergamon Press, New York, pp. 101-107 (1982).
12. Panar, M. and Beste, L.F., *Macromolecules*, *10*, 1401 (1977).
13. Austin, P.L., U.S. Patent 4,059,457 (1977).
14. McCormick, C.L., Callais, P.A., and Hutchinson, B.H., Jr., *Macromolecules*, *18*, p. 2394 (1985).
15. Conio, G., Corazzo, P., Bianchi, E., Tealdi, A., and Citerri, A., *J.*

*Polym. Sci., Polym. Lett. Ed.*, *22*, p. 273 (1984).

16. El-Kafrawy, A., *J. Appl. Polym. Sci.*, *27*, pp. 2435-2443 (1982).

17. Golova, L.K., Kulichikhin, V.G., and Papkov, S.P., *Polymer Science U.S.S.R.*, *28*, p. 1995 (1986).

18. El-Kafrawy, A., *Chem. Abstr.*, *100*, 141033h (1984) .

19. Germain, J.S., Vincendon, M., *Org. Mag. Res.*, *21*, No. 6, pp.371-375 (1983).

20. Hutchinson, B., Ph.D. dissertation, University of Southern Mississippi, Hattiesburg, MS (1985).

21. Turbak, A.F., *TAPPI*, *67*, p.94 (1984).

22. Vincendon, M., *Makromol. Chem.*, *186*, pp. 1787-1795 (1985).

23. McCormick, C.L., Callais, P.A., and Hutchinson, B.H., Jr., *Polym. Prepr., Am. Chem. Soc., Div. Polym. Chem.*, *24*, p. 271 (1983).

24. Turbak, A.F., El-Kafrawy, A., Snyder, F.W., Auerbach, A.B., U.S. Patent 4,352,770 (1982).

25. Nevell, T. and Zeronian, S., in *Cellulose Chemistry and Its Applications*, Nevell, T., and Zeronian, H., (Eds.) John Wiley & Sons, New York, NY, p. 15 (1985).

26. Wadsworth, L., and Daponte, D., in *Cellulose Chemistry and Its Applications*, Nevell, T., and Zeronian, H. (Eds.) John Wiley & Sons, New York, NY, p. 344 (1985).

27. Zeronian, S.H. in *Cellulose Chemistry and its Applications*, Nevell, T.P., and Zeronian, S.H., (Eds.), Ellis Norwood Ltd., New York, pp. 138-158 (1985).

28. Takahasi, S., Fujimoto, T., Miyamoto, T., and Inagaki, H., *J. Polym. Sci.: Part A: Polym. Chem.*, *25*, pp. 987-994 (1987).

29. McCormick, C.L., Anderson, K.W., Pelezo, J.A., and Lichatowich, D.K., in *Controlled Release of Pesticides and Pharmaceuticals*, Lewis, D.H., (Ed.) Plenum Press, New York, pp. 147-158 (1981).

30. McCormick, C.L., Lichatowich, D.K., Pelezo, J.A., and Anderson, K.W., *ACS Symp. Ser.*, *121*, pp. 371-380 (1980).

31. McCormick, C.L., Callais, P.A., *Polym. Prep.*, *27*, pp.91-92 (1986).

32. Diamantoglou, M., Brandner, A., and Meyer, G., *Chem. Abstr.*, *101* (1984) 132370j.

33. Namikoshi, H., Okumwa, Y., and Sei, T., *Chem. Abstr. 100* (1984) 105398x.

34. Diamantoglou, M., and Meyer, G., *Chem. Abstr.*, *102* (1985) 63312p.

35. Daicel Chemical Ind., Ltd., *Chem. Abstr.*, *100* (1984) 87151e.

36. Khalik, E., Gal'braikh, L., Ilieva, N., Meerson, S., and Rogovin, Z., *Chem. Abstr.*, *79* (1973) 67975w.

37. Daicel Chemical Ind., Ltd., *Chem. Abstr.*, *100* (1984) 105397w.

38. Pawlowski, W., Ph.D. Dissertation, North Carolina State University, Raleigh, NC (1986).

39. Okamoto, Y., and Hatada, K., *Chem. Abstr.*, *104* (1986) 188583k.

40. Isogai, A., Ishizu, A., and Nakano, J., *J. Appl. Poly. Sci.*, *29*, pp. 2097-2109 (1984).

41.   Nevell, T., in *Cellulose Chemistry and Its Applications*, Nevell, T., and Zeronian, H. (Eds.) John Wiley & Sons, New York, NY, p. 15 (1985).

42.   Klein, E., and Snowden, J., *Ind. and Eng. Chem.*, 50, pp. 80-82 (1958).

43.   Masnovi, J., Koholic, D., Berki, R., and Binkley, R., *J. Am. Chem. Soc.*, 109, pp. 2851-2853 (1987).

44.   Binkley, R., and Flechtner, T., in *Synthetic Organic Photochemistry*, (Ed. W. Horspool), Plenum Press, New York, NY, p. 377 (1984).

45.   Heuser, E., Heath, M., and Shockley, W., *J. Am. Chem. Soc.*, 72, pp. 670-674 (1950).

46.   Trask, B., Drake, G., Jr., Margavio, M., *J. Appl. Poly. Sci.*, 33, pp. 2317-2331 (1987).

47.   McCormick, C.L., Dawsey, T.R., and Newman, J.K., Carbohydrate Research, 208, pp. 183-191 (1990).

48.   Vigo, T., Daigle, D., and Welch, C., *Polym. Letters*, 10, pp. 397-406 (1972).

49.   Ulery, H., *J. Org. Chem.*, 30, pp. 2464-2465 (1965).

50.   Cuculo, J., *Textile Research Journal*, pp. 321-326 (1971).

51.   Wadsworth, L., Cuculo, J., and Hudson, S., *Textile Research Journal*, pp. 445-455 (1979).

52.   Johnson, E., and Cucolo, J., *Textile Research Journal*, pp. 283-293 (1973).

53.   McCormick, C.L. and Dawsey, T.R., Macromolecules, 23, pp. 3606-3610 (1990).

54.   Carey, F. and Sundberg, R., *Advanced Organic Chemistry, Part B: Reactions And Synthesis*, Plenum Press, New York, NY, p. 103 (1983).

# Chapter 14

# CELLULOSIC GRAFT COPOLYMERS

## VIVIAN T. STANNETT[*]

*Research Triangle Institute*
*P. O. Box 12194*
*Research Triangle Park, NC  27709*

## INTRODUCTION

The author's active interest in extending the then rather new chemistry of graft polymerization to cellulosics began immediately after he joined the New York State College of Forestry faculty in mid-1952.  Burroughs, Waltcher and Jahn had started such a program using the chain transfer method.  The work was presented at the 1953 IUPAC meeting on polymers in Stockholm.[1]  Although never published, this probably represents the first conscious attempt to graft to a cellulosic polymer.  Work has continued since then throughout the world with more than 1,000 papers and patents having been published.  One excellent monograph covering work through 1980 has been published.[2]

It is pertinent to ask the question in 1990, why such enormous interest was generated and sustained.  The main reason was that grafting provided, for the first time in many years, a new technique for modifying cellulose and its derivatives.  Furthermore, graft copolymers were interesting and potentially useful materials in their own right.  In the case of the cellulosics, there was an additional incentive to the study.  Unlike most of the purely synthetic graft copolymers, one could, in the case of cellulosics, destroy the backbone by acid hydrolysis and measure the properties of the isolated side chains.  This technique was exploited by the author and his colleagues, and much new information was obtained.  A particular aspect of this approach was to uncover details of the kinetics and other synthetic details of the various methods of preparing graft copolymers.  The combined knowledge thus obtained was clearly essential for pursuing structure-property relationships per se.

## SYNTHETIC CONSIDERATIONS

Basically, two main routes for the synthesis of graft copolymers have been developed.

---

[*] Camille Dreyfus Professor Emeritus, North Carolina State University

## Method 1

Described below is the principal method overwhelmingly studied.  It requires generating, on an existing polymer chain, an active species, capable of initiating the polymerization of a monomer.  This species is then, simultaneously or consecutively, reacted with the monomer in bulk or solution or as a vapor.  Thus, schematically, we can write:

$$\text{Polymer A } (P_A) \quad \rightarrow \quad P_A^*$$

$$P_A^* + n\,M_B \quad \rightarrow \quad P_A g(P_B)$$

$P_A^*$ is most commonly a macro free radical, although anionic and cationic initiations have also been developed.  This method is still the most popular and is, in general, a practical and economic technique.  Although some studies have been conducted in a homogeneous liquid phase, even with cellulose itself, the main approach has been heterogeneous.  That is, to generate free radicals on the cellulosic, $P_A$, in the solid state and then expose it to monomer B.  There are advantages and disadvantages to the heterogeneous method.  It has enabled cellulose and cellulosic fibers and films to be grafted in their existing state.  Mainly because of the well known gel effect, the yields are high as are the molecular weights of the grafted side chains.  Furthermore, with the heterogeneous method, techniques have been developed to locate the grafted polymer at the film or fiber surfaces, throughout the polymer or in the interior only.  This is clearly a great advantage for many practical applications.

When the polymer $P_A$ is semicrystalline in nature, as is the case of cellulose itself, a number of properties of the base polymer, including the mechanical properties, can be largely maintained, while adding some of the special properties of the grafted side chain polymer.  An obvious example is the grafting of flame-retardant polymers to cellulose fibers such as rayon and cotton.  More recently, in what is, essentially, the reverse situation, synthetic polymers, such as polystyrene, can be grafted onto cellulosics to impart partial biodegradability.  The intrinsic biodegradability of the cellulosic component, plus the fact that biomass in its many forms can be utilized, has led to renewed interest in cellulosic graft copolymers.

In view of one of the major interests of the honoree of the symposium that served as the basis for this book is in biodegradable polymers, this aspect of graft copolymers will be discussed in somewhat more detail, after presenting some further discussion on the synthetic aspects.  In addition, some discussion on the value of graft copolymers to increase the efficiency of polymer blends will be presented.

The general method of grafting outlined above has some limitations that may or may not be important, depending on the application.  These have been discussed by the author and others[3,4] and can be summarized as follows:

(1)     the molecular weights of the grafted side chains are very high with a wide distribution. They are also difficult to control or to modify. As a consequence, only a few grafted chains per cellulosic backbone are formed. Furthermore, many backbone chains are ungrafted.

(2)     Depending on the grafting method and technique used, there are often considerable amounts of homopolymer formed. This is difficult and costly to remove. It should be pointed out, however, that it is not always necessary to remove such homopolymer or that the ungrafted cellulosic material is undesirable.

(3)     There is little or no knowledge of the nature and location of the linkage between the cellulosic backbone and the grafted side chains. These and other limitations of this approach to the synthesis of graft copolymers have been discussed.[3,4]

**Method 2**

A second general method of grafting can be outlined. Schematically, this can be written as follows:

$$P_A + P_B \quad \rightarrow \quad P_Ag(P_B)$$

This would have many advantages; the molecular weights and molecular weight distributions can be predetermined. Furthermore, if there are many reactive centers on the backbone cellulosic polymer and only one reactive end of polymer B, then essentially any number of side chains can be added.

In the early days, there were many attempts, notably in the Soviet Union, to react the cellulosic hydroxyl groups directly with synthetic polymers, prepared with reactive end groups. Again, these were heterogeneous in nature and only partly successful. It is obviously difficult to develop the close proximity of the reacting groups for reaction to take place. Nevertheless, there have been some limited successes, notably by Rogovin and his coworkers[5] and others.[6,7] A particularly interesting study was that by Avny and Schwenker,[8] who used the living polymer technique to synthesize low molecular weight polystyrene, terminated with carbon dioxide to form carboxylic end groups. These were then converted to the acid chlorides with thionyl chloride. Cotton yarn was reacted with the polystyrene acid chloride in benzene solution after treatment with sodium hydroxide, drying and solvent exchanging. The reactions were successful, but only low yields with a maximum of 22% grafting and low degrees of substitution (DS), maximum 0.037 were obtained. Nevertheless, the approach was encouraging and the advantages clearly outlined inasmuch as homopolymer was avoided and control of the DS and side chain molecular weights made possible. This was also pointed out clearly by Hebeish and Guthrie.[2]

It was nearly 20 years later, however, before reactions were developed which gave excellent yields, using the anionic route, albeit with cellulose derivatives rather than cellulose itself. This work, mainly by Narayan and his coworkers, in general has utilized the living polymer technique. Preformed, living polymer such as polystyrene, was prepared with e.g. sec- butyl lithium as the initiator. These polymers have the advantage of narrow molecular weight distribution, and these properties can be determined separately before forming the graft copolymers. Purely synthetic grafts have been previously reported, using somewhat related techniques. Narayan and Tsao[9] reacted the living polymer with secondary cellulose acetate. The reaction can be shown schematically as follows:

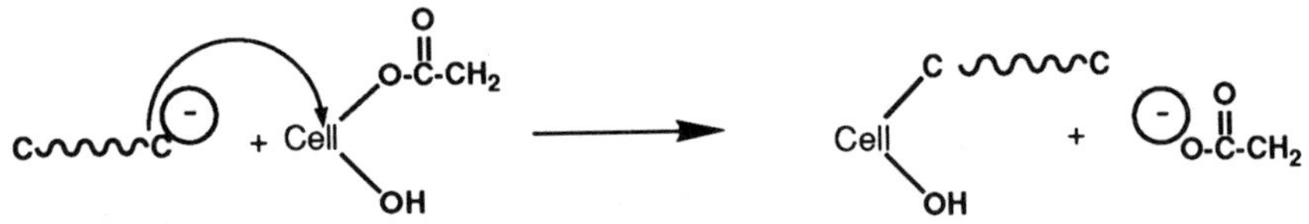

Unfortunately, acetate is a poor leaving group, and the same authors quickly followed by tosylating the remaining free hydroxyl groups.[10,11] Tosyls are better leaving groups, and no free hydroxyl groups were left to deactivate the polyanion. Nevertheless, a considerable amount of homopolymer was formed, plus some degradation of the cellulosic backbone. To prevent these problems, the polyanion (mainly polystyrene anion was studied) was reacted with 1,1-diphenylethylene, a method used earlier in "living polymer" research.[11,12] This gave a less basic and sterically hindered carbanion end group. Excellent yields of graft copolymer were obtained. Mild hydrolysis with ammonia removed all the acetate and remaining tosyl groups, leaving a true cellulose-polystyrene graft copolymer.

In a more recent development, the living polystyrene or other polyanion was reacted with carbon dioxide to form carboxylic end groups.[13] The carboxylate anion is not sufficiently nucleophilic to displace acetate groups but does react efficiently to displace mesylate groups, thus:

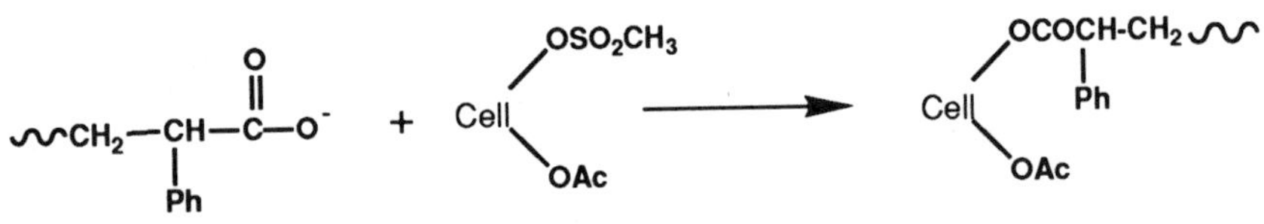

It should be noted that the grafted side chains are attached to the cellulosic backbone by ester linkages, in contrast to the direct reactions referred to earlier. In principle, therefore, the grafted side chains could be removed by alkaline hydrolysis to convert the grafted acetate ester to cellulose itself. A great advantage of the use of the carboxylate polyanion is that the coupling reaction is comparatively insensitive to water. Since there are many ways to add carboxylate groups to synthetic polymers, this considerable advance opens up the way to synthesize many cellulosic graft copolymers. A number of these have already been prepared by Narayan and his group.[14-17] There are two aspects to these interesting developments. One is the original aim to prepare well-defined graft copolymers with various degrees of substitution and side chain molecular weights. This could not be achieved using the current free radical methods, which have dominated the field up to this point. A second aspect is that, using mesylated cellulose derivatives, a large variety of graft copolymers can be prepared, often in a very practical way. A disadvantage, as with the present author's free radical approach, has been that cellulose acetate or other derivatives provide the backbone rather than cellulose itself. It could be, however, that these grafts can be converted to celluloses grafts *in situ*, for example by mild alkaline hydrolysis.

The present author and his colleagues have clearly acknowledged the limitations of the free radical approach to cellulosic graft copolymer synthesis. Nevertheless, the careful removal of the two homopolymers and characterization of the cellulosic backbone and the grafted side chains remaining, did give, for the first time, considerable insight into various aspects of grafting. By fortunate coincidence, this early work and the recent work by Narayan et al. involved the secondary cellulose acetate-polystyrene graft copolymer system. This "special relationship" will be reviewed with respect to two aspects of graft copolymer structure-property relationship, the compatibilization of polymer blends and the imparting of biodegradability to polymers.

## Cellulose Acetate Graft Copolymers as Compatibilizing Agents for Their Blends

Secondary cellulose acetate and polystyrene are highly incompatible polymers. The addition of even a few percent of one polymer to the other leads to white opaque films upon casting, for example, from a common solvent. The addition of a graft copolymer of the two, leads to much clearer films. This was demonstrated and discussed in some detail, using carefully prepared, purified, and characterized grafts, made by radiation-initiated free radical polymerization.[18] A plot of the percent light transmission (normalized) of films versus the composition is shown in Figure 1. The addition of about 25% of a 50:50 graft copolymer rendered a 50:50 level of 75% transparency as indicating compatibility; a phase diagram of such blends is presented in Figure 2. The addition of the graft copolymer shows clear films over a wide range of compositions. The viscosity average molecular weights of the cellulose acetate backbone and the grafted polystyrene side chains were very similar. This gave the best results, compared with widely differing values. However, blends of nonmatching molecular weights were much more efficient in conferring transparency than when used alone. Details are given in full in reference 18.

The clarity of such films is, of course, only one measure of compatibility and is a function more of the size of the domains that may exist in such systems. Evidence that such domains do exist is provided by, for example, the diffusion constants of water in the films. If the mixing was on a truly molecular level, the diffusivities would be roughly intermediate. In fact, they lie close to those of cellulose acetate, whereas polystyrene itself has a diffusion constant for water nearly 16 times larger. The domains are small enough not to scatter light but large enough for the water vapor to diffuse

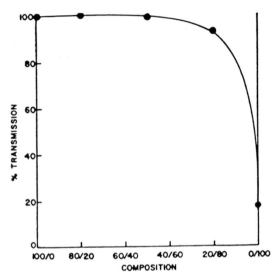

Figure 1.    Light transmission vs composition of films made from a 50:50 blend of secondary cellulose and polystyrene and a 44.1% polystyrene-grafted cellulose acetate.    Reproduced with permission from ref. 18.

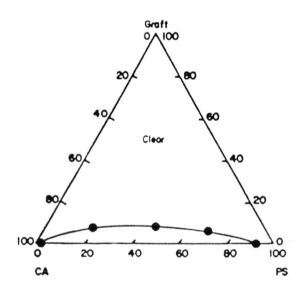

Figure 2.    Phase diagram of blends of secondary cellulose acetate, polystyrene and a 44.1% polystyrene-grafted cellulose acetate. A 75% light transmission of film was selected as an arbitrary compatibility limit. Reproduced with permission from ref. 18.

through the rate determining cellulose acetate domains. Since this early work, a large amount of work has been presented, showing this aspect of the compatibilization of blends by grafts of the parent homopolymers. This has mainly been concerned, however, with purely synthetic polymer systems.

Recently, a brief initial study of the use of cellulose acetate-polystyrene grafts as compatibilizing agents for blends of the two parent homopolymers has been presented by Narayan et al.[19] The anionic approach to grafting was employed, as described earlier. There were differences from the earlier approach of the present author and his colleagues, apart from the actual synthetic method used. The cellulose acetate used in the early studies had a viscosity average molecular weight of 111,000 and a D.S. of 2.25, whereas the anionic synthesis had a D.P. of 110 ($M_w$ ~60,000) and a D.S. of 2.47 with the remaining hydroxyl groups mainly mesylated. The polystyrene side chains had a viscosity average molecular weight of 119,000 in the free radical polymers and 20,000 with the anionic grafts. The blends prepared in solution used dimethylformamide in both investigations. The blends were cast into films and the morphology studied by Narayan et al. by photomicrography. Separate phases always appeared to be present but the size of the dispersed phase was reduced by the addition of graft copolymer. This new study will, undoubtedly, be expanded, including the mechanical behavior of the various blends. Already, the use of the grafts to impart bonding of polystyrene to resin treated wood has been studied.[20] Interesting new composites can be produced in this way and the initial results are encouraging. These and the results of other studies will be eagerly awaited by those interested in graft copolymer technology.[21,22]

## Biodegradable and Biomass Aspects

A brief summary of the areas of application that are in use or under active consideration will follow this more specific aspect. Following the initial euphoria accompanying the development of cellulosic graft copolymers, there was a long period of disenchantment. Some of the reasons for this have been discussed. Apart from a number of promising uses to be discussed later, two considerations have arisen in recent years to bring about renewed interest and optimism in their future. One obvious virtue, which will not be expanded upon here is that such grafts can produce plastics, rubbers, fibers and so on, which are not completely utilizing biomass but are at least up to 50% non-petrochemical based. This aspect, pointed out by Narayan et al. in 1985[10] with respect to grafting to wood, has been followed by a number of publications from the same authors.[22] Their synthetic method is elegant and effective, particularly with cellulose derivatives, and it has recently been extended to lignin. The present author has not specifically addressed this aspect of the possibilities of grafts as biomass-based materials.

The second consideration is that cellulose and its derivatives, particularly at low degree of substitution, are clearly biodegradable by enzymes and, of course, by any type of acid hydrolysis. Although much has been written and discussed about the survival of cellulose in landfills, it is, in fact, highly degradable. A general discussion of the overall problems of degradable plastics, for example, has been published, including an interesting contribution by Andrady,[23] concerning plastics in the marine environment. A more recent review with specific reference to cellulosic grafts has been published by Narayan.[24]

Early in the 1970s, Gilbert and his students and the present author demonstrated that biodegradable block copolymers can be prepared by the incorporation of cellulose or amylose blocks into an otherwise purely synthetic backbone. Briefly, cellulose or amylose is acetylated and acid hydrolyzed to give oligomeric blocks with terminal hydroxyl groups. These can be reacted with, e.g. diisocyanates which can be reacted with various hydroxy terminated synthetic polymers. The resulting block copolymer can be readily hydrolyzed and converted from cellulose acetate to cellulose blocks. These were clearly shown to be degraded by cellulose and amylose enzymes, more efficiently than cellulose or amylose themselves. Furthermore, acid controls did not degrade, so it was purely enzymatic. In addition, films were converted to friable powders, a fact with considerable environmental significance. Work up to 1982 and 1985 has been summarized respectively in two review chapters.[25,26] Gilbert and his colleagues have continued this work and recently extended the nonpolysaccharide blocks to poly(amino acid) blocks. Surprisingly, little degradation with cellulose was reported.[27] No doubt, this work will continue, and completely biodegradable block copolymers will be prepared. Other workers are pursuing methods of preparing peptide and other blocks and grafts of polysaccaride and are included in reference 27.

The elegant and controllable syntheses described above are pioneering and of considerable potential value, but are elaborate and expensive to carry out. A much simpler but less controllable approach would be to simply graft either to cellulose directly or to cellulose acetate, followed by deacetylation.

Penn, Stannett and Gilbert carried out such syntheses, both by condensation[28] and by the more conventional and familiar free radical process.[29] Again, secondary cellulose acetate was used for convenience, followed after grafting by deacetylation. The chain transfer method of grafting was used, wherein the cellulose acetate was partly mercaptoethylated with ethylene sulfide following the procedure of Hermans and Chaudhuri.[30] Styrene was then polymerized in benzene solution, containing small pieces of mercaptoethylated cellulose acetate in film form. Benzoyl peroxide was the initiator. The products were freed from the two homopolymers by extraction with toluene, followed by acetone. The grafts were then deacetylated with sodium methoxide in methanol. Enzymatic hydrolysis of the samples was carried out with 0.5% cellulysin solution at pH 5 for 4 days. Controls at the same pH but without the enzyme showed essentially no degradation. The results are summarized in Table 1. As with the earlier work with the block copolymers, the films were converted to friable powders after degradation, although they survived the deacetylation step in good form. The grafted samples degraded more and faster than cellulose itself.[29]

Table 1.  Enzymatic Degradation of Cellulose-Polystyrene Graft Co-polymers.

| Sample | % Grafting | % Wt Loss* |
|--------|-----------|-----------|
| 1. Enzymatic | 7.2 | 11.4[a] |
| 2. Enzymatic | 14.5 | 40.0[a] |
| 1. Control | 7.2 | 1.6 |
| 2. Control | 14.5 | 0 |

\* Of deacetylated polymers

[a] After 4 days in 0.5% Cellulysin solution pH 5 controls were conducted after 4 days at pH 5 in absence of Cellulysin. Full details may be found in reference 29.

Degradation studies on cellulose-styrene maleic anhydride copolymer graft copolymers have recently been reported by Narayan et al.[31] Direct grafting through esterification was carried out, using cellulose acetate (DP = 2.45) in dimethyl sulfoxide at 108°C. This method is a departure from the anionic technique used previously by this group of researchers. It is more direct and economic and leads to half ester linkages. Amylose grafts were also prepared and studied.

The degradation studies utilized not cellulase but a thermophilic styrene degrading *bacillus* strain. In some cases, the culture medium was supplemented with an organic nitrogen source of yeast extract. Direct metabolism to carbon dioxide was used to monitor the degradation. It was interesting that these styrene-maleic anhydride copolymers alone degraded more than polystyrene itself. The cellulose acetate copolymer also degraded well but less than the corresponding amylose grafts. The work was very preliminary but of considerable interest *per se*. In this approach, not only the biodegradable polymer needs to be designed, but also an appropriate waste disposal scheme.

## OTHER APPLICATIONS

Applications of graft copolymers in general and in cellulose grafts in particular have been less emphasized than the synthesis and characterizations. With cellulose grafts, their possible use for super water-absorbing polymers is still very much a possibility.[32] In the past 5 years, research into possible applications has increased, and close to 40% of papers and particularly patents are concerned with this aspect. This is especially true of Japanese research. A breakdown worldwide shows that the active areas are for membranes and ion exchange materials. Enzyme immobilization, antimicrobial hemostatic and related uses including controlled release and biocompatibilitiy are also being actively researched. Other application studies involve coatings and for diazoprinting, copying and recording applications. Limited research is being undertaken for adhesives, catalyst support systems, oil absorbers, foaming agents, and water treatment agents for the paper and textile industries. It is not clear, however, whether any commercialization of these studies has been achieved at the present time.

## ACKNOWLEDGMENT

Dr. Ramani Narayan is thanked for making available much unpublished work and for many helpful comments and suggestions.

# REFERENCES

1. E. C. Jahn. paper presented at the IUPAC conference on Macromolecules. Stockholm, 1953.
2. A. Hebeish and J. T. Guthrie. *The Chemistry and Technology of Cellulosic Copolymers,* Springer-Verlag, New York, 1981.
3. V. T. Stannett in *Graft Copolymerizations of Lignocellulose Fibers. ACS Symposium Series 157*, 3-20 (1982).
4. R. Narayan in a number of papers starting with reference 11 in this review.
5. Z. A. Rogovin. *J. Polym. Sci.,* 48, 443 (1960) and papers cited therein.
6. C. Bruneau. *Compt. Rendu, 252,* 1413 (1961).
7. C. W. Schroeder and F. E. Condo. *Text. Res. J.,* 27, 135 (1957).
8. Y. Avny and R. F. Schwenker. *Text. Res. J.,* 37, 817 (1967).
9. R. Narayan and G. T. Tsao. *ACS Polym. Preprints, 25(2),* 29 (1984).
10. R. Narayan, M. Krauss, and G. T. Tsao. *ACS PMSE Preprints,* **52** 112 (1985).
11. R. Narayan and M. Shay. *ACS Polym. Preprints, 27,* 204 (1986).
12. R. Narayan and M. Shay in *Recent Advances in Anionic Polymerization,* Ed. T. E. Hogen-Esch and J. Smid, Elsevier Publishing Co., New York, 1987.
13. C. J. Biermann, J. B. Chung, and R. Narayan. *Macromolecules, 20,* 954 (1987).
14. C. J. Biermann and R. Narayan. *Polymer, 28,* 2176 (1987).
15. C. J. Biermann and R. Narayan, *Forest Products J., 38,* 27 (1988).
16. R. Narayan, R. Neu, K. L. Haehl. Proc. of the Tenth Coextrusion Conference "Coex 89", Schotland Business Research, Inc., Princeton, NJ, 1989, pages 89-98.
17. C. J. Biermann and R. Narayan, *Carbohydrate Polymers, 12,* 323 (1990).
18. J. D. Wellons, J. L. Williams, and V. T. Stannett, *J. Polym. Sci., A1(5),* 1341 (1967).
19. R. Narayan and R. P. Neu. *Mat. Res. Soc. Symp. Proc., 197,* 55-66 (1990).
20. R. Narayan, C. J. Biermann, M. O. Hunt, and D. P. Horn in *Adhesives from Renewable Resources,* ACS Symposium Series, 385 , 337-354 (1989).
21. R. Narayan, *Appl. Biochem. and Biotech. J.,* 17, 7 (1988).
22. R. Narayan, work quoted and to be published.
23. A. L. Andrady, *J. Appl. Polym. Sci., 39,* 363 (1990).
24. R. Narayan, *Kunstoffe, 79,* 1022 (1989).
25. R. D. Gilbert, V. T. Stannett, C. G. Pitt, and A. Schindler, *Developments in Polymer Degradation,* N. Grassie (Ed.), 4, 259-293, Applied Science Publishers, 1982.
26. J. W. Carson and R. D. Gilbert in *High Technology Fibers Part A.* Ed. M. Lewin and J. Preston, Marcel Dekker, Inc., New York, 1985, pages 127-168.
27. S. V. Lonikar, R. D. Gilbert, R. E. Fornes, and E. Stejskal, *ACS Polymer Preprints, 31(1),* 640 (1990).

28.  B. G. Penn, V. T. Stannett, and R. D. Gilbert, *J. Macromol. Sci. - Che., A16*, 473 (1981).
29.  B. G. Penn, V. T. Stannett, and R. D. Gilbert, *J. Macromol. Sci. - Chem., A16*, 481 (1981).
30.  D. K. Chaudhuri and J. J. Hermans, *J. Polym. Sci., 51*, 373 (1961).
31.  K. C. Srivastava, Li Nie, Limin Zhang, and R. Narayan, Presented at the AIChE Annual Meeting, Chicago, IL, Nov. 11-16 (1990).
32.  V. T. Stannett, G. F. Fanta, and W. M. Doane. Chapter 8 in *Absorbency,* Ed. P. K. Chatterjee, Elsevier Publishing Co., New York, 1985, pages 257-279.

# Chapter 15

# A NOVEL SYNTHESIS OF MESOGENIC N-METHYL POLYURETHANES, And DEMONSTRATION OF THE EFFECT OF HYDROGEN BONDING ON POPLYURETHANE LIQUID CRYSTALLINE PROPERTIES

F. PAPADIMITRAKOPOLOUS, S. W. KANTOR
AND W. J. MACKNIGHT

*Department of Polymer Science and Engineering*
*University of Massachusetts*
*Amherst, Massachusetts 01003*

## INTRODUCTION

Technological progress is directly linked with the demand for specific and efficient materials to improve mechanical properties and provide unique characteristics. A great number of mesogenic-containing polymers like polyesters, polyethers, polycarbonates, etc, exhibit thermotropic liquid crystalline phases and strongly ordered structures in the solid state.[2] Unlike the above polymers, mesogenic containing polyurethanes are strongly interacting through their hydrogen-bonds. A large body of work details the effect of specific hydrogen-bonding interactions in polyurethane systems.[3-6] The fact that mesogenic containing polyurethanes rarely form stable mesophases,[7-9] is attributed to the strong specific interactions of hydrogen-bonds[1,10-12] and to the poor thermal stability of the urethane bond,[13,14] which decomposes above 200°C. Additionally, the strong intermolecular interactions stabilize the crystalline phase at the expense of the mesophase.

The polyurethane designated 2,4-LCPU-6 has been shown to exhibit monotropic liquid crystalline (LC) behavior.[1]

2,4-LCPU-6

Systematic studies of the structural elements of this system, (H-bonding, asymmetric position of the methyl group in the 2,4-TDI moiety, copolymer effects, etc) and their effects on the phase behavior are lacking due to synthetic difficulties and thermal instabilities. It is of fundamental interest to understand the property-structure relationships of 2,4-LCPU-6 and its family of related mesogenic containing polyurethanes. This is the first of a series of papers in which structure-property relationships of such polymers will be addressed in a systematic fashion. The mesogen of choice for such studies is based on biphenol. Six methylene units on both sides of the biphenol were introduced to provide the appropriate amount of flexibility[15-17] without diluting the mesogen too much. The resulting molecule, 4,4'-bis(6-hydroxyhexoxy)biphenyl (BHHBP or Diol-6), exhibits a highly ordered, stable, smectic mesophase.[18]

$$HO-(CH_2)_6-O-\bigcirc-\bigcirc-O-(CH_2)_6-OH$$

Diol - 6

The introduction of Diol-6 into various polymers, like polyesters[19,20] and polycarbonates[21,22] has resulted in mostly smectic A or C mesophases. The incorporation of Diol-6 into polyurethanes based on 2,4 TDI has also been reported to result in smectic A or C mesophases.[1,10-12] This fact is an indication that the strong intermolecular interactions introduced by hydrogen-bonding do not govern the mesophase morphology. In this chapter we confirm this hypothesis by examining the phase behavior of a polymer similar in structure to 2,4-LCPU-6 but without the hydrogen bonding.

N-Substituted polyurethanes (polycarbamates) have been synthesized in the past, mainly from the condensation of N-substituted diamines with bischloroformates.[23,24] However the N-methyl analogue of 2,4- LCPU-6 is difficult to prepare in this way due to the highly unstable nature of the diamine (2,4 TMA) (**Scheme I**). Here we report the synthesis of high molecular weight N-Methyl 2,4 LCPU-6 (hereafter referred to as NM-2,4-LCPU-6) using a novel high temperature polymerization of a biscarbamoyl chloride with a mesogenic diol (**Scheme II**). This novel high temperature polymerization is closely related to polyester high temperature polymerization from acid chlorides and alcohols, and manifests most of the advantages of such reactions. Bilibin et al.[25-28] as well as other researchers[29] have reported that this polymerization provides high molecular weight polyesters with excellent molecular weight control utilizing monomer stoichiometry. In contrast, normal polyurethane polymerization from the condensation of diisocyanates and alcohols results in moderate molecular weight polymers with poor molecular weight control.[30]

In this chapter we report the detailed thermal and morphological characterization of NM-2,4-LCPU-6. The results obtained for NM-2,4,-LCPU-6 are directly compared with those for 2,4-LCPU-6 previously obtained from our laboratory. The thermodynamic effects of hydrogen bonding on the mesophase are clearly demonstrated on the basis of the observed entropy and enthalpy of fusion,[17,31] as well as X-ray diffraction data. Schematic temperature dependences of the Gibbs free energy[32] for 2,4-LCPU-6 and NM-2,4,-LCPU-6 rationalize the monotropic-enantiotropic behavior as a function of H-bond content. Finally, we discuss some interesting behavior of the NM-2,4,-LCPU-6 near its glass transition temperature.

<u>SCHEME I</u>

SYNTHESIS OF MONOMERS

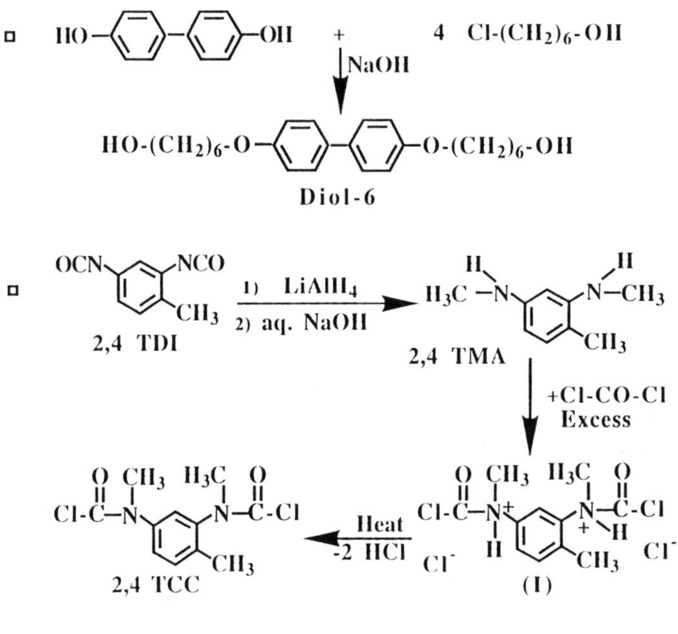

## SCHEME II
### POLYMERIZATION

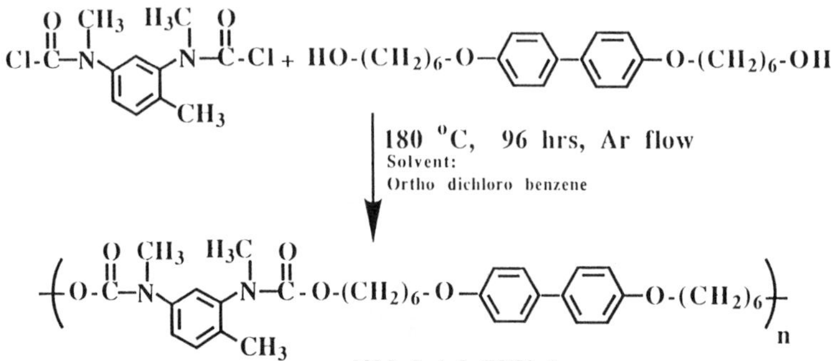

$$180 \ ^{\circ}C, \quad 96 \ hrs, \ Ar \ flow$$
Solvent:
Ortho dichloro benzene

NM-2,4-LCPU-6

## EXPERIMENTAL

### Materials

All chemicals were obtained from Aldrich except for 2,4-toluene diisocyanate (2,4-TDI) which was obtained from Fluka. The reaction solvents were dried and distilled before their use and recrystallization solvents were stored previously over activated 3-4 Å molecular sieves. Prepurified Ar and $N_2$ inert gases were previously passed over BTS catalyst ($O_2$ scavenger)[33] and $CaCl_2$ desiccant.

### Synthesis and Characterization of 2,4-LCPU-6

This is described elsewhere.[11]

### Synthesis of NM-2,4-LCPU-6    (Schemes I &II)

Synthesis of Diol-6. In a 500 mL three-neck, round-bottom flask fitted with a condenser, pressure equalizing dropping funnel, inert gas inlet and magnetic stirrer, was added 250 mL absolute ethanol and 16.0 g (0.4 mol) NaOH. Purified Ar was bubbled through the solution for 15 minutes. Subsequently 18.6 g (0.1 mol) 4,4'-dihydroxybiphenyl was added, maintaining the bubbling of Ar for 15 more minutes at ambient temperature to prevent oxidation of the dianion of biphenol to diphenoquinone (purple color). The resulting slurry was heated to reflux, and 61.1 g (0.44 mol) 6-chlorohexanol, previously purged with Ar, was added slowly. The reaction mixture is refluxed for 24 hours, cooled to room temperature and poured into 1 liter of distilled ice water. The crude solid was washed twice with distilled water, acidified to pH 5-6 with HCl and filtered. The solid product was washed with distilled water to pH 7, dried and recrystallized twice from dioxane-activated charcoal. High purity Diol-6 can be obtained by final recrystallization from ethyl acetate-charcoal and drying under vacuum at 100 °C for a day (mp = 179 °C, yield 70%). *Elemental analysis*; calculated for $C_{24}H_{34}O_4$ : C, 74.57% ; H, 8.87%. found : C, 74.52% ; H, 8.86%.

Synthesis of 2,4 TMA.[34-36] A 2 liter flask was charged with 470 mL of a 1.0 M diethyl ether dispersion of $LiAlH_4$ and 35.0 g (0.201 mol) of freshly distilled 2,4 TDI diluted with 150 mL diethyl ether was added slowly with stirring. The reaction mixture was refluxed for 3 hours before the excess $LiAlH_4$ was decomposed with water. The reaction complex, $LiAl[NR(CH_3)]_4$,[34] was hydrolyzed with 500 mL of a 30% NaOH solution. After the ether was distilled, the resulting mixture was refluxed for 2 hours. The brown oil was extracted with ether, dried with $CaH_2$ and vacuum distilled over $CaH_2$ to give 19.9 g (66% yield) of a colorless liquid (bp = 117-118 °C at 0.7 mmHg). **This compound turns yellow after 5 minutes under vacuum.** The more stable hydrochloride salt decomposes above 200 °C prior to melting. *Elemental analysis* ; calculated for $C_9H_{16}N_2Cl_2$ : C, 48.44% ; H, 7.23% ; N, 12.56% ; Cl, 31.78%. found : C, 48.62% ; H, 7.22% ; N, 12.41% ; Cl, 31.64%.

Synthesis of 2,4 TCC.[36-38] A 1000 ml flask was charged with 300 mL dry, $O_2$ free ethyl acetate, 37.2 g (0.125 mole) of triphosgene[39] and 7.1 g (0.047 mol) of freshly distilled 2,4-TMA was added at 25°C, forming immediately a white precipitate (I). Overnight reflux at 55 °C resulted in a clear solution, which was distilled to dryness. The residue was recrystallized three times from a mixed solvent [$CCl_4$/hexane 3:1, with charcoal] to yield 7.1 g of 2,4-TCC (55% yield), mp = 116.5 °C. *Elemental analysis* ; calculated for $C_{11}H_{12}N_2O_2Cl_2$: C, 48.02% ; H, 4.40% ; N, 10.18% ; Cl, 25.77%. found: C, 47.98% ; H, 4.30% ; N, 10.12% ; Cl, 25.49%.

Polymerization. (**Scheme II**) A dry Schlenk tube with condenser was charged with 0.50000 g (1.8173 mmol) 2,4-TCC, 0.70167 g (1.8154 mmole) Diol-6 (0.1% excess TCC) and 12.0 mL orthodichlorobenzene, freshly distilled over $CaH_2$. The reaction mixture was maintained between 175 and 180 °C under a slow stream of dry, $O_2$ free Ar for 96 hours and precipitated in MeOH. The solid was redissolved in $CH_2Cl_2$ and reprecipitated twice with MeOH, Soxhlet extracted in hot MeOH and vacuum dried to give 0.930 g of NM-2,4-LCPU-6 (yield 87.0%). Its inherent viscosity in $CH_2Cl_2$ at 30.0 °C was 0.82 dL/g. *Elemental analysis* ; calculated for $C_{35}H_{44}N_2O_6$ repeat unit : C, 71.40% ; H, 7.53% ; N, 4.76% ; Cl, 0.00% found : C, 71.38% ; H, 7.51% ; N, 4.68% ; Cl, 0.18%.

## Characterization Techniques

Inherent Viscosities. Inherent viscosities for NM-2,4-LCPU-6 were determined at 30.0°C in $CH_2Cl_2$, using a Cannon-Ubbelohde viscometer. The NM-2,4-LCPU-6 polymers referred as low and as high molecular weight were determined to have an inherent viscosity $[\eta]_{inh}$ of 0.48 and 0.82 dL/g, respectively.

Solution NMR. Solution $^1$H-NMR spectra of NM-2,4-LCPU-6 and 2,4-LCPU-6 polymers were recorded on a Varian XL-200 operating at 200 MHz in deuterated solvents. All spectra were referenced relative to the solvent chemical shift.

Optical Microscopy. Optical microscopy was performed on a Carl Zeiss Ultraphoto II polarizing microscope equipped with a Linkham Scientific Instruments TMS 90 temperature controller and a TMH 600 hot stage. The hot stage temperature was calibrated with vanillin and potassium nitrate melting point standards.

Thermal Analysis. DSC measurements were conducted with a Perkin-Elmer DSC-7 employing a 20 mL/min flow of dry nitrogen as purge gas for the sample and reference cells. The coolant was ice-water bath except for the case of NM-2,4-LCPU-6 polymer, where chopped dry-ice was employed. The temperature and power ordinates of the DSC were calibrated with respect to the known melting point and heat of fusion of a high purity indium standard. For exothermic and endothermic processes the peak temperatures were taken as the transition temperature, while for the glass transition the midpoint of the heat capacity step was taken as the transition temperature. Long term annealing was performed under nitrogen or vacuum to ensure the absence of thermal degradation.

X-Ray Diffraction. Room temperature WAXS patterns were obtained with a Statton X-ray camera using Ni filtered Cu $K_\alpha$ radiation.

## RESULTS AND DISCUSSION

### Scheme I

This scheme illustrates the synthesis of the monomers needed to achieve the synthesis of the NM-2,4-LCPU-6 illustrated in **Scheme II**. The present polymerization scheme was adopted due to the peculiarities of this system which caused several problems with previously reported methylation techniques. Although it was possible to synthesize the $\alpha,\omega$-bis-chloroformate of Diol-6,[23,36,40] the highly unstable nature of the 2,4 toluene-dimethylamime (2,4-TMA) (**Scheme I**) inhibited the formation of high molecular weight NM-2,4-LCPU-6 using either interfacial or homogeneous polymerization in $CH_2Cl_2$. In addition, the metallation methods of Cooper et al.[41-43] using NaH - $CH_3I$ in dry DMF, resulted in severe molecular weight degradation[43] and only partial methylation, as indicated from solution $^1$H- NMR spectra.

Since the salts of 2,4-TMA are more stable than the free amine, they were utilized to improve the yield in the synthesis of the 2,4 toluene-di(N-methylcarbamoylchloride) (2,4-TCC). When the freshly distilled 2,4-TMA was added to the ethyl acetate-triphosgene solution, a white precipitate

(Scheme I)[44] was formed instantaneously and protected the 2,4-TMA from further degradation. The 2,4-TCC is a very stable compound and easy to purify to high purity levels. Although the synthesis of the NM-2,4-LCPU-6 (Scheme II) resembles polyester polymerization[25-29] from acid chlorides and alcohols, it requires special conditions to take place quantitatively and to result in high molecular weight polymer. The fact that no polymer was obtained from solution polymerization carried out at 90°C, using bases such as pyridine as an acid acceptor, is attributed to the stability of carbamoyl chloride/base salts[36,44] similar to (Scheme I). This salt equilibrates with the hydrochloric/base salt altering the stoichiometry and the efficient removal of the by-product HCl. In addition, reaction temperatures higher than 110°C resulted in severe discoloration of the reaction solution probably due to pyridine side reactions. The absence of base was proven necessary in order to obtain high molecular weight polymer. The reaction without an acid acceptor starts around 160°C and was monitored by the evolution of gaseous hydrogen chloride. The optimum reaction temperature was around 180 °C where the reaction proceeded moderately. A slow stream of dry, $O_2$ free argon gas was used to remove the by-product HCl, in order to drive the reaction to completion. Higher temperatures resulted in lower molecular weight due to the instability of the urethane bond[13,14] above 195 °C. Solution polymerization in high boiling solvents resulted in higher molecular weights than melt polymerization. The inherent viscosities, solvent type and experimental conditions for the melt and solution polymerization are summarized in Table 1. The melt polymerization required much greater

Table 1
Polymerization of NM-2,4-LCPU-6
at Various Polymerization Conditions

| Solvent | Polymerization Time (h) * | Temp. (°C) | $\frac{[N(CH_3)COCl]}{[OH]}$ † | Inherent Viscosity (dL/g) |
|---|---|---|---|---|
| 1,2-dichlorobenzene | 96 | 175-180 | 1.0020 | 0.48 |
| 1,2-dichlorobenzene | 96 | 175-180 | 1.0010 | 0.82 |
| 1,2-dichlorobenzene | 96 | 175-180 | 1.0001 | 0.57 |
| 1,2-dichlorobenzene | 72 | 175-180 | 1.0010 | 0.65 |
| 1-chloronapthalene | 72 | 180-190 | 1.0010 | 0.55 |
| 1-chloronapthalene | 40 | 205-215 | 1.0010 | 0.32 |
| Diphenyl-ether | 40 | 190-210 | 1.0010 | 0.29 |
| Melt | 8 - 8 | 170-175 | 1.0200 | 0.17 |
| Melt | 8 - 12 | 170-175 | 1.0100 | 0.38 |

* At the melt polymerization the first number indicates the polymerization time with the Ar flow on, while the second number indicates the time that a vacuum of 5 mmHg was applied.
† Molar ratio.

control over the polymerization conditions mainly due to the high viscosity of the melt, poor stirring and particularly the loss of 2,4-TCC due to sublimation. For this purpose we added a much greater excess of 2,4-TCC compared to the solution polymerization, and lowered the reaction temperature. A small excess of 2,4-TCC (0.1 mol%) proved to be the optimum for the solution polymerization. This polymerization scheme seems to give excellent molecular weight control based on monomer stoichiometry. Thus, the reaction of carbamoyl chloride with an alcohol proceeds quantitatively[45] under these conditions. The only disadvantage of this polymerization scheme is the low reaction rate, because the carbonyl electrophilicity of carbamoyl chlorides is much less than that of acid chlorides, due to the +R resonance effect of the adjacent nitrogen atom. Additionally, the low reaction rate is exacerbated by the narrow temperature window available (160 °C reaction begins-195°C urethane degradation begins). The polyester of isophthaloyl dichloride and Diol-6 with similar inherent viscosity to the high molecular weight NM-2,4,-LCPU-6 was obtained in 8 hours compared to 4 days for the latter. Nevertheless the molecular weights that can be achieved (at long times) with this method for N-substituted polyurethanes are much higher than those reported for conventional polyurethanes as well as for N-substituted polyurethanes.

The solution $^1$H-NMR spectra of NM-2,4-LCPU-6 and 2,4-LCPU-6 are shown in Figure 1. The molecular weight of both polymers was the highest obtained and the concentration was 5% w/v. The high spectral resolution of both samples indicate a relatively stiff backbone chain. In agreement with the structures, both downfield peaks of N-H protons of 2,4-LCPU-6 (at 9.29 and 8.52 ppm) have been removed, and two new peaks are present in the spectra of NM-2,4-LCPU-6 (at 3.16 and 3.25 ppm) due to the N-CH$_3$. The rest of the spectra[11] for both samples are virtually the same.

Representative 10 °C/min DSC heating and cooling scans of the NM-2,4-LCPU-6 are presented in Figure 2. Curve (A) is usually observed upon heating a sample that has been left at room temperature for more than 5 minutes. The first peak at 45.6°C is associated with the enthalpic relaxation[46-48] at the glass transition temperature. The second endotherm around 56°C is weaker and much broader than the first one, and is associated with the melting of a mesophase. Curve C, the representative cooling scan, shows a very broad exothermic region associated with the formation of this mesophase, which is interrupted by the glass transition. The enthalpy relaxation peak is enhanced because room temperature is 10 to 15°C below the Tg and this is the optimum range for physical aging. In order to avoid the enthalpy relaxation which interferes with the second endotherm, the samples were maintained at temperatures below -20°C after the cooling cycle and the heating cycle was started from -30°C. In this case, curve B was obtained, showing a sharp glass transition at 38.6°C and a broad region of exothermic behavior associated with the isotropization of the mesophase. The molecular weight seems to have very little effect on the transition temperatures and enthalpies for both NM-2,4-LCPU-6 samples. Annealing at 50°C, which is in the middle of the narrow temperature window between solidification and

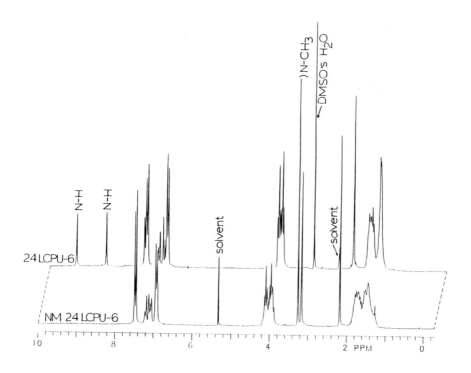

Figure 1.        $^1$H NMR spectra of a 5% solution of 2,4-LCPU-6
$[\eta] = 0.60$ dL/g$^3$ in d$_6$-DMSO and NM-24-LCPU-6 $[\eta]_{inh} = 0.82$ dL/g in
CD$_2$Cl$_2$ at room temperature.

isotropization, increased the amount of the mesophase as well as perfecting it, as shown in Figure 3.

In order to understand the temperature-phase behavior of NM-2,4-LCPU-6 better and to provide a basis for the interpretation of the effect of hydrogen-bonding in the LCPU systems, the optical textures obtained by cooling samples of high and low molecular weight from the melt to the annealing temperature of 50 °C were studied by polarized light microscopy. The results are shown in Figure 4. Rapid cooling from the melt to temperatures below the Tg resulted in a "glassy" appearing material with no

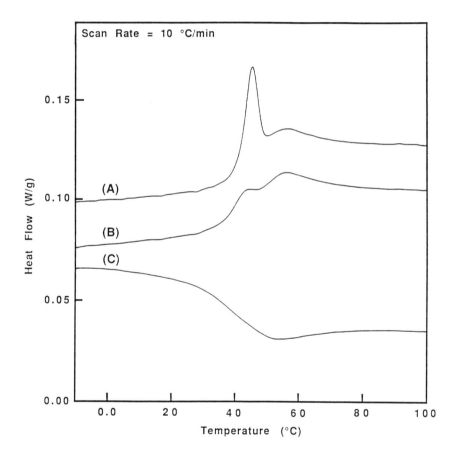

Figure 2. DSC heating (A) and (B) and cooling (C) traces of NM-24-LCPU-6 [η]$_{inh}$=0.48 dL/g. (A) Heating trace of a sample cooled at 10°C/min from the melt and aged around 15 minutes at room temperature, (B) Heating trace of a sample cooled at 10°C/min from the melt, and maintained at -20°C before the heating scan. See text for details.

texture. On the other hand, cooling rates of 10°C/min and lower generated around 52-55°C a very fine texture (white region on the right and left of Figure 4A). It was observed that long before this fine texture formed, the sample could not be sheared because of high viscosity, due to the close proximity of the glass transition. Annealing at 50°C resulted in a threaded texture formed around air bubbles or dust particles shown clearly in Figure 4A, indicating a nucleation type of growth. Further annealing spread and enhanced the threaded textures (Figures 4B and 4C) which are more akin to that reported for a variety of liquid crystalline polymers.[49,50] Further annealing did not change the texture and no evidence of banded spherulitic textures appeared. On heating, these perfected mesophases melted around 60-65°C. The molecular weight affects the coarsening rate of the threaded structure and since all this happens near Tg, the low molecular weight sample developed the morphology more quickly.

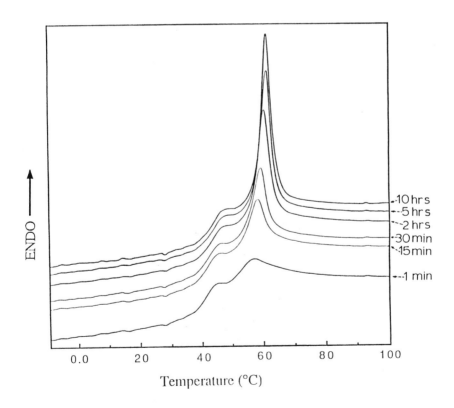

Figure 3.    DSC heating traces at 10°C/minute of NM-2,4-LCPU-6 $[\eta]_{inh} = 0.48$ dL/g previously annealed at 50.0°C for various annealing times.

(A)

(B)

(C)

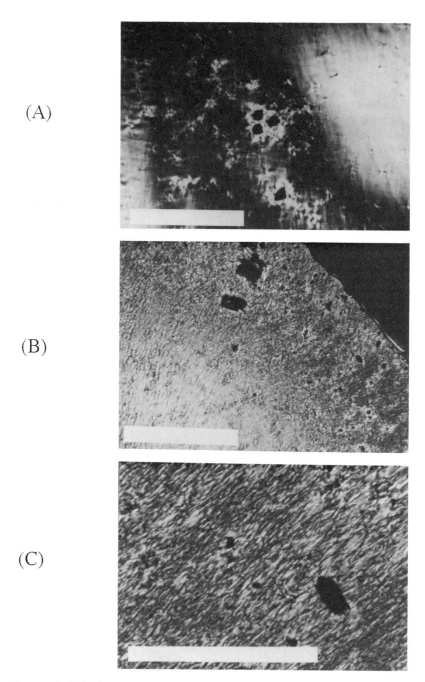

Figure 4. Polarized optical micrographs of NM-2,4-LCPU-6 [$\eta$]$_{inh}$ = 0.48 dL/g, displaying the evolution of a mesophase upon annealing at 50°C. (A) after 10 minutes, (B) after 2 hours, (C) after 6 hours. Bar represents 100 -$\mu$m marker.

(A)

(B)

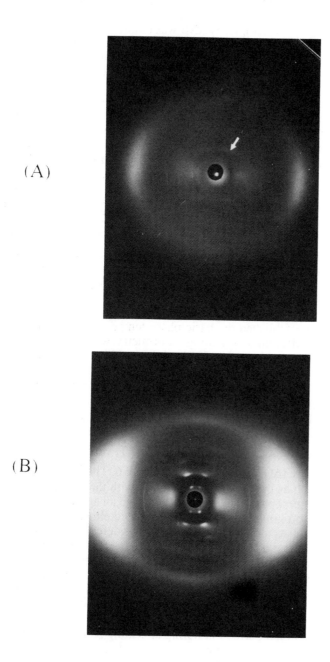

Figure 5. WAXS patterns of a fiber of NM-2,4-LCPU-6 $[\eta]_{inh} = 0.82$ dL/g
(A) drawn from the melt and (B) annealed for 2 days at 50°C. Fiber
axis is oriented vertically.

The observations of differential scanning calorimetry and polarized light microscopy suggest the presence of an enantiotropic mesophase very close to the Tg of the sample. WAXS was employed in order to establish the nature of this mesophase. Fibers of the high and low molecular weight NM-2,4-LCPU-6 were drawn from the melt and quenched in air. The X-ray diffraction pattern for a single fiber of the high molecular weight sample is shown in Figure 5A. The pattern exhibits diffuse equatorial reflections corresponding to a spacing of about 4.5 Å. In addition one can observe four weak off-meridional reflections. These types of patterns have been observed for a number of main-chain liquid crystalline polymers[51-53] and have been attributed to the so-called "Cybotactic" nematic structure, a morphology intermediate between nematic and smectic C phases. The off-meridional reflections correspond to a spacing of approximately 15 Å. The fiber, upon annealing at 50°C, develops a smectic C phase[54,55] embedded in the residual nematic phase (Figure 5A). The pattern exhibits again the diffuse equatorial reflections corresponding to a spacing of about 4.5 Å, while the diffuse off-meridional reflections have increased in number as well as sharpening considerably. The tilt angle is $\beta_t = 20\pm1°$ with layer spacing d = 29.8±0.4 Å and repeat length l = 31.7±0.5 Å. The molecular weight seems to have no effect on the spacing distances and tilt angle, although the higher molecular weight resulted in fibers with higher orientation.

Having established the basic phase-temperature behavior of the NM-2,4-LCPU-6, differential scanning calorimetry was used to provide more detailed information about the effects of thermal history. It is noteworthy that since the isotropic-to-mesophase transition occurs so close to the Tg, lengthy annealing has to be employed in order to achieve an equilibrium morphology. Figure 3 displays the 10°C/min DSC heating traces of the low molecular weight sample previously annealed at 50.0°C for various annealing times. The measurements from Figure 3 of the Tg, the change in the heat capacity $\Delta C_p$ at Tg and the isotropization temperature $T_i$ versus mesophase-isotropic melting enthalpy ($\Delta H_{m,i}$) are shown in Figure 6. The increase of the melting enthalpy as well as the clearing temperature with the annealing time indicate that the initial mesophase is very disordered and an appreciable fraction of the polymer is amorphous. This material is located either in disordered regions between mesophase domains or in the various defects within the mesophase domains.[15] The fact that a completely amorphous material is not obtained, no matter how fast the quench was, indicates that the isotropic-mesophase transition occurs instantaneously,[56] with a cooperative mechanism involving aligning of the mesogenic units. This mechanism leads to a very fine dispersion of the mesophase with a large quantity of grain boundaries and defects. A secondary mechanism of the mesophase perfection is observed, similar to the crystal perfection during annealing resulting in higher melting enthalpy and melting temperature.

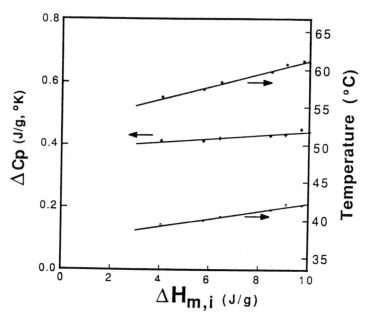

Figure 6. Heat capacity $\Delta C_p$ (♦), clearing temperature $T_i$ (•) and glass transition temperature $T_g$ (o) vs mesophase-isotropic melting enthalpy ($\Delta H_{m,i}$) of NM-2,4-LCPU-6 $[\eta]_{inh}$ = 0.48 dL/g. Data from Figure 3.

As has been previously reported[15,57] low order mesophases such as nematic and smectic A and C exhibit glass transitions due to their inherent disorder. There is an increasing interest in the behavior of the glass transition as a function of phase disorder. Wunderlich et al.[57-59] have addressed this problem from both theoretical and experimental points of view. The dependence of glass transition temperature and change in heat capacity of thermotropic nematic liquid crystal azoxy polyesters on their spacer length and molecular weight has been reported recently by Blumstein et al.[60,61] Zachmann et al.,[62,63] investigating thermotropic liquid crystal copolyesters of ethylene terephtalate (ET), ethylene naphthalene-2,6-dicarboxylate (EN) and oxybenzoate (HB), observed by DSC and DMTA different glass transition temperatures for the liquid-crystalline phase ($T_g^{LC}$) and amorphous phase ($T_g^i$). Characteristically, for the copolyester of ET/EN/HB (35/35/30) the $T_g^{LC}$ was about 40°C lower than the $T_g^i$ and the relative amounts of the two phases was governed by the annealing temperature that produced the mesophase. The free volume of the chains in the isotropic state, below or above the $T_g^i$, will always be higher than their free volume in the mesophase, below or above the $T_g^{LC}$, respectively, due to the loss of two translational modes in the mesophase. Zachmann's interpretation was based on the relative difference of the volumes of the glass

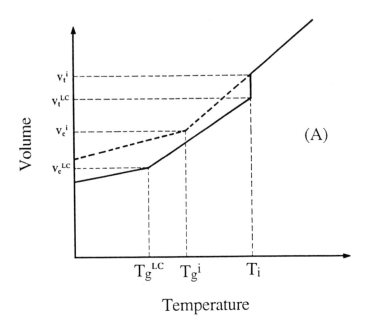

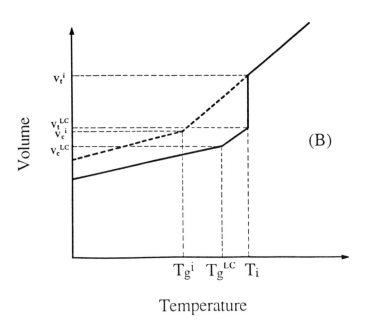

Figure 7. Schematic volume dependence as a function of temperature for $T_g^{LC}$ < $T_g^i$ (A) and for $T_g^{LC}$ > $T_g^i$ (B) (See text for details).

transition $\Delta V_c$ ($\Delta V_c = V_c^i - V_c^{LC}$) versus the change in volume at the mesophase-isotropic transition $\Delta V_t$ ($\Delta V_t = V_t^i - V_t^{LC}$) (Figure 7). This approach explains both possible cases, for $T_g^i$ higher or lower than $T_g^{LC}$. The criteria for $T_g^{LC} < T_g^i$ is $\Delta V_t < \Delta V_c$ while for $T_g^{LC} > T_g^i$ is $\Delta V_t > \Delta V_c$. The magnitude of $\Delta V_t$ corresponds to the mesophase order, while the $\Delta V_c$ is related to the difference in the free volume between the isotropic phase at its glass transition and the mesophase at the glass transition temperature.

The fact that the glass transition temperature of the NM-2,4-LCPU-6 is only 20 °C below the mesophase-isotropic transition affords the possibility to vary systematically the fraction of mesophase present. In addition, because the mesophase transition occurs so fast and so close to the Tg, phenomena like "cold crystallization" of the mesophase from the amorphous phase immediately above Tg, do not interfere with its determination as they do in Zachmann's system. The increase of the Tg with the mesophase melting enthalpy (mesophase perfection) of Figure 6 shows that the mesophase has a higher glass transition ($T_g^{LC}$) than the amorphous ($T_g^i$). It is noteworthy that $\Delta H_{m,i}$ less than 4 J/g was not obtainable due to the fast rate of the mesophase formation. The increase of Tg is linear with $\Delta H_{m,i}$ and on the basis of extrapolation the $T_g^{LC}$ seems to be about 6 °C higher than the $T_g^i$. This can be easily explained with Zachmann's model, assuming that $\Delta V_t > \Delta V_c$. The above assumption is reasonable since Zachmann's copolyesters exhibit less ordered nematic mesophases than the more ordered smectic mesophase of NM-2,4-LCPU-6.

If the increase of isotropization temperature $T_i$, with the mesophase melting enthalpy (Figure 6) is due to an increase of the mesophase order, then the following explanation is proposed. Assuming that $\Delta V_c$ remains more or less constant, then $\Delta V_t$ is the only variable. The $\Delta V_t$ increases with the mesophase order, leading to a linear increase of $T_g$ (Figure 8), in accordance with the model of Zachmann's.

In addition to the linear increase of $T_g$, a subtle increase in the change in $\Delta C_p$, within the experimental error, is also observed (Figure 6). Presently we cannot propose a model to explain this phenomenon which will imply knowledge of the differences of the translational, vibrational, and rotational modes[59] of the present system in its mesophase and amorphous phase.

## COMPARISON OF 2,4-LCPU-6 AND NM-2,4-LCPU-6

Utilizing previously published data[1] from our laboratory, we will attempt to draw conclusions about the effect of the hydrogen bonding on the 2,4-LCPU-6 liquid crystalline properties based on data comparison with the methylated version, NM-2,4-LCPU-6. Representative 10°C/min, normalized DSC heating and cooling scans of the two polymers are presented in Figure 9.

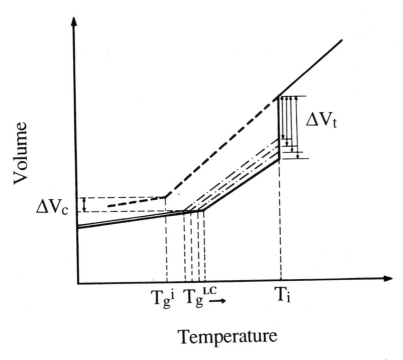

Figure 8.    Schematic volume dependence as a function of temperature for NM-2,4-LCPU-6.  See text for details.

For the low molecular weight ($[\eta]$=0.46 dL/g) 2,4-LCPU-6, upon heating the glass transition is observed at approximately 85°C. This is followed by a region of "cold crystallization" peaking at about 11°C but continuing until the onset of melting as characterized by the endothermic peaks at 158 and 170°C. Since the polymer has been identified as monotropic[1], these two endotherms are associated with crystal melting transitions. Upon cooling, two exotherms at 142 and 138°C are observed. The high temperature exotherm is associated with the isotropic-mesophase transition while its lower temperature counterpart is associated with the mesophase-imperfect-crystal transition. These crystals are associated with the 158°C endotherm of the heating cycle. On the basis of WAXS of fibers drawn from the melt, the 2,4-LCPU-6 shows a smectic A mesophase[11] with a layer spacing of about 31 Å, while a smectic C mesophase[10] can be obtained by shearing the melt between glass plates, probably due to the slower cooling rates of the films compared to those of the fibers. The repeat length of the 2,4-LCPU-6 in both the smectic A and C mesophase is about 31 Å, slightly shorter than in the NM-2,4-LCPU-6, indicating that, although the hydrogen-bonding does not affect the mesophase morphology, it results in a less extended chain conformation (Table 2).

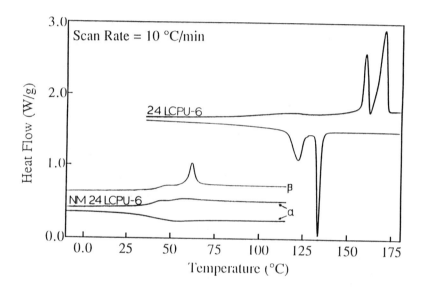

Figure 9. Normalized DSC heating and cooling traces of 2,4-LCPU-6 [η] = 0.46 dL/g[1] and NM-2,4-LCPU-6 [η]$_{inh}$ = 0.48 dL/g. α normal cyclic scan, β after 10 hours annealing at 50.0°C.

Utilizing DSC annealing data from both polymers and the familiar relationship ($T_m = \Delta H_m/\Delta S_m$), the calculated mesophase-isotropic change in entropy for NM-2,4-LCPU-6 is $\Delta S_{m,i} = 0.16$ J/g,°C. In the case of the monotropic 2,4-LCPU-6[1] the calculated isotropic-mesophase change in entropy is $\Delta S_{i,m} = 0.15$ J/g,°C while the calculated crystal-isotropic change in entropy is $\Delta S_{c,i} = 0.24$ J/g,°C. Comparing the $\Delta S_{m,i}$ for the NM-2,4-LCPU-6 with the $\Delta S_{i,m}$ of the 2,4-LCPU-6, and assuming the same degree of disorder for both the molten phases, confirms the conclusion that hydrogen bonding has little effect on the mesophase morphology (Table 2). In other words, in these systems the main effect of H-bonding on the mesophase-isotropic transition is primarily enthalpic in nature[64] and is easily visualized in Figure 9. The transitions in the 2,4-LCPU-6 are much larger than in the NM-2,4-LCPU-6, but the mesophase-isotropic $\Delta S$ is almost the same. On the other hand the $\Delta S_{c,i}$ of the 2,4-LCPU-6 is almost 1.5 times greater than the $\Delta S_{i,m}$, once again indicating the monotropic nature of the

Table 2.
Comparison of 2,4-LCPU-6 and NM-2,4-LCPU-6

| Property * | 2,4-LCPU-6 $[\eta] = 0.46$ dL/g † | NM-2,4-LCPU-6 $[\eta]_{inh} = 0.48$ dL/g |
|---|---|---|
| $T_g$ | 85 °C | 39 °C |
| $T_m$ | 158 °C, 170 °C | ---- |
| $T_{m,i}$ | ---- | 56 °C |
| $T_{i,m}$ | 142 °C | 53 °C |
| $\Delta S_{c,i}$ | 0.24 J/g,°C † | ---- |
| $\Delta S_{m,i}$ | ---- | 0.16 J/g,°C |
| $\Delta S_{i,m}$ | 0.15 J/g,°C † | ---- |
| Mesophase type | $S_a$, $S_c$ | $S_c$ |
| Repeat length | 30-31 Å ‡ | 31.7±0.5 Å |

* Transition temperatures correspond to 10 °C/min heating or cooling scans, while transition entropies were calculated from annealing experiments using the plateau $\Delta H$ values.
† From ref. 1.
‡ From refs. 10 and 11.

sample (Table 2). To elucidate better the monotropic-enantiotropic character of both samples, the schematic temperature dependence of the Gibbs free energy is presented in Figure 10.[32] For the sake of simplicity the phases are represented as solid lines while below the glass transition, where equilibrium

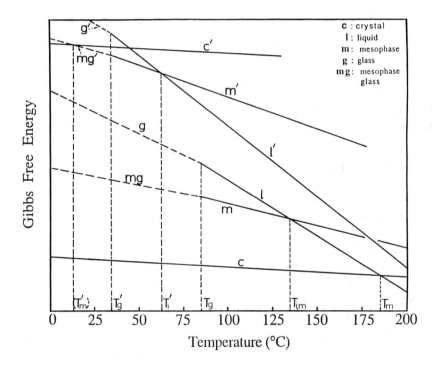

Figure 10. Schematic temperature dependence of Gibbs free energy for 2,4-LCPU-6 and for NM-2,4-LCPU-6 (primed).  See text for details.

cannot be attained, the lines are broken. The above $\Delta S$ values correspond to the slope changes between phases at the transition. In the case of 2,4-LCPU-6 the mesophase is less stable than the crystalline phase mainly due to the strong, highly directional H-bonds in the crystal lattice. Upon cooling from the melt, the material supercools along the liquid phase line until it meets the mesophase. The spontaneous transition to the mesophase provides the necessary nuclei for crystallization, and a transformation takes place to the stable crystal phase. Upon heating, the crystal phase free energy is always less than the mesophase and it melts before it transforms to a mesophase. In the case of NM-2,4-LCPU-6 the mesophase is stable relative to the crystalline phase because H-bonding is lacking. Unfortunately the glass transition temperature lies so close to the mesophase-isotropic transition that it is impossible to observe a crystal phase in NM-2,4-LCPU-6. The relative positions of the various phases in Figure 10 are schematic as data are not available to provide quantitative information.

The absence of hydrogen bonding results also in the enhancement of the solubility of NM-2,4-LCPU-6 in most organic solvents, except saturated hydrocarbons and alcohols. On the other hand, 2,4-LCPU-6 is soluble only in polar aprotic solvents (DMF, DMAC, DMP, DMSO, HMPA, etc).

## CONCLUSIONS

- Scheme II illustrates a novel general route for the preparation of N-substituted polyurethanes with high molecular weights and excellent molecular weight control .

- 2,4-LCPU-6 and NM-2,4-LCPU-6 exhibit similar mesophase morphologies, showing that the presence or absence of H bonding is not important in determining the nature of the mesophase (Table 2).

- H-bonding affects the temperatures of the various transitions, primarily through an enthalpic effect (Table 2).

- The NM-2,4-LCPU-6 is an excellent model to provide understanding of the glass transition temperature as a function of mesophase perfection. The $T_g$ of the mesophase is estimated to be $6^\circ C$ higher than the amorphous phase $T_g$, on the basis of extrapolation to zero fraction of mesophase.

## ACKNOWLEDGMENTS

The authors wish to thank Dr. K. Sanui, Dr. R.W. Lenz, Dr. P. Bhowmik and Dr. H. Fischer for helpful discussions. We are grateful to the Center for UMass-Industry Research in Polymers (CUMIRP) and the Army Research Office, ARO 23941-CH, for support of this research.

## REFERENCES

1.  Smyth, G.; Valles, E. M.; Pollack, S. K.; Grebowicz, J.; Stenhouse, P. J.; Hsu, S. L.; MacKnight, W. J. *Macromolecules*, 1990, 23, 3389.
2.  Chapoy, L. L. *Recent Advances in Liquid Crystalline Polymers*, Elsevier Applied Science Publishers, New York, 1985.
3.  Brunette, C. M.; Hsu, S. L.; MacKnight, W. J. *Macromolecules* 1982, 15, 71.
4.  (a) West, J. C.; Cooper, S. L. *J. Polym. Sci., Polym. Symp.* 1977, 60, 127. (b) Seymour, R. W.; Estes, G. M.; Cooper, S. L. *Macromolecules* 1970, 3, 579.
5.  (a) Koberstein, J. T.; Gancarz, I.; Clark, T. C. *J. Polym. Sci., Polym. Phys. Ed.* 1986, 24, 2487. (b) Christenson, C. P.; Harthcock, M. A.; Meadows, M. D.; Spell, H. L.; Howard, W. L.; Crestwick, M. W.; Guerra, R. E.; Turner, R. B. ibid 1986, 24, 1401.
6.  Coleman, M. M.; Lee, K. H.; Skrovanek, D. J.; Painter, P.C. *Macromolecules* 1986, 19, 2149.
7.  Iimura, K.; Koide, N.; Tanabe, H.; Takeda, M. *Makromol. Chem.* 182, 2569 (1981).
8.  (a) Tanaka, M.; Nakaya, T. *J. Macromol. Sci.-Chem.*, A 24(7), 777 (1987). (b) Tanaka, M.; Nakaya, T. *Macromol. Chem.* 187, 2345 (1986).
9.  Mormann, W.; Brahm, M. *Macromol. Chem.* 190, 631 (1989).
10. Pollack, S. K.; Shen, D. Y.; Hsu, S. L.; Wang, Q.; Stidham, H. D. *Macromolecules*, 1989, 22, 551.
11. Stenhouse, P. J.; Valles, E. M.; MacKnight, W. J.; Kantor, S. W. *Macromolecules*, 1989, 22, 1467.
12. Pollack, S. K.; Smyth, G.; Stenhouse, P. J.; Papadimitrakopoulos, F.; Hsu, S. L.; MacKnight, W. J., submitted for publication.
13. Dyer, E.; Hammond, R. J. *J. Polym Sci.*: A, 2, 1(1964).
14. Yang, W. P.; Macosko, C. W.; Wellinghoff, S. T. *Polymer*, 27, 1235, (1986).
15. Blumstein, R. B.; Blumstein, A. *Mol. Cryst. Liq. Cryst.*, 165, 361 (1988).
16. Saotome, K.; Komoto, H. J *J. Polymer Sci.: Part A-1.* 5, 119 (1967).
17. (a) Macknight, W. J.; Yang, M. *J. Polymer Sci., Polym. Symp.* 1973, 42, 817. (b) Macknight, W. J.; Yang, M.; Kajiyama, T., in *Analytical Calorimetry*; R.S. Porter, J.F. Johnson, Ed. Plenum, New York, (1977)
18. Smyth, G.; Pollack, S. K.; Hsu, S. L.; MacKnight, W. J., submitted for publication.
19. (a) Reck, B.; Ringsdorf, H. *Makromol. Chem., Rapid Commun.* 6, 291 (1985). (b) Zentel, R.; Reckert, G. *Makromol. Chem.*, 187, 1915 (1986). (c) Bualek, S.; Zentel, R. *Makromol. Chem.*, 189, 791 (1988).
20. Fischer, H.; Karasz, F. E.; MacKnight, W. J., to be published.

21.  (a) Sato, M.; Nakatsuchi, K.; Ohkatsu, Y. *Makromol. Chem., Rapid Commun.* 7, 231 (1986). (b) Sato, M.; Nakatsuchi, K.; Ohkatsu, Y. *Makromol. Chem., Rapid Commun.* 8, 383 (1987).

22.  Sato, M.; Kurosawa, K.; Nakatsuchi, K.; Ohkatsu, Y. *J. Polym. Sci., Part A: Polym. Chem.* 26(11), 3077 (1988).

23.  Harrell, L. L.; *Macromolecules*, 1969, 2, 607.

24.  Sorenson, W. R.; Campbell, T. W., *Preparative Methods in Polymer Chemistry*, 2nd ed. Wiley Interscience, New York

25.  Bilibin, A. Yu.; Ten'kovtsev, A. V.; Piraner, O. N.; Skorokhodov, S. S. *Pol. Sci. U.S.S.R*, 26, 12 pp. 2882 (1984).

26.  Bilibin, A. Yu.; Pashkovsky, E. E.; Tenkovtsev, A. V.; Skorokhodov, S. S. *Makromol. Chem., Rapid Commun.*, 6, 545 (1985).

27.  Bilibin, A. Yu.; Zuev, V. V.; Skorokhodov, S. S. *Makromol. Chem., Rapid Commun.*, 6, 601 (1985)

28.  Skorokhodov, S. S.; Bilibin, A. Yu. *Makromol. Chem., Macromol Symp.*, 26, 9 (1989).

29.  Melendez, E.; Navarro, F.; Pinol, M.; Rodriguez, J. L.; Serrano, J. L. *Mol. Cryst. Liq. Cryst.*, 1988, 155, 83.

30.  P. J. Manno, *Urethane Chemistry and Applications (K. N. Edwards)* ACS Symposium Series 172, 5.

31.  Bechtoldt, H.; Wendorff, J. H.; Zimmerman, H. J. *Makromol. Chem.* 1987, 188, 651.

32.  Keller, A. ; Ungar G., To be published.

33.  Shriver, D. F.; Drezdzon, M. A. *The Manipulation of Air-Sensitive Compounds.* 2nd Edition, Wiley-Interscience, New York, 1986.

34.  Finholt, A. E.; Anderson, C. D.; Agre, C. L. *J. Org. Chem.*, 18, 1338 (1953).

35.  Ellzey, S. E.; Mack, C. H. *J. Org. Chem.*, 28, 1600 (1963).

36.  Smith P. A. S. *The Chemistry of Open-Chain Organic Nitrogen Compounds*, vol 1. W. A. Benjamin Inc., New York (1965).

37.  Raiford, L. C.; Alexander, K. *J. Org. Chem.*, 5, 300 (1940).

38.  Weygand, F.; Mitgau, R. *B.* 88, 301 (1955).

39.  Eckert, H.; Forster, B. *Angew. Chem. Int. Ed. Engl.*, 26 (1987) No. 9, 894.

40.  Saotome, K.; Komoto, H. J *J. Polymer Sci.: Part A-1.* 5, 107 (1967)

41.  Hwang, K. K. S.; Speckhard, T. A.; Cooper, S. L. *J. Macromol. Sci.-Phys.*, B 23(2), 153 (1984).

42.  Srichatrapimuk, V. W.; Cooper, S. L. *J. Macromol. Sci.-Phys.*, B 15, 267 (1978).

43.  Adibi, K, George, M. H.; Barrie, J. A. *Polymer*, 20, 483 (1979).

44.  Babad, H.; Zeiler, A. G. *Chemical Reviews*, 73(1), 75 (1973).

45.  Odian, G. *Principles of Polymerization*, 2nd Edit. Wiley-Interscience, New York, (1981).

46.  Struik, L. C. E. Physical Aging in Amorphous Polymers and Other Materials, Elsevier Scientific Pub. Co., New York (1978).

47.  Prest, W. M. Jr.; Chow, T. S. *J. Appl. Phys.*, 53, 6568 (1982).

48.  Kovacs, A. J.; Aklonis, J. J.; Hutchinson, J. M.; Ramos, A. R. *J. Polymer Sci., Polym. Phys. Ed.* 1979, 17, 1079.

49.  Demus, D.; Richter, L. *Textures of Liquid Crystals* ; Verlag Chemie: Weiheim, 1978.
50.  Noel, C. in *Recent Advances in Liquid Crystalline Polymers*; Chapoy, L. L., Ed.; Elsevier Applied Science Publishers, New York, 1985.
51.  Azaroff, L. V.; Schuman, C. A. *Mol. Cryst. Liq. Cryst.*, 1985, 122, 309.
52.  Leadbetter, A. J.; Norris, E. K. *Mol. Phys.* (1979), 38, 3, 669.
53.  Blumstein, A.; Thomas, O.; Asrar, J.; Makris, P.; Clough, S. B.; Blumstein, R. B. *J. Polym. Sci., Polym. Lett. Ed.* 1984, 22, 13.
54.  Azaroff, L. V. *Mol. Cryst. Liq. Cryst.* 1987, 145, 31.
55.  Roviello, A.; Sirigu, A. *Eur. Polym. J.* 1979, 15, 61.
56.  Cheng, S. Z. D. *Macromolecules* 1988, 21, 2475.
57.  Grebowicz, J.; Wunderlich, B. *J. Polymer Sci., Polym. Phys. Ed.* 1983, 21, 141.
58.  Wunderlich, B.; Loufakis, K. *J. Phys. Chem.*, 1988, 92, 4205.
59.  Wunderlich, B. *Polymer Preprints* , 31 (1), 272 (1990).
60.  Kim, D. Y.; Blumstein, R. B. *Polymer Preprints*, 30 (2), 472 (1989).
61.  McGowan, C. B.; Kim, D. Y.; Blumstein, R. B. *Polym. Prep.*, 31 (1), 261 (1990).
62.  Chen, D.; Zachmann, H. G. Submitted to *Polym.*.
63.  Zachmann, H. G.; Chen, D.; Nowacki, J.; Olbrich, E.; Schulze, C. Submitted to *Integration of Fundamental Polymer Science and Technology*, Eds. Kleintjents, L. A.; Lemstra, P. J.; Elsevier, New York, (1990).
64.  Van Krevelen, D. W.; Hoftyzer, P. J. *Properties of Polymers, Their Estimation and Correlation with Chemical Structure*, Elsevier Amsterdam, (1976).

# Chapter 16

# SYNTHESIS AND AGGREGATION PROPERTIES OF IONIC AMPHIPHILIC SIDE CHAIN SILOXANE POLYMERS

DAVID R. ZINT AND PETER K. KILPATRICK

*Department of Chemical Engineering*
*North Carolina State University*
*Raleigh, North Carolina 27695*

## INTRODUCTION

Amphiphilic polymers have attracted considerable attention of late for their application as associative thickeners.[10,22] Comb or regular side chain polymers offer a variety of uses in thermotropic liquid crystalline form as media for optical data storage, electrooptic display devices, nonlinear optical devices, photoconductors, solid polymer electrolytes and stationary phases in chromatography applications.[15]

Recently, efforts to synthesize side chain lyotropic liquid crystal-forming polymers with either amphiphilic or rigid thermotropic mesogenic side chains have revealed that lyotropic liquid crystallinity can be greatly enhanced by attaching the lyotropic LC-forming group to a flexible polymeric backbone. This enhanced stability manifests itself in terms of both enlarged compositional and thermal extent of individual LC phases.[8,9,13,14] Using polymethylhydrosiloxane (PMHS) as a derivatizable hydrophobic backbone, Finkelmann and coworkers have attached a variety of both rigid and flexible amphiphilic side chain groups to this polymer by hydrosilylation.

Finkelmann and Rehage[7] reported the synthesis of several liquid-crystal-forming polysiloxanes with rigid mesogenic side-chains. The mesogenic group was either a substituted biphenyl or a substituted phenyl ester of benzoic acid. These were attached by propyloxy or butyloxy groups to the siloxane backbone, and the end of the side chains consisted of either a methoxy, a hexyloxy or a nitrile group. These side chain polymers all exhibited thermotropic liquid crystallinity, with lower glass transition temperatures than their hydrocarbon analogues because of the flexibility in the backbone.

Finkelmann *et al.*[8] first reported on solvent-induced or lyotropic liquid crystallinity of polysiloxane with amphiphilic side chains. The amphiphilic side-chain was the ester of 10-undecenoic acid and either tetra-, hexa-, or octaethylene glycol monomethyl ether. This nonionic amphiphile was added to the PMHS backbone by hydrosilylation. The monomer undecenoyl-octaethylene glycol monomethyl ether (UOG) exhibits a hexagonal lyotropic LC phase at concentrations from 49-70 wt % surfactant in water and at temperatures from -10 to 20°C. The amphiphilic side-chain polymer in which UOG is coupled to the PMHS backbone of degree of polymerization (DP) 95 exhibits a large region of lyotropic hexagonal phase from 40-75 wt % polymer and from -10 to 50°C. In addition, at higher concentrations (70-90%) and over a wider temperature range (-10 to 65°C), there exists a stable lamellar LC phase. Thus, the polymeric amphiphile was observed to have lyotropic LC phases of enlarged compositional and thermal extent than the analogous monomeric amphiphile. Finkelmann *et al.*[8] speculated that this enhanced stability may be due to restricted translational and rotational mobility of the amphiphilic groups when they are anchored to a polymeric backbone.

Luhmann and Finkelmann[14] documented a similar enhancement of lyotropic liquid crystallinity upon attachment of nonionic amphiphiles to PMHS in which the hydrophobic portion of the amphiphile consisted of a propyloxy or hexyloxy group linked to a biphenyl moiety. The hydrophilic portion of the amphiphilic side-chain was a monomethyl ether of nonaethylene glycol or undecaethylene glycol. As with the flexible hydrophobic side chain, greatly enhanced thermal and compositional extents of the hexagonal and lamellar LC phases were observed. Moreover, this effect was apparent even at very small DPs of 3-6.

Based on the successful attemptsof Finkelmann and coworkers to synthesize nonionic polymeric surfactants that exhibit lyotropic mesomorphism with enhanced stability over the corresponding monomer, it seemed reasonable that ionic amphiphilic side chain polymers based on PMHS could also be synthesized that might exhibit comparable enhanced lyotropic mesomorphism. In this chapter, we describe the synthesis of polymeric surfactants comprised of PMHS functionalized with undecenoic acid side chains, which are subsequently neutralized to yield the corresponding cesium salts of the polymeric carboxylic acid. These materials have been characterized by $^1$H-NMR spectroscopy to determine the number average molecular weight and the degree of side chain functionalization. The aggregation properties of these materials in water have been studied indirectly by surface tensiometry and directly by quasi-elastic light scattering (QLS). The lyotropic liquid crystallinity has been probed by making aqueous solutions with deuterated water ($D_2O$) and observing the resulting deuterium NMR spectra and their dependence on composition. The phase behavior at high concentration and the aggregation behavior at low concentration of the polymeric surfactants are compared to those of the monomeric surfactant, cesium undecenoate.

# EXPERIMENTAL PROCEDURES

## Materials
The hydrophobic polymer backbone, polymethylhydrosiloxane (PMHS) was obtained from Aldrich Chemicals (catalog no. 17,620-6, DP = 30) and Petrarch Systems (catalog no. PS119, $M_w$ = 1500). 10-Undecenoic acid, sodium hexachloroplatinate, and dicyclopentadiene were the purest grades available from Aldrich Chemicals. The reaction and separation solvents tetrahydrofuran, toluene, methanol and 1-propanol were HPLC grade materials from Fisher Scientific. Cesium hydroxide was from Carus Chemicals, $D_2O$ was 99.8% isotopic purity from Wilmad Glass and $CDCl_3$ and $CD_3OD$ were from MSD Isotopes. Undeuterated water was drawn through a four-stage Nanopure filtration system and stored in glass bottles.

## Methods
### Polymer Synthesis
Hydrosilylation of the PMHS backbone by 10-undecenoic acid was performed according to the procedure described by Apfel *et al.*[1]. PMHS and 10-undecenoic acid (4.7-11.6 % excess) were dissolved in one of the three solvents, THF, toluene or 1-propanol. The solution was warmed to the reaction temperature (50-100°C), and the catalyst added (150-6600 ppm). For four of the reactions, the catalyst components (NaPtCl•6HO and dicyclopentadiene) were added directly in solution form to the reaction mixture. For the other two reactions, the catalyst product (dicyclopentadienylplatinum chloride) was synthesized prior to being introduced to the reaction mixture[1]. Active catalyst was prepared by dissolving 0.33 g of NaPtCl•6HO in 7 mL of glacial acetic acid, diluting with 14 mL of water, adding 0.38 g of dicyclopentadiene, heating to 70°C and stirring vigorously for 24 hours. The product catalyst was centrifuged, decanted and washed three times with THF. The hydrosilylation reaction, which ran for 2-24 hours with a nitrogen sparger to agitate the mixture and prevent oxidation, should form predominantly the anti-Markovnikov product. Hydrosilylation was carried out in a 500-mL three-necked round bottom flask. Inserted into the vertical neck of the flask was a 50 cm Allihn drip tip condenser with glass wool in its outlet to reduce the escape of solvent. The reaction mixture was sparged continuously with nitrogen and the reaction temperature was continuously monitored. A Glas-Col heating mantle, controlled by a Staco 120 volt variable transformer, was used to heat the reaction mixture.

## Polymer Isolation
Following reaction, solvents were replaced with methanol by diafiltration of the reaction mixture through a YM-5 cellulose acetate membrane (5000 MW cutoff) in an Amicon 8400 stirred cell ultrafiltration system. The acidic form of the polymer - poly(methyl undecanoic acid siloxane), or PMUS - was converted to the cesium salt form by stoichiometric addition of cesium hydroxide in methanol. The neutralized cesium salt form of the polymer (PMCsUS) was converted to an aqueous solution by diafiltration with either deuterated or undeuterated water.

## Gel Permeation Chromatography

Gel permeation chromatography (GPC) was used to monitor the course of the reaction and reaction solvent replacement during diafiltration. The setup included a Perkin-Elmer Series 10 liquid chromatography pump, supported by a Perkin-Elmer LC-25 differential refractive index (RI) detector, and a Fisher Recordall Series 5000 chart recorder. One GPC column used was a Waters Associates μStyragel™ linear gel permeation column with a reported molecular weight range of 500-1,000,000. The second GPC column used was a DuPont ZORBAX porous silica microsphere (PSM) 60S high performance size exclusion column (HPSEC) with a molecular weight range of 100-10,000.

## Nuclear Magnetic Resonance Spectroscopy

Two nuclear magnetic resonance (NMR) spectrometers were used to characterize the polymer and to probe for evidence of liquid crystallinity: a General Electric GN-300 Omega spectrometer and an IBM-CX 100 spectrometer. The GE GN-300 Omega spectrometer was equipped with a 5 mm dual probe with a proton resonance frequency of 300.522 MHz. Chemical shifts of protons are reported relative to TMS. The IBM-CX 100 spectrometer was equipped with an IBM VTU temperature controller, an Aspect 3000 software package and a drum plotter. A 10 mm deuterium probe with a resonance frequency of 15.371 MHz was used with this spectrometer. Typical spectral conditions in obtaining deuterium spectra were a 5000 Hz spectral width, 12 μs pulse width (corresponding to about 45°), 0.5-1.0 seconds acquisition time and 500-1000 transients.

## Surface Tensiometer

Surface tensions were measured by the DuNuoy Ring detachment method with a Fisher 215 Autotensiomat Surface Tension Analyzer and a 6.005 cm circumference platinum-iridium DuNouy ring. A Fisher Recordall Series 5000 chart recorder was used to display the results of the analyses. A Lauda RC 20 refrigerating circulator was used to maintain constant temperature.

## Quasielastic Light Scattering

The quasielastic light scattering (QLS) Instrument used was comprised of an Innova 70-3 argon ion laser (Coherent) equipped with a Brookhaven Instruments (BI)- 200 SM Automatic Goniometer, a BI-2030 digital correlator, a BI computer, a BI high voltage power supply and a Newport 815 power meter. The system temperature was controlled by a BI RTE-5DD refrigerated circulating bath.

# RESULTS

## Surface Tension Measurements of Cesium Undecenoate-Water Solutions

In order to determine the efficacy of lowering surface tension and the critical micelle concentration(CMC) of the monomer cesium undecenoate, surface tension measurements were made on aqueous solutions of cesium undecenoate in the concentration range from 80 μM to about 0.1 M. The surface tension decreases monotonically (Figure 1) with increasing concentration below about 0.06 M. Above this concentration, the surface tension is virtually independent of concentration. We attribute this leveling of the surface tension to micellization of the surfactant. Application of the Gibbs adsorption isotherm to the surface tension data[2] yields a surface excess concentration for cesium undecenoate just below the CMC of 0.239 nmol/cm$^2$. This corresponds to an area/molecule of approximately 69 Å$^2$.

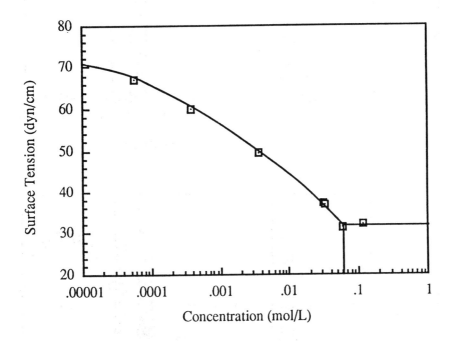

Figure 1.  Dependence of surface tension on surfactant concentration of cesium 10-undecenoate-water solutions.

## $^2$H NMR Spectroscopy of Cesium Undecenoate-D$_2$O Mixtures

Evidence for lyotropic liquid crystallinity of concentrated (50-90 wt %) samples of cesium undecenoate in deuterated water was obtained by deuterium quadrupole NMR spectroscopy.[4,11,17,19] By measurement of the net anisotropy of D$_2$O molecules bound to the surface of the anisotropic liquid crystalline aggregates as gauged by the quadrupole splitting, it is possible to construct a plot of quadrupole splitting with increasing concentration which delineates liquid crystalline phase boundaries. This is shown in Figures 2 for cesium undecenoate-water mixtures at 298and 333 K. Below approximately 50 wt % surfactant, the deuterium nmr spectrum consists of a narrow (< 15 Hz line width) isotropic peak and the mixture is fluid and freely flowing. We

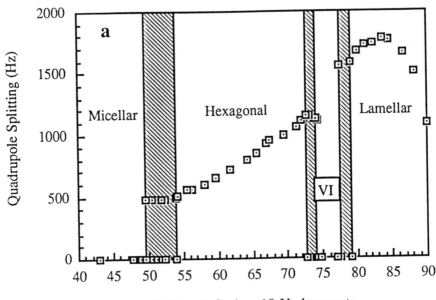

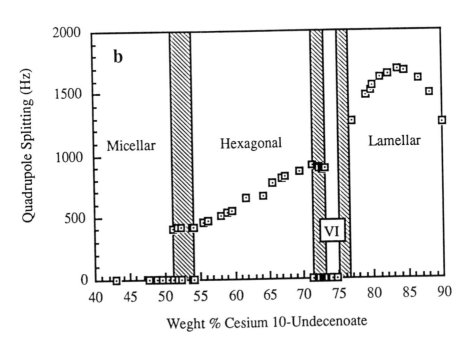

Figure 2. Dependence of $^2$H NMR quadrupole splitting in Hz on concentration of cesium-10-undecenoate-D2O solutions at (a) 298 K and (b) 333 K. VI denotes the region of stable viscous isotropic liquid crystal. Crossed-hatched regions denote two-phase regions.

interpret these observations as evidence of isotropic micellar solutions. At concentrations from 50 to 54 wt %, the deuterium nmr spectrum consists of a superposition of an isotropic peak and a Pake pattern of quadrupole splitting of 500 Hz. Beyond 54 wt %, the deuterium NMR spectrum is a Pake pattern with quadrupole splitting, which increases monotonically with increasing concentration. The visual appearance of the sample is that of an extremely viscous birefringent phase. This phase exhibits striation textures typical of hexagonal LC phases when viewed between crossed polarizers in an optical microscope. This hexagonal phase is stable up to concentrations of 72 wt %. Between 72 to 74 wt %, deuterium NMR spectra consist of a superposition of an isotropic peak and a Pake pattern with a quadrupole splitting equal to that of the hexagonal phase at 72%. The phase which is stable beyond 74% is isotropic, extremely viscous, and exhibits strain birefringence. We interpret these observations as evidence of a viscous isotropic liquid crystalline phase. It is stable to concentrations of 77%, whereupon an additional anisotropic phase with quadrupole splittings of 1500 Hz is observed. This highest concentration phase is less viscous than hexagonal or viscous isotropic, and exhibits textures typical of a lamellar LC phase when viewed under crossed polarizers in the optical microscope. Using these NMR data and visual observations, a partial phase diagram of cesium undecenoate-$D_2O$ was constructed and is shown in Figure 3.

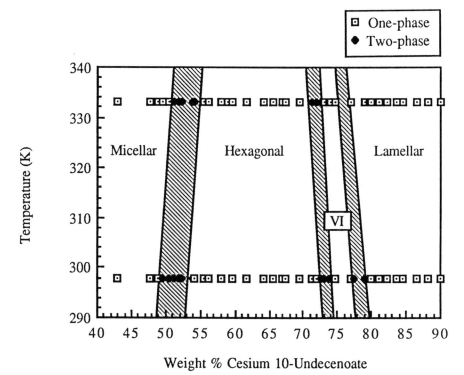

Figure 3.    Partial binary phase diagram of cesium 10-undecenoate-water mixtures as determined by $^2H$ NMR spectroscopy.

## $^1$H NMR Analysis of PMHS Backbone

$^1$H-NMR spectra of the parent backbone PMHS polymers from Aldrich and Petrarch were acquired to determine their number average molecular weights ($M_n$) and to aid in identification of $^1$H-NMR spectral peaks of the functionalized polymer. Two readily resolved resonances were observed: a peak at 4.76 ppm corresponding to protons that are directly attached to the silicon atoms in PMHS and a peak at 0.13 ppm corresponding to methyl protons directly attached to the backbone and capping the ends of the polymer. A linear PMHS backbone of infinite molecular weight would possess a ratio of methyl to silyl protons of 3. The degree of polymerization (DP) can thus be determined by forming the ratio of methyl proton peak area to backbone proton peak area, subtracting the contribution due to backbone methyls and dividing into the relative peak area of the end cap methyls:

$$DP = \frac{18}{(\frac{MP}{BP} - 3)} \tag{1}$$

where MP and BP denote the integrated peak areas of methyl and backbone silyl protons, respectively. Applying equation (1) to the peak areas obtained for the two PMHS samples gives degrees of polymerization of 25 and 28 for the Aldrich and Petrarch materials, respectively. The number average molecular weights, determined from the structures corresponding to these degrees of polymerization, are 1666 and 1846, respectively.

## Synthesis and Isolation of Side Chain Polymers

Four different sets of reaction conditions were employed to determine the effects of reaction temperature, catalyst concentration and reaction time on the degree of functionalization of the side chain polymer. These reaction conditions are summarized in Table 1. The determination of the degree of functionalization of each of these polymers by $^1$H NMR is described in the following section. After completing the reaction, the reaction mixtures were cooled to ambient temperature and converted to methanolic solutions by dilution and rotary evaporation. Residual undecenoic acid was removed by diafiltration in methanol. The disappearance of the acid monomer was monitored by GPC. The acid side chain polymer (PMUS) was converted to the saponified form by stoichiometric neutralization with CsOH in methanol. The solvent was then replaced by diafiltration with either water or $D_2O$ through a YM5 ultrafiltration membrane (5000 MW cutoff). Complete retention of the polymer was taken as indirect evidence of resistance of the polymer backbone to chain scission upon contact with the basic CsOH solution.

Table 1.

Reaction Conditions for Acid Side-Chain Functionalized Polysiloxanes
(PMCsUS)

| | PMCsUS 1 | PMCsUS 2 | PMCsUS 3 | PMCsUS 4 |
|---|---|---|---|---|
| PMHS (gms) | 2.61 | 13.7 | 9.98 | 8.00 |
| Excess 10-undecenoic acid (%) | 6.6 | 4.7 | 9.0 | 11.6 |
| Catalyst (ppm) | 150 | 6600 | 340 | 150 |
| Reaction time (hrs) | 5 | 24 | 12 | 24 |
| Reaction temp. (°C) | 50 | 50 | 100 | 90 |
| Solvent | THF | THF | Toluene | i-Propanol |
| Percent functionalization | ----- | 41.1 | 55.3 | 28.2 |
| Molecular feight | ----- | | 4885 | 6154 4354 |

PMCsUS 2 and PMCsUS 3 were synthesized using the Aldrich Chemicals
PMHS (DP 25). PMCsUS 4 was synthesized using the Petrach Chemicals
PMHS (DP 28). PMCsUS 3 and PMCsUS 4 were synthesized by preparing
the catalyst according to the method of Apfel et al. [1]

## [1]H NMR Analysis of Functionalized Polymer PMCsUS

In order to identify the peaks in the [1]H NMR spectra for the side chain
functionalized polymers, it was first necessary to make spectral assignments
for the 10-undecenoic acid side chain. This compound is indexed in the
Sadtler library of proton NMR spectra and these assignments were taken
directly from this source. The acidic proton was not observed experimentally
due to exchange with the deuterated methanol solvent. [1]H-NMR spectra of
the acidic (PMUS 2) and saponified polymers (PCsUS 2) were also obtained.
The residual unfunctionalized backbone protons were not observed in these
spectra as they are unresolved from the methanol solvent peak. Evidence of
the reaction is provided by the appearance of a peak at 0.59 ppm which
represents protons on the eleventh carbon of the acid side chain and which is
directly attached to the siloxane backbone. This assignment was made by
comparing the chemical shift to that of methylene protons directly attached to
siloxane silicon in tetraethylsilane indexed in the Sadtler library. The other
peaks in the PMCsUS spectra are common to the 10-undecenoic acid or
PMHS spectra.

The percent functionalization (PF) (i.e., the percent of the siloxane backbone protons functionalized by 10-undecenoic acid) can be determined from the peak areas of protons in the PMCsUS spectrum. By taking the ratio of the area of the peak at 0.13 ppm (the methyl protons, MP) to the area of the peak at 0.59 ppm (the protons on the eleventh carbon of the attached acid, $C^{11}H_2$), the percent functionalization is obtained from:

$$\% \text{ Functionalization} = \frac{3(C^{11}H_2) \, (DP + 6)}{2(MP)(DP)} \tag{2}$$

In equation (2), DP is the degree of polymerization obtained from analysis of the $^1$H-NMR spectra of the parent PMHS backbone polymers. Applying equation (2) to the spectra of the functionalized polymers, PMCsUS 2, 3 and 4 were found to have percent side-chain functionalizations of 41.1, 55.3 and 28.2%, respectively. These results are reported along with the reaction conditions in Table 1.

The primary difference in the spectra of the acidic and ionic forms of the functionalized polymers appears to be broadening of the peaks of the saponified polymer, which is most likely due to aggregation of the polymer in aqueous solution. It should be noted that the acidic side-chain polymers are not water-soluble to any appreciable extent. This line broadening was quantified by noting that the peaks corresponding to the eleventh carbon on the undecanoate chain directly attached to the siloxane backbone and the unresolved methylenes corresponding to carbons 4 through 10 have half-line widths of 21.6 and 8.7 Hz for the acidic polymer PMUS 2 and 39.3 and 19.8 Hz, respectively, for the saponified polymer.

### Surface Tension Measurements of PMCsUS-Water Solutions

In an effort to monitor aggregation of saponified polymer in aqueous solution, surface tensions were measured for isotropic solutions of PMCsUS 2, 3 and 4 at varying concentrations ($10^{-5}$-$10^{-1}$ M) (Figure 4). As Figure 2 indicates, the polymers are about equally effective in lowering solution surface tension. Solutions of the two functionalized polymers with the lower % sidechains, PMCsUS 2 and 4, were observed to be translucent at concentrations of 0.01 M and greater. Solutions of PMCsUS 3, which had a % sidechain of 55%, were transparent at all concentrations. Solutions of PMCsUS 2 and 3 exhibited concentration-independent surface tensions at sufficiently high concentrations. Surface excess concentrations ($\Gamma$) and area per molecule ($\sigma$) were calculated for the three saponified polymers by applying the Gibbs adsorption isotherm[2] to the surface tension data in the concentration regime over which the surface tension was linear with the logarithm of the polymer concentration. These results are reported in Table 2.

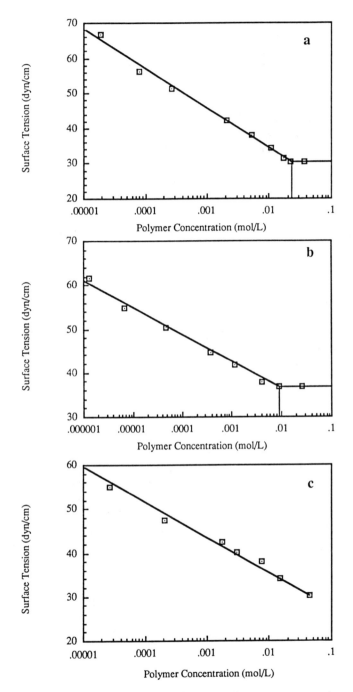

Figure 4.   Dependence of surface tension on polymer concentrations at 25°C of  PMCsUS-D2O solutions on polymer concentration for (a) PMCsUS 2, (b) PMCsUS 3 and (c) PMCsUS 4.

Table 2.

Size Transition Concentration, Surface Excess Concentration, and Area per Molecule at Air-Water Interface for the PMCsUS Polymers

|  | PMCsUS 2 | PMCsUS 3 | PMCsUS 4 |
|---|---|---|---|
| Size transition concentration | | | |
| (mol/L polymer) | 0.022 | 0.009 | 0.045 |
| (weight % polymer) | 1.07 | .534 | 1.95 |
| $\Gamma$ (pmol/cm$^2$) | 197 | 113 | 129 |
| $\sigma$ (Å$^2$/molecule) | 84.3 | 147.3 | 128.5 |
| Molecular weight | 4885 | 6154 | 4354 |

## QLS Measurements of PMCsUS-Water Solutions

QLS measurements were performed on samples of varying concentrations of aqueous solutions of the saponified PMCsUS polymers. The mean diffusion coefficient D was obtained directly from a cumulant fit to the scattered light intensity autocorrelation function.[12,21] This diffusion coefficient was converted to an effective hydrodynamic diameter using the Stokes-Einstein equation.

Shown in Figure 5 are polymer aggregate diameters as a function of concentration for solutions of the three saponified polymers PMCsUS 2, 3 and 4. QLS measurements could not be performed on the most concentrated samples of the PMCsUS 2 and 4 polymers because these solutions were turbid. Upon centrifugation, these translucent solutions yielded an observable precipitate. Solutions of the polymer with the highest degree of side chain functionalization, PMCsUS 3, were transparent at high concentrations (> 0.01 M). The apparent polymer aggregate size varied continuously (from ≈ 200 Å at PMCsUS concentrations of 0.3-0.4 mM to ≈ 800-900 Å at concentrations of 10 mM). With the most soluble polymer, PMCsUS 3, aggregate sizes of ≈ 4000 Å were measured at 20 mM. The approximate size of monomeric functionalized PMCsUS polymer should be no larger than 50 Å in diameter based on an end-to-end length of the PMHS backbone of 60-65 Å[3]. It thus seems clear that the sizes observed by QLS correspond to aggregates of polymer rather than to monomeric polymer at all concentrations studied.

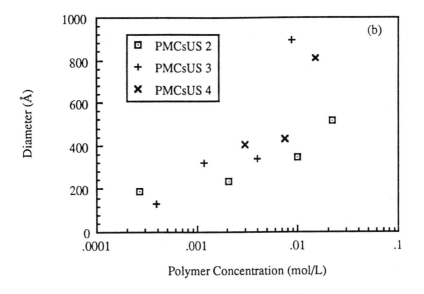

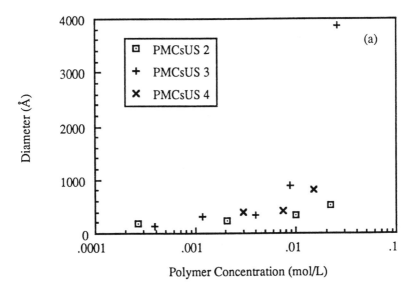

Figure 5.    Variation in Stokes diameter as measured by quasi-elastic light
scattering of PMCsUS polymeric aggregates.    Plot (b) is an
expanded version of plot (a).

## $^2$H NMR Spectroscopy of PMCsUS-D$_2$O Samples

Binary PMCsUS-D$_2$O samples were prepared from each of the three polymers as described previously. Samples made from PMCsUS 2 and 4 remained macroscopically phase separated from the water after vigorous mixing. This is consistent with the visible precipitation observed in the light scattering samples commented on above. Samples made with PMCsUS 3 equilibrated rapidly (a few days) to yield optically clear homogeneous phases which were then analyzed by $^2$H NMR spectroscopy. These experiments were performed at 298 and 333 K and from 10 to 80 wt % PMCsUS 3. From 10-32 wt % polymer, the samples were observed to be fluid and isotropic. Above 32% polymer, the samples became visibly viscous and ceased to flow under their own weight at concentrations greater than 36 wt %. All the deuterium NMR spectra on samples varying in composition from 10-70% polymer were isotropic peaks; there was no evidence of anisotropic LC formation from either the NMR experiments or from visual observations between crossed polarizers. However, the line shape was observed to change between 32 and 36% from a single Lorentzian to a superposition of two Lorentzian lineshapes: a broader Lorentzian ($\Delta v_{1/2} = 20$ Hz) and a narrower Lorentzian ($\Delta v_{1/2} = 10$ Hz). This is illustrated in Figure 6, a sample consisting of 34.25 % PMCsUS 3 and D$_2$O. The lineshape is clearly super-Lorentzian, and one interpretation of this is that of a phase transition from a less viscous fluid phase to a more viscous isotropic phase.

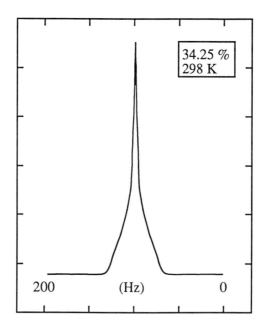

Figure 6. $^2$H quadrupole NMR spectrum of mixture of 34.25 wt % PMCsUS 3 and 65.75 wt % D$_2$O.

## DISCUSSION

The monomer cesium undecenoate lowers surface tension in a manner similar to other alkali metal carboxylates. Its CMC of 60 mM is lower by a factor of 2 than the comparable CMC of 117 mM determined for sodium undecenoate by Durairaj and Blum.[5] This is consistent with the observed depression of the CMC of alkali metal carboxylates as the size of the bare counterion is increased from sodium to cesium.[16] The monomer cesium undecenoate also exhibits lyotropic mesomorphism at higher concentrations which is typical of alkali metal carboxylates of short alkyl chain lengths. Sodium n-octanoate is observed to exhibit hexagonal, viscous isotropic and lamellar LC phases with increasing surfactant concentration[6]. We contrast this behavior with that of the functionalized polymers PMCsUS.

As is evident from the reaction data presented in Table 1, the conditions of the hydrosilylation reaction have a strong effect on the percentage of the PMHS backbone that is functionalized. The saponified polymer with the largest degree of undecanoate side chain functionalization, PMCsUS 3 (55% side chain), also exhibited the greatest water solubility (32 wt %), the largest aggregate size by QLS ($\approx$ 4000 Å) and a viscous isotropic liquid crystalline phase at high concentrations in water (36-80 wt %). Apparently, the degree of functionalization plays a large role in solubility of the polymer, and this is likely attributable to the hydrophobicity of polymer.

All three functionalized polymers that were studied (PMCsUS 2,3 and 4) were effective in lowering surface tension of aqueous solutions. Polymers 2 and 4 were slightly more effective than polymer 3 as gauged by the surface excess concentrations (see Table 2). Again, this is likely attributable to the greater hydrophobicity of polymers 2 and 4, relative to polymer 3, which may lead to increased surface activity at comparable concentrations. With typical aqueous solutions of ionic surfactants, surface tension is observed to decrease in a nearly linear fashion with the logarithm of the surfactant concentration until the CMC, or critical micelle concentration, is reached. Beyond the CMC, additional surfactant participates in the formation of micellar aggregates and the surface tension is observed to change little with increasing concentration. The surface tension behavior of aqueous solutions of polymers 2 and 3 is typical of micelle-forming ionic surfactants and it is tempting to construe the break points in surface tension as $\approx$ 0.02 and $\approx$ 0.01 M, respectively, as CMCs. However, the QLS data indicate that at concentrations well below this breakpoint, there are aggregates of diameter 200 Å and greater. Moreover, the size of these aggregates grows continuously with increasing concentration for all three polymers. These observations suggest that there are a distribution of polymer molecular weights and degree of ionic surfactant side-chain functionalization. At low concentrations, the polymer molecules with lowest CMC values (presumably those with greatest hydrophobicity) aggregate, while that fraction of the distribution with highest CMC values adsorb to the air-water interface and lower surface tension. Interestingly, the distribution of surface excess concentrations of the individual polymer species averages to yield a linear surface tension plot with logarithm of concentration. At sufficiently high concentrations, even the most functionalized of polymer species participate in aggregate formation.

At elevated concentrations of polymer 3 in water (10-30 wt %), solutions of the polymer remain fluid and isotropic. Beyond 32 wt % polymer, mixtures of polymer and water become viscous and at 36 % polymer, they gel into a viscous isotropic phase. The $^2$H nmr lineshape of polymer 3-$D_2O$ mixtures changes in this transition regime (32 to 36%) from a simple Lorentzian line shape of 15 Hz linewidth to a superposition of two Lorentzian line shapes in which the broader component is ca. 30-40 Hz in width. This suggests that a first-order phase transition separates low and high concentration polymer-$D_2O$ mixtures. While we provide no evidence here of three-dimensional order, the high viscosity and optical isotropy of the high concentration polymer phase suggest that this is a viscous isotropic lyotropic liquid crystalline phase.

Unlike the nonionic amphiphilic side chain polysiloxanes synthesized and studied by Finkelmann et al.[8] and by Luhmann et al.,[13,14] the ionic amphiphilic side chain polymers prepared here show no anisotropic liquid crystallinity at high concentrations (>30 wt%) of polymer. This may be due to incomplete functionalization of the PMHS backbone; the largest degree of side chain functionalization obtained was only 55%. With only partial functionalization, one might expect some intramolecular hydrophobic interactions, which may lead to globular aggregates. With more complete functionalization, the electrostatic repulsion between adjacent carboxylate moities may favor a more distended polymer chain. This in turn might favor the formation of anisotropic aggregates. Clearly, future studies on ionic amphiphilic side chain polysiloxanes should address the issues of backbone chain size distribution and the distribution of side chain functionalization as determinants of aggregation in aqueous solution and possible lyotropic liquid crystal formation.

## CONCLUSIONS

Cesium undecenoate (CsU) exhibits a CMC of 60 mM at 25°C and lyotropic LC formation typical of other short-chain alkali metal carboxylates. Two anisotropic phases are stable in the monomer-water system: a hexagonal and a lamellar phase. These phases have broad compositional and thermal extent. In addition, a narrow concentration range is observed in CsU-D2O mixtures (72-76%) in which a viscous isotropic LC phase is stable. Polymethylhydrosiloxane (PMHS) polymers of DP 25-30 have been functionalized by hydrosilylation with the amphiphilic side chain: 10-undecenoic acid. The degree of side chain functionalization obtained varied from 30 to 55%. The acidic form of the side chain was neutralized with CsOH to yield the ionic carboxylate form. This polymer was water soluble while the acid form and the PMHS backbone were not. The carboxylated PMCsUS polymers were effective in lowering surface tension and associated in solution to yield aggregates of 200-800 Å in diameter. The surface excess concentrations of the three PMCsUS polymers increased monotonically with decreasing ionic side chain percentage. Only PMCsUS 3 with the highest side-chain percentage (55%) was soluble in water above a concentration of 0.01 M (1-2 wt%). This polymer formed isotropic solutions in water up to concentrations of 32 wt %, beyond which a viscous isotropic phase was observed. The tendency of these ionic side-chain polysiloxane polymers to form isotropic rather than anisotropic phases may be due to a distribution in molecular weights and side-chain percentages.

## ACKNOWLEDGMENT

This work was supported in part by grants from the 3M Company and Colgate-Palmolive.

## REFERENCES

1.    Apfel, M.; H. Finkelmann, G. Janini, R. Laub, B. Lühmann, A. Price, W. Roberts, T. Shaw, and C. Smith, "Synthesis and properties of high-temperature mesomorphic polysiloxane solvents: biphenyl- and terphenyl-based nematic systems", *Anal. Chem.*, 57, 651-658 (1985).
2.    Adamson, A. W., Physical Chemistry of Surfaces, 4th edition, Wiley, New York, 1982.
3.    Billmeyer, F. W., Textbook of Polymer Science, 2nd edition, Wiley-Interscience, New York, 1971.
4.    Davis, J. H., "The description of membrane lipid conformation, order and dynamics by $^2$H-NMR", *Biochimica et Biophysica Acta*, 737, 117-171 (1983).
5.    Durairaj, B.; and F. D. Blum, "Synthesis and characterization of oligomeric micelles of sodium carboxylates", *ACS Polym. Prepr.*, 26, 239-240 (1985).
6.    Ekwall, P., "Composition, properties, and structure of liquid crystalline phases in systems of amphiphilic compounds", *Advances in Liquid Crystals*, 1, 1-130 (1975).
7.    Finkelmann, H.; and G. Rehage, "Investigations on liquid crystalline polysiloxanes, 1 synthesis and characterization of linear polymers", *Makromol. Chem., Rapid Commun.*, 1, 31-34 (1980).
8.    Finkelmann, H.; B. Lühmann, and G. Rehage, "Phase behavior of lyotropic liquid crystalline side chain polymers in aqueous solutions", *Coll. Polym. Sci.*, 260, 56-65 (1982).
9.    Finkelmann, H., and M. A. Schafheutle, "Lyotropic liquid crystalline phase behavior of a monomeric and a polymeric monosaccharide amphiphile in aqueous solution",*Coll. Polym. Sci.* , 264, 786-790 (1986).
10.   Glass, J. E., editor, *Polymers in Aqueous Media, American Chemical Society Advances in Chemistry Series V. No. 223*, Washington, D.C., 1989.

11. Johansson, Å.; and T. Drakenberg, "Proton and deuterium magnetic resonance studies of lamellar lyotropic mesophases", *Mol. Cryst. Liq. Cryst.*, 14, 23-48 (1971).

12. Koppel, D. E., "Analysis of macromolecular polydispersity in intensity correlation spectroscopy: the method of cumulants", *J. Chem. Phys.*, 57, 4814-4820 (1972).

13. Lühmann, B.; H. Finkelmann, and G. Rehage, "Phase behavior and structure of polymer surfactants in aqueous solution", *Makromol. Chem.*, 186, 1059-1073 (1985).

14. Lühmann, B.; and H. Finkelmann, "Lyotropic Liquid Crystalline Phase Behavior of Bmphiphilic Monomers and Polymers having a Rod-like Hydrophobic Moiety", *Colloid & Polymer Sci.*, 265, 506-516 (1987).

15. McArdle, C. B., Side Chain Liquid Crystal Polymers, Blackie and Son, London, 1989.

16. Mukerjee, P.; and K. J. Mysels, "Critical Micelle Concentrations of Aqueous Surfactant Systems", *Nat. Stand. Ref. Data Ser.*, Nat. Bur. Stand. (U.S.), (1971).

17. Persson, N-O; and B. Lindman, "Deuteron Nuclear Magnetic Resonance in Amphiphilic Liquid Crystals. Alkali Ion Dependent Water and Amphiphile Orientation", *J. Phys. Chem.*, 79, 1410-1418 (1975).

18. Sadtler Research Laboratories, Inc., 1085, 464, (1978).

19. Seelig, J., "Deuterium Magnetic Resonance: Theory and Application to Lipid Membranes", *Quarterly Reviews of Biophysics*, 10, 353-418 (1977).

20. Siderer, Y.; and Z. Luz, "Analytical expressions for magnetic resonance lineshapes of powder samples", *J. Mag. Res.*, 37, 449-463 (1980).

21. Stock, R. S.; and W. H. Ray, "Interpretation of photon correlation spectroscopy data: A Comparison of Analysis Methods", *J. Poly. Sci.: Poly. Phys. Ed.*, 23, 1393-1447, (1985).

22. Wang, Y., and M. A. Winnik, "Onset of Aggregation for Water-Soluble Solymeric Associative Thickeners: A Fluorescence Study", *Langmuir*, 6, 1437 (1990).

# Chapter 17

# PROCESSIBLE HEAT-RESISTANT RESINS BASED ON NOVEL MONOMERS CONTAINING *O*-PHENYLENE RINGS

J. PRESTON[a], J. W. TREXLER, JR[b] AND Y. TROPSHA[a]

*[a]Research Triangle Institute*
*P. O. Box 12194*
*Research Triangle Park, NC 27709*

*[b]Tennessee Eastman Company*
*Kingsport, TN 37662*

## INTRODUCTION

For more than three decades there has been a great effort to prepare heat resistant polymers for the preparation of films, fibers and resins for composites. (The literature in this area is too voluminous to cite here. For anyone who may be interested in learning more on this subject, two reviews,[1,2] and two books[3,4] that may be of help in gaining an insight to the field are referenced.) The routes to the synthesis of these polymers have been to prepare polyheterocycles of many types either by direct synthesis or *via* a soluble precursor.

Because of the requirement that heat resistant polymers be processible, research efforts have been directed toward balancing optimum heat resistance against those structural modifications that will make processing feasible. For example, in polymers of type I,

I                    where X = -O-, -S-, and -NH-

only polymer containing X = -NH- is sufficiently soluble in common organic solvents (e.g., dimethylacetamide) to be processed to film and fiber or to be used in the preparation of composites. However, the long-term thermooxidative stability at an elevated temperature (e.g., ~300°C in air) is low. Nevertheless, because processibility is so important, only this variant is made commercially, despite its moderate heat resistance.

Because polymers of I where X = -O- and -S- are so very heat resistant, attempts have been made to use a precursor, II,

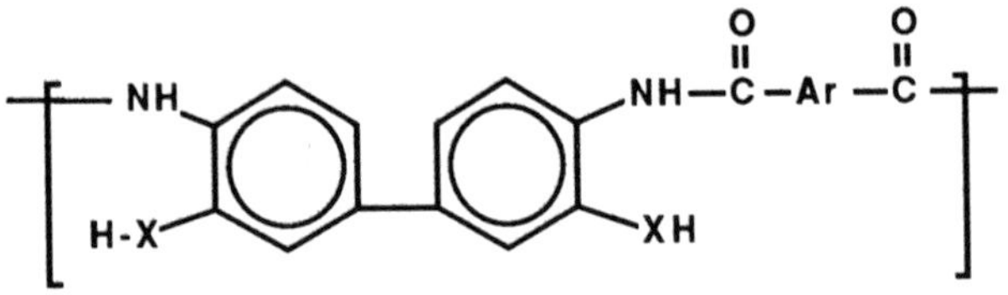

II

for the fabrication step to prepare films, fibers and composites. Attempts are then made to form polymer I in an afterstep. However, the conditions for converting II to I are so severe (long-term heating at ≥350°C in the absence of oxygen) that this route is impractical, especially since complete conversion is never achieved.

A rather successful route to a polyheterocycle employing a precursor polymer is available in the case of the preparation of polyimides. (Again, there are innumerable references that could be cited but only two reviews[5,6] are referenced here.) A typical preparation of a polyimide *via* a soluble precursor polymer, a polyamic-acid, is illustrated in equation (1).

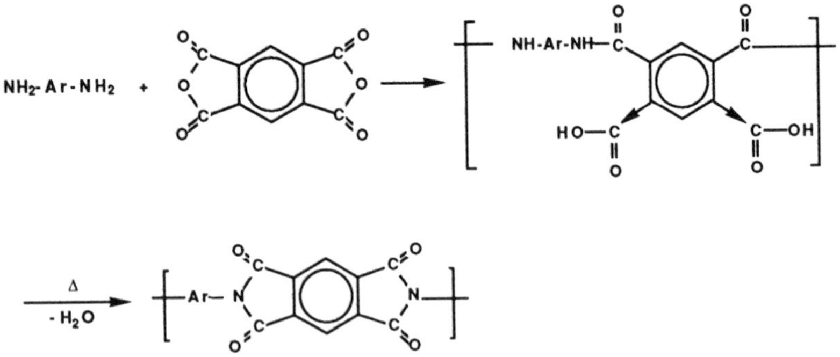

(1)

Another route to polyimides that is particularly useful for resins is that employing a mixture of monomers instead of a precursor polymer. This route (Equation 2), for lack of a better term, could be called the diacid-diester route because the dianhydride employed is first reacted with an alcohol (alternatively, a glycol) prior to reaction with a diamine.[7]

$$ \tag{2} $$

Despite the very good thermal stability of polyimides in general, there are some problems:

1. During cyclodehydration of the polyamic-acid precursor, a large amount of water is released, a property that is particularly bad for the processing of composites because voids may be formed.

2. The resulting cured polyimide has a fairly large water regain and is relatively sensitive to hydrolysis.

In attempts to make various polyheterocycles (including polyimides) more soluble, use has been made of linkages such as -O-, -S-, and -C(CH$_3$)$_2$- in monomers (e.g., diamines and dianhydrides). The use of such groups generally produces a considerable enhancement of polymer solubility, but the penalty in thermooxidative stability in the resultant polymers is high. Use of C(CF$_3$)$_2$ linkages in polyheterocycles has been more successful in maintaining thermooxidative stability and has even lowered moisture regain, but monomers containing such groups are quite expensive and when degradation of resins containing such linkages does occur, highly toxic HF is eliminated.

In the present work, an attempt has been made to develop a new heat resistant resin having maximum processibility without sacrificing heat resistance (especially thermooxidative stability) due to the use of solubilizing groups or precursors that may not yield fully cyclized heterocyclic groups. This approach uses no solubilizing groups such as -O-, -S-, -C(CH$_3$)$_2$-, -C(CF$_3$)$_2$-, etc. Moreover, a fully preformed heterocyclic group known for its high thermooxidative stability is employed. This group also is not as prone to high moisture regain as an imide group alone. Processability is assured by use of polyimide type chemistry, making use of either the polyamic-acid precursor route or the diacid-diester route.[7] (The latter route is possible only when the diamine monomer selected is highly soluble in the reaction mixture employed.)

Earlier, one of the present authors published data on fibers from ordered polybenzoxazole-imide copolymers that were exceedingly heat resistant at an elevated temperature in air.[8] Data for fiber from one of these polymers, III, are given in Table 1. Note that fiber from the control, IV, used in the test embrittled very rapidly, probably because the polyimide used for this fiber contains an ether linkage, -O-, which made processing easier but served as a site for initiation of oxidation.

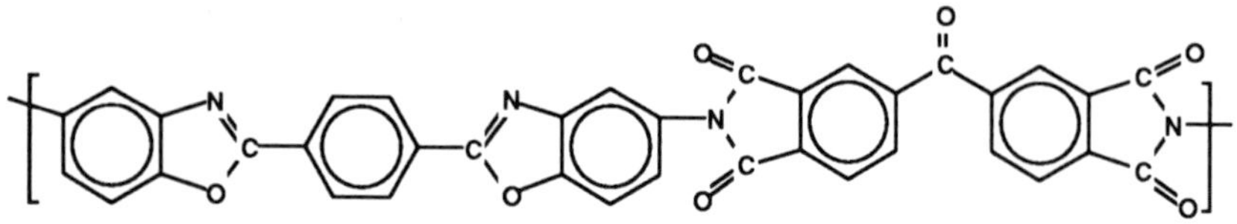

III

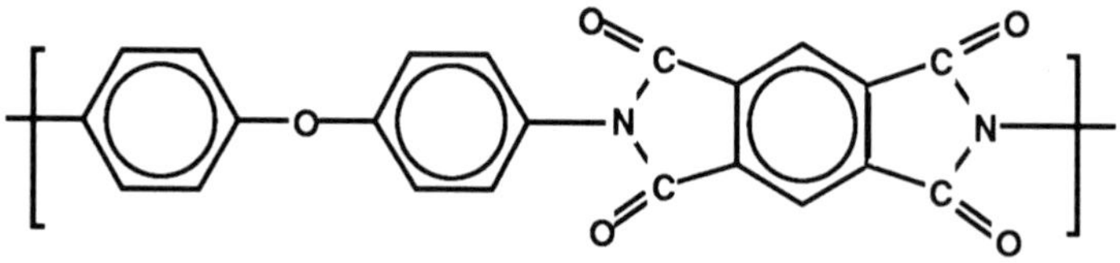

IV

Table 1.    Thermooxidative Stability Comparison of Fibers from Polymers III and IV

| Fibers made from: | Number of Days at 300°C in Air | | |
|---|---|---|---|
| | 10 | 49 | 70 |
| *Tenacity, % Retention* | | | |
| IV | 0.0 | - | - |
| III | 94 | 83 | 30 |
| *Elongation, % Retention* | | | |
| IV | 0.0 | - | - |
| III | 78 | 62 | 26 |

Because of the excellent heat resistance of fibers of III, it was thought that III might be an excellent candidate for use as a heat resistant resin, especially if processed via the diacid-diester route to a polyimide. However, the solubility of the benzoxazole diamine used to make III (i.e., A, in amide type solvents) was only about 6%, a value much too low for the high solids systems required for good resin systems.

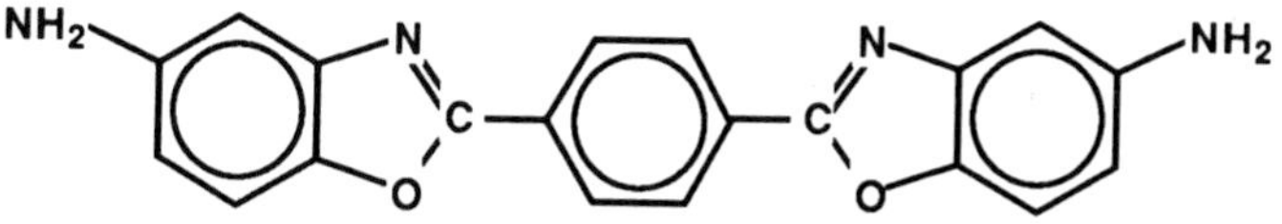

A

Another benzoxazole diamine that would yield just as thermooxidatively resistant a resin as III is that based on B, which is structurally similar to A, but contains an orthophenylene ring instead of a *para*-phenylene ring as the bridging group between the benzoxazole moieties of the diamine (cf. A and B).

Polybenzoxazole-imides based on B were reported[9] earlier by one of the

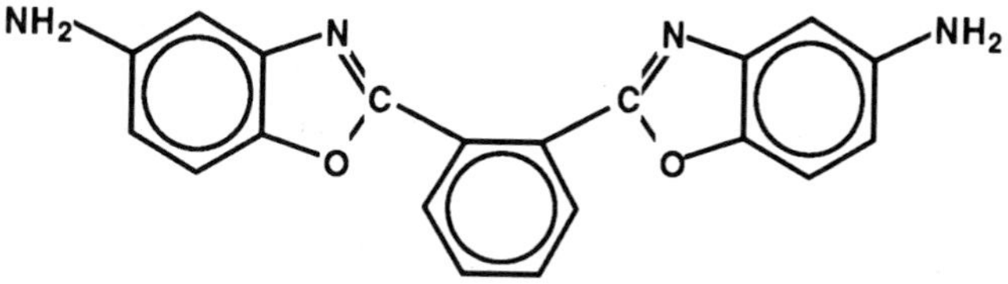

B

authors, but very few physical data were given for such polymers. In the present work, physical data on polymer V (Equation 3) are given in some detail.

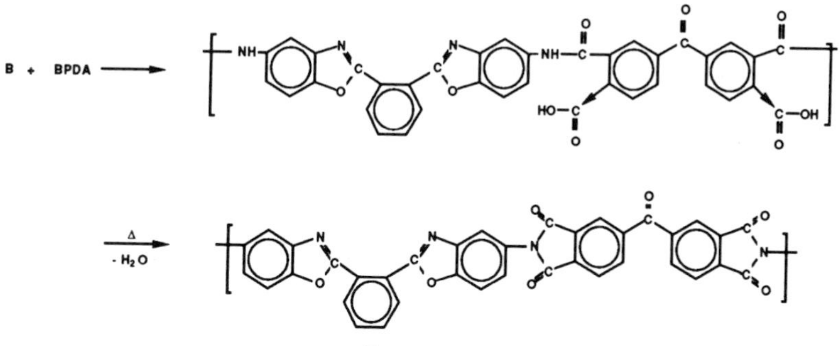

$$(3)$$

V

# EXPERIMENTAL

## Preparation of Diamine B

Diamine **B** was prepared as described in a patent[10] and in a paper[11] by one of the present authors. Numerous experiments by the authors of the present work confirm the high crude yield of product claimed in the patent, but the yield of pure product falls to only about 6%. No differences were found in yields of crude or purified product using the two published procedures, both of which are based on the method of Hein.[12]

## Preparation of Diamine C

Diamine **C** was prepared by the same procedure used for diamine **B**, using instead of phthalic anhydride 1,10-decanedicarcarboxylic acid. After treatment of the crude product in ethanol with decolorizing charcoal, a white material with a melting point of 117.5-117.8°C (determined by means of DSC) was obtained. Unlike the wholly aromatic diamines (e.g. **B**), **C** could not be sublimed without decomposition.

## Polymerization

Polymerizations were carried out as reported previously.[9] Films were cast from the reaction mixtures. The films were converted from the polyamic-acid precursor form to the polyimide form by heating at ~110°C to remove solvent, followed by heating for 3 hours at 150°C and at 170°C overnight. The mostly imidized films were heated at ~300°C for about 5 minutes to fully "cure" the film and to assure that any iso- imide groups are converted to imide ones.

## Determination of Glass Transition Temperature

The glass transition temperature ($T_g$) for polymer V was determined by means of torsion braid analysis (TBA) performed by Professor J. K. Gillham of Princeton University and Plastics Analysis Instruments, Inc. of Princeton, New Jersey. In this test, a 25% solids solution of the polyamic-acid precursor to V in dimethylacetamide (DMAc) was spread on a glass braid and mounted in the TBA unit at 30°C in a dry helium atmosphere. The polymer was cured according to the following regimen:

> 30-80°C over the course of 30 minutes
> 80-100°C over the course of 60 minutes
> 100°C for one hour
> 100-150°C at 10°C/hour
> 150°C for one hour
> 150-180°C at 10°C/hour
> 180°C for 12 hours

TBA data were obtained on the sample, which had been heated/cooled as follows:

> 180-300°C at 5°C/minute
> 300°C for 5 minutes
> 300°C to -180°C
> - 180°C to 350°C
> 350°C to -180°C
> - 180°C to 350°C
> 350°C to 30°C

The specimen was further cycled 63 times between 200 and 350°C at 1.5°C/minute under an atmosphere of helium.

## Pressing of Films

Several layers of thin films of V were placed in a Pasadena Hydraulics press at 310°C and pressed together for one hour at 10,000 psi. When the material was removed from the press, a single film was obtained. When thin films of III were treated in similar fashion, no indication of any bonding together was observed; the individual films hardly even flattened out and were no smoother after the pressing than before.

Interleaving films of V between thin films of IV and carrying out the pressing operation as described above gave a consolidated film.

## RESULTS AND DISCUSSION

### Structure/Property Relationships

For heat resistant polymers, it has been known for some time that those containing *para*-oriented rings generally give higher melting polymers than similarly structured ones containing *meta*-oriented rings. Extrapolating this trend to polymers containing *ortho*-oriented rings would lead to the prediction that such polymers would give still lower melting polymers. However, data on heat resistant polymers containing *ortho*-oriented rings are very meager and the data to be found are highly suspect in light of the discussion following. Consequently we are in the process of reexamining some of the published experiments to see if high molecular weight polymer indeed was prepared or if merely heat resistant oligomers in fact were produced.

The reason for our skepticism concerning the formation of heat resistant polymers containing *ortho*-oriented rings is the fact that when difunctional groups are placed *ortho* to one another, they tend to form simple heterocyclic rings instead of participating in chain extending reactions. Because yields for the formation of chain extending reactions must be ≥99.99% in order to achieve a degree of polymerization of about 100, monomers containing *ortho*-oriented functional groups cannot be expected to yield high molecular weight polymers except in special cases where only one reaction can occur, e.g., the formation of a polyester or the formation of a polyamide from a secondary amine.

Because it is a mute point at present whether heat-resistant polymers have been prepared heretofore containing *ortho*-oriented rings, structure property relationships cannot be drawn with certainty. However, model compounds can be used to draw conclusions concerning the effect of ring orientations and the trends of the melting points are quite clear in the series (Table 2): *para- > meta- >> ortho-*. Because the trends for solubility generally follow melting point inversely, the order for solubility for polymers in this series should be: *ortho- >> meta- > para-*. This is precisely what we have found and are attempting to exploit in the present work.

Table 2    Melting Points of *bis*-Benzheterocyclic Model Compounds
Containing Phenylene Rings

| Structure | mp (°C)[a,b] | |
|---|---|---|
| | X = S | X = O |
| | 115.5[14] (177)[10,13] | 179[14] (111.5-112.0)[15] |
| | 234.5 (229)[13] | 229[14] |
| | (355-356)[13] | 264.9[14] (257-258)[15] |

[a]   Determined from the peak of the DSC curve for a sublimed sample.

[b]   Values in parentheses are for recrystallized samples reported in the
literature.

The contribution of an *ortho*-oriented ring to melting point in a heat
resistant polymer, based on the above points, might be expected to be great
and indeed was found to be so. In order to get some idea of how profound an
effect is obtained, we prepared a polymer similar to V but containing an
alkylene chain derived from the diamine C. The polymer, VI, was prepared

C

in the usual manner via the polyamic acid precursor, which was converted to
the cured polymer by means of heat. It is of interest to note that VI, with a

$$(4)$$

VI

rather long, flexible alkylene chain, has a melting point of ~360°C while V, a wholly aromatic polymer, has a melting point and/or flow point of ~310°C.

Some idea of the effect of the length of alkylene chains on the melting points of polymers containing *bis*-benzoxazole units may be obtained from an examination of corresponding model compounds (Table 3). From these data we might expect that the effect on an *ortho*-oriented ring would be comparable to that of an ethylene unit (cf. first entry of Table 2 and Table 3). The fact that polymer VI has a melting point of ~360°C would suggest that the melting point for polymer V should be much higher still. (If the observed $T_g$ of ~298°C is used to calculate a melting point for V using the two-thirds rule, then the melting point of V should be ~570°C, or about the same as that observed for IV.[9])

## Polymer Syntheses

Although monomers with *ortho*-oriented functional groups cannot be expected to yield high molecular weight polymer (see above), the yields for the preparation of monomers containing *o*-phenylene rings can be within acceptable limits. This is the approach taken in the present work: to prepare a monomer (e.g., a diamine) containing an *ortho*-oriented ring and to polymerize it by known methods to a high molecular weight polymer (e.g., a polyimide).

Such a polymer, V, has been reported[9] previously by one of the present authors. However, in the previous publication, little more than a softening point was given by way of physical data.

## Characterization of Polymer V

We have reconfirmed the previously reported softening point of ~290°C for a thin film of V when it is placed on a heated surface. However, after pressing several thin films together, the consolidated film obtained was found to withstand much higher temperatures without softening. Thus, the consolidated film from pressing was placed over two metal bars (spaced approximately 6 cm apart) in a furnace at 450°C in air. The consolidated film did not flow or sag under these conditions even though some softening might have been expected in light of the fact that the specimen was heated in excess of its $T_g$ (see section below).

Table 3    Melting Points of *bis*-Benzheterocycles Containing Alkylene Chains[13]

| Structure | m.p. (°C) |
|---|---|
| | 192-193 |
| $-(CH_2)_4-$ | 126-127 |
| $-(CH_2)_6-$ | 96-98 |
| $-(CH_2)_8-$ | 87-88 |

The $T_g$ of V was determined by means of DSC, thermomechanical analysis (TMA) and TBA. The values of Tg obtained by these methods were, respectively, 290, 283, and 298.3°C (value after 63 cycles from 200 to 350°C). The fact that the $T_g$ changes so little on such a large number of recycles led Professor Gillham to state, "This is a remarkably stable polymer".[16] It should be noted that most simple polyimides derived from BPDA have an initial Tg on the order of 289°C. In addition to the Tg, a transition below glass transition, $T_{sec}$, of 122°C was obtained by means of TBA.

Because of the precision of the TBA method, additional data can be given.

Increasing temperature $T_g$ (180 to 300°C at 5°C/minute)  =  196.0°C

Decreasing temperature $T_g$ (300 to -180°C)          = 286.9°C
Decreasing temperature $T_{sec}$ (300 to -180°C)         = 126.3°C

Increasing temperature $T_{sec}$ (-180 to 350°C)         = 125.0°C
Increasing temperature $T_g$ (-180 to 350°C)          = 285.6°C

| | |
|---|---|
| Decreasing temperature $T_g$ (350 to -180°C) | = 288.6°C |
| Decreasing temperature $T_{sec}$ (350 to -180°C) | = 130.0°C |
| | |
| Increasing temperature $T_{sec}$ (-180 to 350°C) | = 126.6°C |
| Increasing temperature $T_g$ (-180 to 350°C) | = 288.8°C |
| | |
| Decreasing temperature $T_g$ (350 to 30°C) | = 289.3°C |
| Decreasing temperature $T_{sec}$ (350 to 30°C) | = 131.8°C |

## Configuration of Polymer V

The fact that polymer V initially shows a softening point at 290-310°C is of considerable advantage for a resin because it permits consolidation after solvent and water (from cyclodehydration of the precursor polymer) have been expelled. Processing of V is similar to that of a thermoplastic resin but after the application of heat and pressure, the resin behaves much like a thermoset. In addition to use as a conventional resin to bind to itself (e.g., the consolidation of thin films and potentially, prepregged fiber), V has shown the ability to bind other films together. Thus films of IV, which cannot be pressed together under any known set of conditions can readily be formed into a consolidated structure when interleafed with thin films of V or coated with a film of V and pressed together at suitable temperatures and pressures, e.g. ~310°C and 10,000 psi.

The mechanism by which this polymer passes from a thermoplastic to a thermosetlike resin is unclear at present. One could speculate that the "floppy" structure of V may order to some extent under heat and pressure to give a more regular "coiled" structure. Computer simulations of the structures of III and V show that the main difference between extended chains of these polymers is the pitch of the coils that they make - a long, looping coil for III and a relatively tight, "corkscrew" coil for V. A computer model of V also reveals the tight "kink" introduced by the *ortho*-phenylene ring into the polymer chain.

The "thermosetting" like behavior of V could be accounted for by crosslinking because this phenomenon is always possible for a polyimide that has not been end-capped. However, the remarkable stability of the $T_g$ of V on recycling some 63 times from 200 to 350°C would argue against severe crosslinking.

## CONCLUSIONS

Heat-resistant resins containing *ortho*-phenylene rings can be prepared in high molecular weight, and one such resin has been shown to exhibit processibility similar to that of thermoplastics but to show other properties similar to those of thermosets. Despite the low initial softening point of the new resin, a $T_g$ of 298°C is exhibited, a value slightly higher than that of polyimides containing the benzophenone-imide moiety ($T_g \cong 289°C$) used in the new resin. Other new polymers (e.g., polyamides and polyheterocycles of many types) can undoubtedly be prepared using the method shown here for the incorporation of *ortho*-oriented rings into reactive monomers.

## ACKNOWLEDGMENT

This work was supported by DARPA and administered by the Office of Naval Research under Contract N0014-88-C-0672.

## REFERENCES

1.  J. Preston, "Heat Resistant Polymer" in *Kirk-Othmer Encyclopedia of Chemical Technology,* Vol. 12, 3rd ed., John Wiley & Sons, New York, 1980, pp. 203-225.
2.  P. M. Hergenrother, "Heat-Resistant Polymes in *Encyclopedia of Polymer Science and Engineering,* Vol. 7, John Wiley & Sons, New York, 1987, pp. 639-665.
3.  A. H. Frazer, High Temperature Resistant Polymers, Interscience Publishers, New York, 1968.
4.  P. E. Cassidy, Thermally Stable Polymers, Marcel Dekker, Inc., New York, 1980.
5.  C. E. Sroog, Polyimides in *Encyclopedia of Polymer Science and Technology,* Interscience Publishers, New York, New York, 1969, pp. 247-272.
6.  J. Preston, "Polyimides" in *Kirk-Othmer Encyclopedia of Chemical Technology,* Wiley Interscience, New York, Supplement Vol., 1971, pp. 746-773.
7.  E. Lavin, A. H. Markhart and R. E. Kass, U. S. Pat. 3,190,856 (1965).
8.  J. Preston, W. B. Black, and W. DeWinter, *Appl. Polym. Symp.,* 9, 145 (1969).
9.  J. Preston, W. F. DeWinter, and W. B. Black, *J. Polymer Sci.,* 7, 283 (1969).
10. E. Nyilas and I. L. Pinter, U. S. Patent 3,314,894 (1967).
11. J. Preston, W. F. DeWinter, and W. L. Hofferbert, Jr., *J. Heterocyclic Chem.,* 5, 269 (1968).
12. D. W. Hein, R. S. Alheim, and J. J. Leavitt, *J. Am. Chem. Soc.,* 79, 427 (1957).
13. E. Nyilas and I. L. Pinter, *J. Am. Chem. Soc.,* 82, 609 (1960).
14. J. Preston and J. W. Carson, Jr., unpublished results.
15. C. Rai and J. B. Braunworth, *J. Org. Chem,* 26, 3434 (1961).
16. Private communication.

# Chapter 18

# LATEX BLENDS AS MODELS OF RUBBER-TOUGHENED PLASTICS: POLY(METHYL METHACRYLATE)

MAURICE MORTON AND FRANK GRANT*

*Institute of Polymer Science
The University of Akron
Akron, Ohio 44325-3909
*2600 Windy Hill Drive
Cleveland, OH 44124*

## INTRODUCTION

Over the past 10 years, blends of plastic and rubber latices have been used in our laboratories as models for the study of the factors governing the properties of high impact plastics. It is well known that the inclusion of a dispersion of fine rubbery particles, as the minor component in a plastic, substantially enhances its impact resistance.[1] However, the complex method of preparation of such heterophase polymer blends has made it difficult to establish unequivocal relationships between their properties and morphology. This is due to the fact that, in most cases, such blends are prepared by the polymerization of the plastic-producing monomer containing some dissolved rubber. During such a polymerization, the rubber components, being incompatible with the plastic polymer, precipitate as a dispersion at some stage of the reaction, forming a separate dispersed phase. It is not surprising, there-fore, that these systems make if difficult to relate the morphology to the physical properties, although some such conclusions have been reached in a number of investigations.

Our approach to elucidating such relationships has been to use blends of plastic and rubber latices, of known rubber particle size, as a means of establishing such relationships more unequivocally. This involved the blending of a latex of the plastic with a lesser amount of the rubber latex of known particle size and coagulating the blend in an alcoholic medium (to remove any surfactant from the rubber particles), thus leading to a plastic matrix containing a rubber dispersion of known particle size. Previous work[2] has thus led to the following conclusions:

1. A plastic such as polystyrene, where fracture is preceded by formation of "crazes," shows an optimum in impact strength at a rubber particle size in the 2-4 $\mu$m range. This corroborates previous findings on such systems.[3-5]

2. A plastic such as poly(vinyl chloride), which is more ductile than polystyrene and therefore fractures by a "shear-yielding" (cold drawing) mechanism, shows an optimum in impact resistance at a rubber particle size of only 0.2-0.3 μm. Thus increasing ductility of the plastic requires a smaller rubber particle size.[1,6]

3. In all cases studied, the presence of rubber-to-plastic chemical bonds (grafting) leads to an increase in impact resistance.

The plastic-rubber blends used in the work above were all relatively compatible with each other. It was thought of interest to study some systems that could show a variation in compatibilities, i.e., blends of poly(methyl methacrylate) (PMMA) with various rubbers. The rubber latices used for this purpose were:

> Polybutadiene (PB)
> Styrene/butadiene copolymer (SBR)
> Poly(butyl acrylate) (BA)
> Butyl acrylate/allyl acrylate copolymer (BAA)
> Styrene/butyl acrylate copolymer (SBA)

These rubbers offered a range of compatibilities with the PMMA. The experimental details of latex preparation, blending and coagulation are described elsewhere.[7] It should be noted, however, that the PMMA prepared in our laboratory in latex form had an $M_v$ value of 118,000, melt index 2.1, suitable for molding. The BAA latex was prepared in order to provide a rubber particle containing some unsaturation, making it suitable for staining with osmium tetroxide for electron microscopy of blend films. This was the usual precaution taken to ensure that the polymer blends contained rubber particles corresponding to the original rubber latex particles, even after latex coagulation, and drying and molding of the composites.[2]

## RESULTS AND DISCUSSION

### Blends with Polybutadiene
A series of five polybutadiene latices was prepared, having particle sizes ranging from 80 to 700 nm, to be used for blending with the PMMA base latex. As shown in Table 1, it required a rubber content of about 40% was required to show any effect in increasing the impact resistance. This is unusually high when compared to the 10-15% rubber content generally used in other systems,[2] but is apparently the accepted practice for PMMA-based plastics. Naturally, as expected, this results in a substantial drop in the flexural modulus of the materials, as shown in Figure 1. (The normal drop in modulus, e.g., of polystyrene containing 10% rubber,[2] is about 25%).

The effect of the rubber particle size on the impact strength is plotted in Figure 2, where an optimum can be seen at about 200 nm. The occurrence of this optimum at that level of particle size indicates that PMMA fractures by shear yielding,[1,6] What is also noteworthy, however, is the relatively small

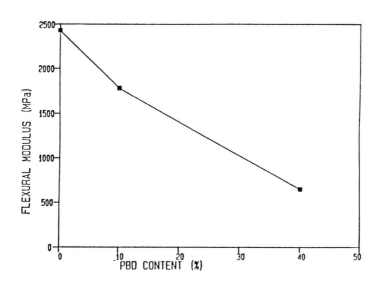

Figure 1.     PMMA/PB blends:  flexural modulus vs. rubber content

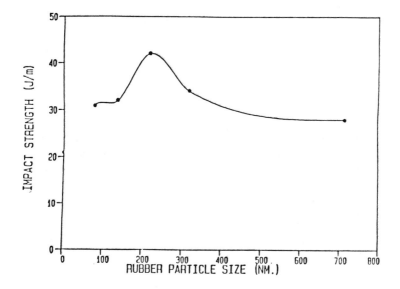

Figure 2.  PMMA/PB blends:  izod impact vs. particle contact  (% rubber)

effect that such a large proportion of the PB rubber (40%) has on the impact strength, raising it from 21 to 42 J/m, even at the optimum rubber particle size. In contrast, the impact strength of polystyrene is generally tripled or quadrupled even at 10% rubber content.[2]

TABLE 1
Izod Impact of PMMA/PB Blends

| Rubber content (%) | 0 | 10 | 10 | 40 | 40 |
|---|---|---|---|---|---|
| PB particle size (nm) | - | 80 | 320 | 80 | 320 |
| Izod impact (J/m) | 21 | 21 | 21 | 31 | 34 |

### Blends with Styrene-Butadiene Copolymer (SBR)

A commercial SBR latex of large particle size (865 nm) was used in this case. The effect of this inclusion is shown in Table 2 for both 10 and 40% rubber content. Although the SBR had a higher Tg (-63°C) than the PB rubber (-83°C), it showed a greater effect in raising the impact strength of the PMMA, at comparable particle sizes (see Figure 2). This was the opposite effect to that previously found for polystyrene,[2,8] and provided the first hint that there is an additional characteristic of the rubber, other than the glass transition temperature, that may affect the impact strength.

TABLE 2
Izod Impact of PMMA/SBR Blends
Rubber Particle Size 865 nm

| Rubber content  (%) | 0 | 10 | 40 |
|---|---|---|---|
| Izod impact (Jm-1) | 21 + 1 | 26 + 1 | 32 + 1 |

### Blends with Poly(butyl acrylate) (BA)

In view of the rather modest increase in impact strength obtained even at a rubber content of 40% with the hydrocarbon-type polybutadiene or SBR, it was thought of interest to examine the effect of a more compatible rubber, such as poly(butyl acrylate) (BA). For this purpose, two BA latices were used. One was prepared in our laboratory and had a particle size of 80 nm while the other was a commercial latex, Rhoplex H-582 (Rhom & Haas), with a particle size of 330 nm. The results are shown in Table 3. It can be seen that the acrylate rubber, BA, has a noticeably greater effect than the hydrocarbon rubber, polybutadiene (PB), at comparable particle size.

### Blends with Poly(butyl acrylate-co-allyl acrylate)(BAA)

As stated previously, the original intent of preparing a butyl acrylate/allyl acrylate copolymer was to introduce some unsaturation into the rubber, so that osmium tetroxide staining could be used for accurate particle size measurement of the rubber particles in the molded material by electron microscopy. The use of these blends for impact strength measurements then led to the surprising results shown in Figure 3, i.e., the inclusion of the rubber particles in the PMMA did not improve the impact strength at all, but in fact, resulted in a decrease! This unexpected result will be discussed later.

TABLE 3
Izod Impact of PMMA/BA Blends
(PMMA = 21 J/m)

| Rubber content (%) | 25 | 40 | 40 |
|---|---|---|---|
| BA particle size (nm) | 80 | 80 | 330 |
| Izod impact (J/m) | 29 | 35 | 41 |

## Blends with Poly(butyl acrylate-co-styrene) (SBA)

Since the SBR rubber (i.e., a copolymer of styrene with butadiene) showed a greater effect on the impact strength than the polybutadiene alone, it was thought of interest to see if a copolymer of styrene and butyl acrylate would show the same effect. Three SBA copolymers were prepared in latex form, with a particle size of about 80 nm and varying styrene content, from

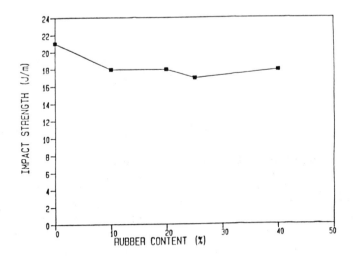

Figure 3. PMMA/BAA blends: izod impact vs. rubber content (particle size = 80 nm).

10 to 20%. The impact strengths of the resulting blends with PMMA are shown in Table 4. There seems to be an optimum in styrene content of the SBA at about 15%, and, at this level, the impact strength is considerably higher than with the other rubbers discussed heretofore, even at a relatively

small rubber particle size. Figure 4 shows that, at the optimum particle size (ca. 200 nm), an even greater impact strength can be achieved.

TABLE 4
Izod Impact of PMMA/SBA Blends:
(40% Rubber Content, 80 nm.)

| % Styrene | Impact Strength (J/m) |
|-----------|----------------------|
| 10 | 45 +/- 1 |
| 16 | 51 +/- 1 |
| 20 | 48 +/- 1 |

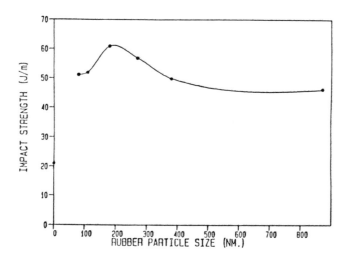

Figure 4. PMMA/SBA blends: izod impact vs. particle size (40% rubber)

## Impact Strength and Rubber-Plastic Compatibility

The results obtained with the PMMA/BAA blends (Figure 3) and the PMMA/SBA blends (Table 4 and Figure 4) indicate that the chemical nature of the rubber can play a strong role in enhancing impact strength of the plastic. It was thought that it might be instructive to attempt a correlation between the impact strength of the blends used in this study and the rubber/plastic compatibility, using the well-known polymer solubility parameter $\delta$ as a guide. Such a correlation is demonstrated in Table 5. The values of $\delta$ for PMMA, PB, SBR and BA were obtained from the literature,[9] while the values for BAA and SBA were calculated, using the "group molar attraction constants" method.[9] The impact strengths of the PBA, SBA and BAA systems were all compared at the same rubber particle size of 80 nm, even though this did not represent the optimum.

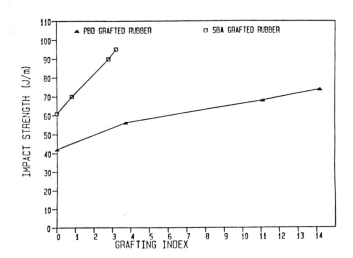

Figure 5. IZOD impact vs. grafting index
(40% rubber: particle size 220 nm).

TABLE 5

Solubility Parameters (δ) vs.Impact Strength
(PMMA = 18.8; Rubber Content = 40%)

| Rubber | Solubility Parameter (δ) $(J/m^3)^{1/2}$ | Particle Size (nm) | Izod Impact (J/m) |
|--------|------|------|------|
| PB | 17.2 | 715 | 28 |
| SBR | 17.8 | 865 | 31 |
| BA | 18.0 | 80 | 35 |
| SBA | 18.2 | 80 | 51 |
| BAA | 18.6 | 80 | 18 |

As can be seen from Table 5, there appears to be a good correlation between rubber/plastic compatibility and impact strength. In other words, when the δ values of the rubber and the PMMA are too far apart (e.g., PB and SBR), there is less reinforcement of impact strength. On the other hand, when the δ values are too similar (e.g., BAA), there appears to be no reinforcement at all, i.e., a homogeneous blend is obtained with no rubber phase. This was actually confirmed by scanning electron microscopy. The optimum case appears to be when the rubber and plastic phases exhibit a good physical adhesion because of their compatibility (e.g., SBA and PMMA).

Similar findings with regard to rubber-plastic compatibility have previously been reported for rubber-toughened poly(vinyl chloride)[10,11].

### Effect of Rubber-Plastic Grafting

Previous work in these laboratories[2,8] has demonstrated that the introduction of chemical bonds between the plastic matrix and the rubber particles always enhances the impact strength. The same effect is demonstrated in Figure 5 for both PB and SBA systems. Here again the "grafting index" expresses the percent, by weight, of rubber bound to the plastic.

## CONCLUSIONS

From the results reported herein, it would appear that, in the case of rubber-toughened plastics, the maximum enhancement of impact strength occurs when a) the rubber used is highly compatible with, but not soluble in, the plastic; b) when the particle size of the rubber is at its optimum, and c) when there is substantial chemical bonding between the two phases.

## REFERENCES

1.    C.B. Bucknall, *Toughened Plastics*; Applied Science Publishers, Ltd: London, 1977.
2.    Maurice Morton, M. Cizmecioglu and R. Lhila, Chap. 14 in *Polymer Blends and Composites in Multiphase Systems*, C.D. Han, Ed., *Adv. in Chem. Ser. No. 206, ACS*, Washington, D.C.,1984,  221.
3.    F.H. Merz, G.C. Claver and M. Baer, *J. Polym. Sci.*,22,325 (1956).
4.    S.L. Rose, *Polym. Eng. Sci.*, 7, 115 (1967).
5.    S. Newman, in *Polymer Blends*, D.R. Paul and S. Newman, Eds., Academic Press, New York, 1978; Vol. 2, Chap. 13.
6.    C.B. Bucknall and D.G. Street, *SCI Monog., 26*, 272 (1967).
7.    F. Grant, Ph.D. dissertation, University of Akron, 1988.
8.    Maurice Morton, N.K. Agarwal and M. Cizmecioglu, in *Polymer Alloys, Vol. III*, D. Klempner and K.C. Frisch, Eds., Plenum Press, New York, 1983, pp 11-18.
9.    H. Burrell, in *Polymer Handbook, 2nd Ed.*, J. Brandrup and E.H.Immergut, Eds., Wiley-Interscience, New York, 1975, pp IV-338, 339.
10.    R. Bauer and M. Guillod, in *Copolymers, Polyblends and Composites*, N. Platzer, Ed., Adv. Chem. Ser. No. 142, ACS, Washington, D.C., 1975, p 231.
11.    J.E. Bramfitt and J.A. Heaps, in *Advances in PVC Compounding and Processing*, M.Kaufman, Ed., Macharen & Sons, London, 1962, 41.

# Chapter 19

# RESIDUAL POROSITY IN POLYMERIC LATEX FILMS

C. M. BALIK

*Department of Materials Science and Engineering*
*North Carolina State University*
*Raleigh, NC 27695-7907*

## INTRODUCTION

A continuing practical problem with latex-based coatings is their high permeability (compared with solvent-based coatings) to water vapor and potentially harmful pollutant gases. Some studies have shown that substrates protected with latex coatings are more susceptible to deterioration than those protected with solvent-based coatings, when exposed to corrosive gases and vapors.[1] Chainey *et al.* [2] have measured permeabilities of helium in films cast from homopolymer latex suspensions and in solvent-cast films of the same polymers. The permeability of the latex films decreased with time but never reached the values found for the solvent-cast films, even for polymers that were well above their glass transition temperatures ($T_g$) and were aged for as long as 30 days. It has also been pointed out that latex film properties continue to slowly change beyond the point where the film is dry, continuous, and transparent.[3-8] In this stage, further gradual coalescence of the latex particles is presumed to occur.

The observations above are highly suggestive of the presence of residual porosity in latex films. A pore network could account for increased permeability and gradual aging effects noted for latex films that are already dry and transparent. The pores or voids must be smaller than the wavelength of light, perhaps as small as several nanometers. They may be uniformly distributed, or they may exist in clusters or channels. The pore morphology must depend on various film-formation conditions such as temperature, drying time, humidity, presence and type of coalescing solvents and initial latex particle size.

The purpose of this chapter is to present additional evidence for the existence of this residual porosity, and to measure the kinetics of its removal at various aging temperatures. This has been accomplished by studying the sorption and diffusion behavior of a probe gas ($CO_2$) in the latex films. A convenient framework for describing the sorption behavior of amorphous polymers is the dual-mode model originally suggested by Barrer.[9] Below $T_g$, the gas is assumed to exist in two distinct sorbed sites or modes. The

Henry's law mode involves dissolution of the gas between polymer chains, analogous to dissolution of a gas in a liquid. The second mode involves sorption into preexisting microvoids, or regions of essentially frozen-in free volume. This is known as the Langmuir mode, since the isotherm for a gas sorbing completely in this mode would resemble a Langmuir adsorption isotherm. The total concentration of gas in the polymer C is given by the sum of the Henry's law $C_D$ and Langmuir mode $C_H$ concentrations:

$$C = C_D + C_H = k_D P + C'_H \frac{bP}{1 + bP}$$

(1)

where $k_D$ is the Henry's law constant, $C'_H$ is the Langmuir capacity, b is the affinity of the gas for the Langmuir sites and P is the gas pressure. $C'_H$, $k_D$, and b are the dual-mode sorption parameters characteristic of a given polymer/gas pair at a fixed temperature. They are obtained by fitting equation (1) to the experimental isotherms of C vs. P, which for glassy polymers are concave to the pressure axis. It has been demonstrated that there is a quantitative connection between $C'_H$ and the excess volume associated with the glassy state.[10,11] As the temperature rises toward $T_g$, $C'_H$ approaches zero. Above $T_g$, the Langmuir mode disappears, and the sorption isotherms become straight lines having a slope equal to $k_D$. The shape of the sorption isotherm is therefore quite sensitive to the presence of microvoids or excess volume in the polymer.

We have used this aspect of gas sorption to characterize the residual porosity in polymer latex films above $T_g$. There is a morphological similarity between a homogeneous glassy polymer containing excess volume, and a polymer latex film above $T_g$ containing some residual porosity. Both are in a nonequilibrium state and should exhibit dual mode sorption behavior. The Langmuir capacity constant $C'_H$ can be quantitatively associated with the amount of porosity in the latex films. If the latex film is above $T_g$, the porosity should disappear with time. All these features have been observed and are discussed herein. $CO_2$ has been chosen as the probe gas, because of its low solubility in the polymer. The total sorbed concentration is therefore dominated by the pores ($C_H$ in equation 1). As will be seen, the total sorbed concentration is quite low, indicating that a very small amount of porosity is present. A more soluble gas (high $C_D$) would tend to make the experimental isotherms more linear and less sensitive to small changes in porosity.

## EXPERIMENTAL

### Samples
The latex was generously provided by UCAR Coatings and Emulsions (Cary, NC). It contained 49.5% solids with both ionic and nonionic surfactants. The polymer was reported to be a terpolymer of vinyl acetate, vinyl chloride and butyl acrylate, present in approximately a 1:1:1 mole ratio. The average particle size was 0.5 μm.

Latex films were cast directly from the suspension on a clean glass plate and allowed to dry for 6 hours, then were placed in a desiccator for aging at 22°C. For the latex films, film age refers to the time between removal of the film from the glass plate and initiation of a sorption run. Solvent-cast films were prepared by dissolving the latex films in chloroform, recasting using a similar procedure, then annealing at 55°C for 4 days, followed by slow cooling in the oven. Film thickness averaged 250 μm for all samples. The $T_g$ of the dried polymer film as measured by DSC, was 10.3°C.

## Apparatus

Sorption and diffusion measurements were made with a Cahn RG recording electrobalance, enclosed in a glass vacuum system. An 8 liter reservoir was attached to maintain constant pressure in the system as the polymer sorbed the penetrant. Polymer films were suspended directly from the balance with thinly drawn glass fibers. The electrobalance was housed in a plywood box thermostated at 30 ± 0.5°C. The sample weight was continuously monitored with a chart recorder. Similar systems have been described by others.[12,13]

To begin a run, a sample was attached to the electrobalance and the system was evacuated to $10^{-3}$ torr and maintained for a fixed degassing time of 38 hours. All samples achieved a constant weight within this time. The system was then backfilled with $CO_2$ at the desired pressure, and mass vs. time readings were initiated on the chart recorder. Plots of mass uptake vs. $(time)^{1/2}$ were initially linear, indicating a Fickian diffusion mechanism.

## RESULTS AND DISCUSSION

### Sorption

Sorption isotherms at 30°C for $CO_2$ in latex films aged from 62 to 480 hours are shown in Figure 1, along with the isotherm for the solvent-cast film. Similar data taken from a second set of samples indicated good reproducibility. The solvent-cast film exhibits the expected Henry's law behavior for a polymer above its $T_g$. The latex films show a much larger sorptive capacity, and their isotherms exhibit the downward curvature normally expected for a glassy polymer. The curvature in this case is attributed to the residual porosity in the polymer. The excess sorptive capacity of the latex samples decreases with aging time and approaches, but never quite reaches, the values for the solvent cast film. This was found to be true for samples aged for as long as 1800 hours (75 days). The decrease in sorptive capacity is attributed to further gradual coalescence of the latex particles, and removal of residual porosity.

Some simplifying assumptions can be made when fitting these data to the dual mode model (equation 1). The Henry's law constant $k_D$ might be expected to have the same value for all samples, since this parameter depends only on the polymer, penetrant, and temperature, all of which are constant for the data in Figure 1. The Langmuir affinity parameter b depends on the same factors and is also not expected to vary for the latex samples. Only the Langmuir capacity parameter $C'_H$ should vary among the samples; furthermore, $C'_H$ should be zero for the solvent-cast sample. These simplifying assumptions lead to the use of five adjustable parameters to describe all four isotherms in Figure 1.

It was also recognized that a significant amount of particle coalescence may occur for the 62 hour sample during acquisition of the isotherm, which took about 12 hours. $C'_H$ for this sample cannot be considered constant, whereas for the other samples very little coalescence occurs during isotherm data collection. This was accounted for by independently measuring the variation of $C'_H$ with time and correcting the data in the 62 hour isotherm to reflect a constant value of $C'_H$. The details of this correction procedure are outlined in the next section. The data shown in Figure 1 for the 62 hour sample have been plotted as obtained, however, without applying any correction for coalescence.

Initially, each isotherm was individually fitted to equation 1, making no assumptions whatsoever concerning the constancy of any of the dual mode parameters among the samples, except for setting $C'_H = 0$ for the solvent cast sample. The parameter values obtained from this fit are displayed in Table 1. The dotted lines in Figure 1 represent the plots of equation 1 using the parameters in Table 1. While each isotherm can be fitted quite well, $k_D$ for

TABLE 1

Dual Mode Parameters : Individual Regressions

| Sample | $k_D \times 10^3$ $cm^3\ CO_2$ (STP)/$cm^3$ polymer-torr | $b \times 10^2$ torr$^{-1}$ | $C'_H$ $cm^3\ CO_2$ (STP)/$cm^3$ polymer |
|---|---|---|---|
| Solvent-cast | 0.99 | --- | --- |
| Latex, aged 62 h | $-0.48 \pm .08$ | $0.68 \pm .01$ | $3.01 \pm .05$ |
| Latex, aged 230 h | $0.85 \pm .03$ | $1.06 \pm .04$ | $0.67 \pm .02$ |
| Latex, aged 480 h | $0.73 \pm .02$ | $0.50 \pm .01$ | $0.46 \pm .01$ |

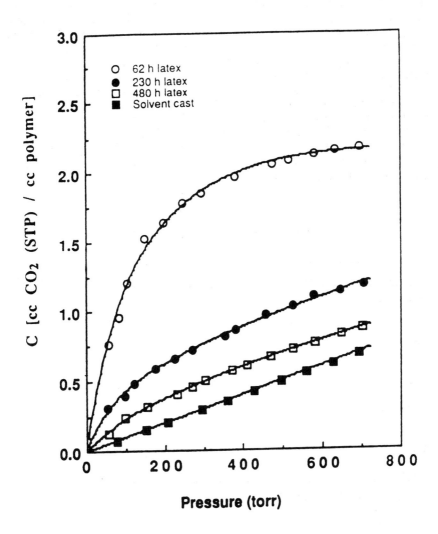

Figure 1.    Sorption isotherms for $CO_2$ in latex and solvent cast samples at 30°C. Each isotherm was individually fitted to equation (1). The lines are based on the parameter values in Table 1.

the 62 hour sample is negative. This is physically unrealistic, and probably is a result of not correcting the data for coalescence during isotherm acquisition. Neither the $k_D$ nor the b values show any definite trend with aging time. The $k_D$ values for the 230 and 480 hour latex films are of the same order as the solvent-cast film. $C'_H$ decreases with aging time, as expected.

The data were next fitted using the assumption of a single value of $k_D$ for all of the samples, and a single value of b for the latex samples. $C'_H$ was again fixed at zero for the solvent cast sample, and the data for the latex samples were corrected for particle coalescence during isotherm acquisition. These results are listed in Table 2. These simplifying assumptions do not appreciably degrade the regression, as evidenced by the similarity in the standard deviations listed in Tables 1 and 2. The value of $k_D$ found in the overall regression is nearly equal to the value found for the solvent-cast sample alone, which lends some validity to the physical significance usually ascribed to this parameter. The values of $C'_H$ for the overall regression are slightly different from those in Table 1, but the decreasing trend is preserved. The $C'_H$ values actually change during isotherm acquisition, especially for the 62 hour sample. Those listed in Table 2 are the values that existed at the beginning of the acquisition of the isotherm.

Table 2

Dual Mode Parameters: Overall Regression

| Sample | $k_D{}^* \times 10^3$ | $b^* \times 10^2$ | $C'_H{}^*$ | Pore fraction |
|---|---|---|---|---|
| Solvent-cast | | --- | --- | --- |
| Latex, aged 62 h | 0.95 | 1.09 ±.03 | 2.24 ±.05 | 0.0049 |
| Latex, aged 230 h | | | 0.67 ±.02 | 0.0015 |
| Latex, aged 480 h | | | 0.27 ±.02 | 0.0005 |

*Units same as in Table 1.

Lines representing the parameters in Table 2 are plotted in Figure 2. It is not possible to plot a single line representing this model through each isotherm. This is due to the variation of $C'_H$ during acquisition of the isotherms (especially for the 62 h sample). Although we know how $C'_H$ varies with time, the relationship between the time each data point was taken and the pressure used is not a smooth function, making a smooth (cross)plot of C vs. pressure from the model impossible. Instead, two lines have been drawn for each isotherm. The upper line uses the (constant) value of $C'_H$ that existed at the beginning of isotherm data collection, and the lower line uses the $C'_H$ value that the sample had at the end of the data collection process (about 12 h later). These lines therefore represent hypothetical samples that do not undergo coalescence, and they should bracket the actual data. This is the case, and the data points move accordingly from near the upper line at the beginning of the isotherm to near the lower line at the end. The spread between each pair of lines indicates how much coalescence occurs during isotherm data collection; this is clearly highest for the sample aged for the least amount of time.

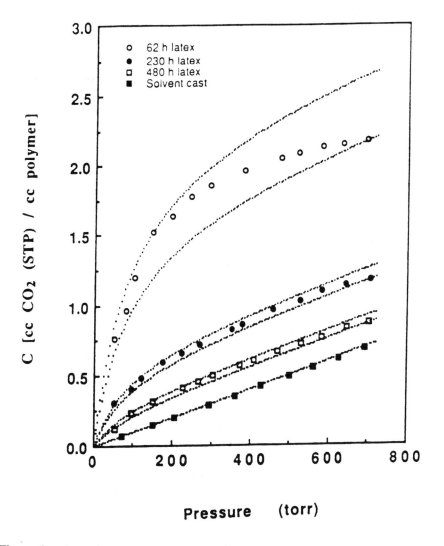

Figure 2.  Sorption isotherms for $CO_2$ in latex and solvent-cast samples at 30°C. An overall regression was used to fit the data, using assumptions listed in the text. The dotted lines are based on parameter values in Table 2 and represent hypothetical samples that do not coalesce during isotherm acquisition.

By associating the Langmuir capacity constant with the amount of residual porosity in the latex samples, it is possible to calculate the pore fraction in these samples, using the known sorbed-state molar volume of $CO_2$. This number has been determined for $CO_2$ in a number of environments, including organic solvents and zeolites,[11] and has been found to be relatively constant at 49 cc/mole between 25 and 50°C. This is also close to the molar volume of liquid $CO_2$ just below the critical point. Assuming that the same value can be applied here, the pore fractions listed in the last column of Table 1 were calculated. The pore fraction is very small; yet the presence of even a minute amount of porosity has a large effect on the sorption behavior. Consider that the difference between the 62 h latex isotherm and the solvent cast isotherm is a pore fraction of about 0.5%. The high sensitivity of this method for characterizing residual porosity can be attributed to the low solubility of $CO_2$ in the polymer itself. Most of the $CO_2$ that is sorbed by the latex samples apparently exists in the pores (especially for samples aged for shorter times).

## Aging Kinetics

The kinetics for removal of the residual porosity can in principle be determined over a relatively long time scale by periodically recording and analyzing sorption isotherms as in Figure 2. From these plots, the time dependence of $C'_H$ can be deduced. However, for shorter time scales (on the order of minutes or hours), a continuous method for monitoring pore removal is desirable. A second type of sorption experiment was designed for this purpose, and the resulting data were used to obtain the time dependence of $C'_H$. This information was used to correct isotherm data for particle coalescence that occurred during isotherm acquisition.

Samples that had been aged for 62 hours were exposed to a constant $CO_2$ pressure of 700 torr at 30°C, and sample weight was monitored continuously over periods of up to 26 days. Typical results are shown in Figure 3, in which the concentration of $CO_2$ sorbed in the Langmuir mode ($C_H$ in eq. 1) is plotted against $t^{1/2}$ (i.e., the constant Henry's law concentration has been subtracted from the total sorbed concentration). An initial uptake of $CO_2$ to an "equilibrium" level close to that shown in Figure 2 at 700 torr is observed, followed by a gradual decrease with time. The decrease in the amount of $CO_2$ in the pores is attributed to the gradual loss of porosity with time. Interestingly, these data level off at a nonzero value, suggesting that pore removal stalls and the sample never reaches zero pore volume over this time period. This is consistent with the fact that the latex film isotherms in Figures 1 and 2 never coincide with the solvent cast film as aging time increases. It is also consistent with observations by others[2] that latex films always have a higher permeability than solvent-cast films of the same polymer, even after several months of aging.

Similar data were obtained at other temperatures ranging from 25 to 50°C; some of which are displayed in Figure 4. The rate at which the porosity was removed increased with increasing temperature; at 50°C $C_H$ leveled out within one hour. In each case, $C_H$ leveled off at the same nonzero value, about 0.2 cm$^3$ $CO_2$ (STP)/cm$^3$ polymer. Apparently, the factor(s) responsible for stopping the pore removal process at this level is not temperature dependent over this range of temperatures. This is surprising, since the rate at which these samples relax varies by several orders of magnitude over the same temperature range, as shown in Figures 3 and 4.

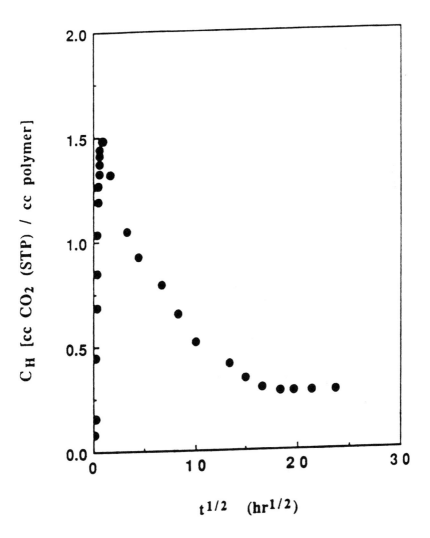

**Figure 3.** Sorption-relaxation plot for $CO_2$ in a 62 h latex sample at 30°C and 700 torr. The ordinate represents the difference between the total sorbed concentration and the constant Henry's law concentration at 700 torr.

It is possible to superimpose all of the sorption-relaxation data from 25 to 50°C on a single master curve, using the shift factors $a_T$ calculated from the WLF equation:

$$\log_{10}(a_T) = \frac{A(T - T_o)}{B + (T - T_o)}$$

(2)

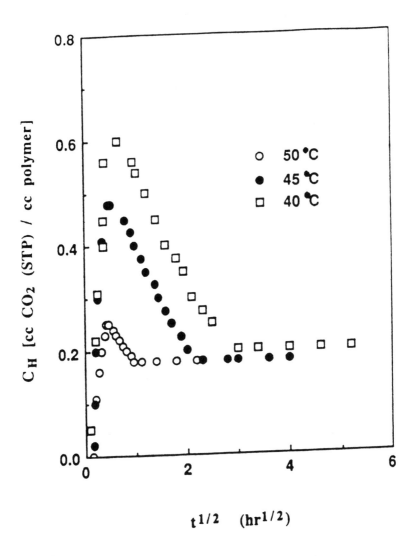

Figure 4.    Sorption-relaxation plots for $CO_2$ in 62 hour latex samples at 700 torr and various temperatures.

Here $T_0$ is the reference temperature, taken as the aging temperature (22°C), and T is the experimental temperatue. The WLF constants evaluated for this choice of $T_0$ were A = 14.174 and B = 64.583. Only data from the decaying portions of each curve were used (*i.e.*, after diffusion equilibrium was attained). The master curve is plotted in Figure 5 at the reference temperature, and the data show good superposition. The shift factors associated with each temperature are listed in Table 3.

Table 3
Sorption-Relaxation Curve Shift Factors

| T (°C) | 25 | 28 | 30 | 33 | 35 | 40 | 45 | 50 |
|--------|-----|-----|-----|-----|-----|-----|-----|-----|
| $a_T \times 10^3$ | 234.9 | 62.39 | 27.40 | 8.653 | 4.217 | 0.8140 | 0.1896 | 0.05168 |

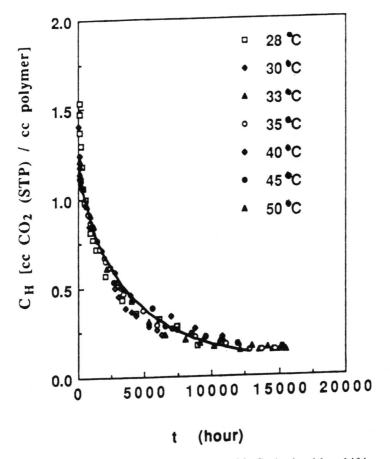

Figure 5. Master sorption-relaxation curve at 22°C obtained by shifting all the individual plots using the WLF equation.

For the purpose of determining the time dependence of $C'_H$, the data on the master curve were represented by the following empirical equation:

$$C_H(t) = w + x \exp(-yt^z) \tag{3}$$

with $w = .0459$, $x = 1.1505$, $y = 0.00139$ and $z = 0.7997$. In equation (3), $t = 0$ corresponds to a film age of 62 hours. To incorporate this time dependence into the dual mode equation, a correction factor $f(t)$ was defined as:

$$f(t) = \frac{C_H(t)}{C_H(t_r)} = \frac{C'_H(t)}{C'_H(t_r)} \tag{4}$$

in which $t_r$ is a reference time, taken as the film age at the beginning of isotherm data collection. $C_H(t_r)$ is the value of equation (3) at $t = 0$. The last equality in equation (3) holds, since pressure is constant at 700 torr for all sorption-relaxation curves. The modified dual mode equation then becomes:

$$C = k_D P + f(t) C'_H \frac{bP}{1 + bP} \tag{5}$$

To calculate $f(t)$, the elapsed time from the beginning of isotherm data collection must be obtained for each data point in the isotherm. To utilize the master curve, these times must be shifted to the corresponding times at the reference temperature of 22°C, using the shift factors in Table 3. The shifted times are inserted into equation (3) to calculate the corresponding values of $C_H(t)$ and $f(t)$. The equation finally used in the nonlinear regression analysis of the sorption data in Figure 2 was therefore equation (5), which incorporates the time dependent correction factor to account for the changing value of $C'_H$.

### Diffusion

A typical sorption–desorption kinetics curve for $CO_2$ in a 62 hour latex sample at 30°C is displayed in Figure 6. $M_t$ represents the amount of penetrant sorbed or desorbed at time $t$, and $M_\infty$ represents the total amount sorbed or desorbed. For Fickian diffusion into or out of a thin film at short times, $M_t/M_\infty$ is given by:

$$\frac{M_t}{M_\infty} = \frac{4}{1} \left( \frac{Dt}{\pi} \right)^{.5} \tag{6}$$

where $D$ is the concentration-independent diffusion coefficient and $l$ is the film thickness. Diffusion coefficients are calculated from the slope of the initial linear region of plots as in Figure 6. The overlapping of the linear regions of plots as in Figure 6 indicate that $D$ is not concentration dependent in this case. This is further substantiated in Figure 7, which shows almost no variation of

D with pressure from 0 to 700 torr. The solid horizontal lines represent the average D for each sample over this range of pressures. Figure 7 also shows that D decreases with increasing aging time for the latex samples and has its lowest value for the solvent cast film. This trend in D is consistent with gradual pore removal and coalescence of the latex films. As was the case for the solubility data, D for the latex samples never reaches the solvent-cast value, once again suggesting that coalescence stops short of producing a pore-free film. The overlapping of the data for samples aged 480 and 1800 hours suggests that very little coalescence occurs beyond 480 hours of aging at 22°C. The diffusion coefficient for $CO_2$ is significantly higher (by about 60%) in the 62 hour latex sample than it is in the solvent cast sample, yet the pore fraction in the 62 hour sample is only 0.49%. A change in D of this magnitude cannot be accounted for by discrete voids, which comprise only 0.49% of the sample, even if unrealistically large values are assumed for the pore diffusivity.

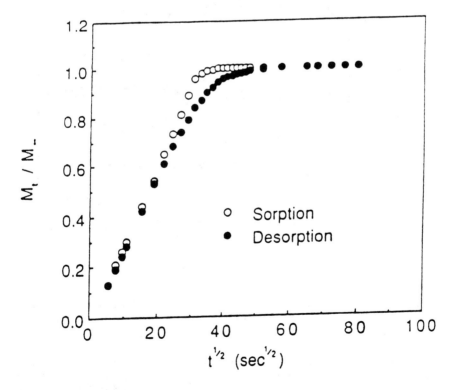

Figure 6.      Typical sorption-desorption kinetics plot obtained at 30°C and 200 torr for $CO_2$ in a latex sample aged 62 hours.

These facts can only be reconciled by assuming that the pores are at least partially connected in some sort of a continuous network that allows relatively fast diffusion into the sample. Indeed, if the pore diffusivity was smaller, a significant deviation from Fickian kinetics would be observed in plots such as Figure 6.

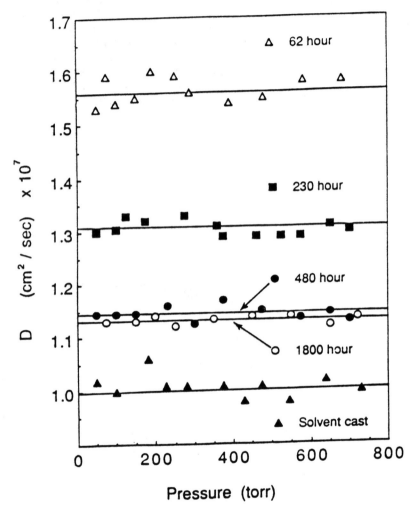

**Figure 7.** Dependence of the $CO_2$ diffusion coefficient on pressure at 30°C for latex films aged for various times and a solvent-cast film. The solid lines represent the average value of D for each sample.

The effect of such a pore network on the overall diffusivity of a gas in a polymer can be quite significant, even for pore fractions as small as 0.49%. The Knudsen (pore) diffusivity for $CO_2$ in a single, cylindrical pore having a diameter of 20 Å at 30°C is estimated to be $2.5 \times 10^{-3}$ cm²/s.[14] This is 4 orders of magnitude higher than the diffusivity of $CO_2$ in the bulk polymer, as measured for the solvent cast sample, which has no porosity (see Figure 7). A continuous network of cylindrical pores 20 Å in diameter can easily be constructed under the constraint of a pore fraction of 0.49%. This suggests that changes in D of 60% resulting from the presence of such a pore network are not unrealistic. A more exact estimate of the effect this type of pore network has on the overall diffusivity requires adoption of a model for the topology of the network; these calculations are beyond the scope of this chapter. A pore morphology on this scale would be very difficult to detect and characterize by other direct methods such as microscopy.

The permeability of a polymer to a penetrant is defined as the product of the diffusion coefficient and the solubility. The solubility of $CO_2$ in the 62 hour latex film is much larger than the solvent-cast film, especially at low pressures, where they differ by almost an order of magnitude (Figure 1 or 2). In contrast, the diffusion coefficients for $CO_2$ in these samples differ by only 60%. Previous permeability measurements for helium in latex and solvent-cast films of other acrylic latex polymers having similar glass transition temperatures show differences of up to a factor of 5 between the two.[2] Our results suggest that it is the excess solubility associated with residual porosity in the latex films that accounts for this, and not diffusivity differences. The effect is most pronounced at low pressures and short aging times. A gas more soluble in the polymer than $CO_2$ or helium would probably exhibit a smaller permeability difference between latex and solvent-cast films, since the pore contribution would be a smaller fraction of the overall solubility. In the limit of high penetrant solubility, the pore contribution would be negligible, and the permeabilities of solvent-cast and latex films should be essentially equal.

## CONCLUSIONS

The residual porosity in latex films of a vinyl chloride/vinyl acetate/butyl acrylate terpolymer has been measured using sorption and diffusion measurements with $CO_2$ as the probe gas. Pore fractions less than 0.49% can easily be measured with this technique. The low solubility of $CO_2$ in the polymer is the key to the high sensitivity of this method. The sorption isotherms obtained for the latex films resemble those of a glassy polymer, and the dual-mode sorption model can be used successfully to describe the data. Pore removal and further gradual coalescence of the latex particles over periods of about one month are accompanied by decreasing $CO_2$ solubility and diffusivity, with solubility changes being more pronounced. The solubility and diffusivity remain above the values measured for $CO_2$ in solvent-cast films of the same polymer, even after 75 days of aging at room temperature.

The pore removal kinetics can be continuously followed by exposing latex films to $CO_2$ at constant pressure and monitoring weight loss as sorbed gas is "squeezed out" during film coalescence. The pore removal rate increases with increasing temperature, and several such "relaxation kinetics" plots could be superimposed onto a single master curve using the WLF equation. This indicates that the mechanism of further gradual coalescence is viscous flow.

The relatively large changes in $CO_2$ diffusivity with aging time, given the very small amount of porosity, suggests that the pores are at least partially connected in a continuous network, which allow rapid diffusion of $CO_2$ into the latex films. This is followed by a slower, Fickian diffusion process, which is experimentally measurable and presumably results from the gas permeating the polymer regions between pore channels.

## ACKNOWLEDGMENTS

The efforts of M. A. Said in performing the experiments cited herein are gratefully appreciated. Acknowledgment is made to the Donors of the Petroleum Research Fund, administered by the American Chemical Society, for the partial support of this work.

## REFERENCES

1.    G. G. Campbell, G. G. Schurr, D. E. Slawikowski, and J. W. Spence, *J. Paint Tech.*, 46, 59 (1974).
2.    M. Chainey, M. C. Wilkinson, and J. Hearn, *Macromol. Chem. Suppl.*, 10/11, 435 (1985).
3.    J. W. Vanderhoff, E. B. Bradford, and W. K. Carrington, *J. Polym. Sci., Polym. Symp.*, 41, 155 (1974).
4.    J. W. Vanderhoff, H. L. Tarkowski, M. C. Jenkins, and E. B. Bradford, *J. Macromol. Chem.*, 1(2), 361 (1966).
5.    E. B. Bradford and J. W. Vanderhoff, *J. Macromol. Chem.*, 1(2), 335 (1966).
6.    R. E. Dillon, L. A. Matheson, and E. B. Bradford, *J. Colloid Sci.*, 6, 108 (1951).
7.    G. L. Brown, *J. Polym. Sci.*, 22, 423 (1956).
8.    S. S. Voyutskii, *J. Macromol. Chem.*, 32, 528 (1958).
9.    R. M. Barrer, J. A. Barrie, and J. Slater, *J. Polym. Sci.*, 27, 177 (1958).
10.    W. J. Koros and D. R. Paul, *Polym. Eng. Sci.*, 20, 14 (1980).
11.    W. J. Koros and D. R. Paul, *J. Polym. Sci., Polym. Phys. ed.*, 16, 1947 (1978).
12.    A. R. Berens and H. B. Hopfenberg, *Polymer*, 19, 489, (1978).
13.    D. J. Enscore, H. B. Hopfenberg, and V. T. Stannett, *Polymer*, 18, 793 (1977).
14.    J. M. Smith, *Chemical Engineering Kinetics*, 2nd ed., McGraw-Hill, New York, 1970, p. 406.

# Chapter 20

# A SIMPLE METHOD FOR LATEX
# AGGLOMERATION AND CONCENTRATION

CARLTON G. FORCE

*Charleston Research Center*
*Westvaco Corporation*
*North Charleston, SC 29411-2905*

## INTRODUCTION

The topic of this chapter involves the agglomeration of latex. Almost all latexes used for adhesives and rubber foam are agglomerated or otherwise increased in particle size from that produced in their polymerization. The primary need for a larger particle size is to permit concentration of the latex to high solids. A year ago Westvaco was granted a patent based on this work involving a new process that allows agglomeration to either narrow or wide particle size distributions. The process also permits the concentration of the latex to its maximum solids content by centrifugation rather than the expensive atmospheric pressure water evaporation methods now used.

For many commercial applications such as the manufacture of foam rubber, the higher the latex solids content and the lower the latex viscosity, the more desirable is the latex composition. Latexes for coating and adhesive applications show the best properties when their particle size is near 200 nanometers (nm) in diameter. At much above 200 nm, the surface tension forces developed in drying the polymer particles are not adequate to produce rapid coalescence into a continuous film. These surface tension forces increase for smaller polymer particle size; however, the smaller the size, the lower the maximum solids that can be achieved before the latex goes from water like viscosity to a thick gel over a 1-2% increase in solids, because of the secondary electroviscous effect.

The larger a latex particle size, the higher the solids achievable before becoming too viscous for satisfactory flow. A broad particle size distribution extending from less than 100 to several hundred nanometers enhances the achievable solids up to about 70%. Applications of synthetic rubber latex in foam rubber articles and carpet backing require high solids content to prevent synersis. Rubber for such articles must coalesce rapidly during drying to optimize cushion characteristics in the finished product.

Large particle size latexes also are required for other applications, such as achieving high impact strength in hard polymers. If a hard brittle polymer receives a sharp blow, it will be apt to shatter unless latex particles are randomly dispersed in the polymer continuous phase. The dispersed latex will absorb the energy and stop the crack.

Fatty acid salt emulsifiers sometimes are used to coat latex particles and maintain the colloidal stability of the emulsion particles in suspension. These soap ions are adsorbed to the latex particles to produce electrically negative surfaces. These charges attract positive cations (counter ions) into the aqueous regions surrounding the particles and produce a diffuse electrostatic double layer around each latex particle. If these diffuse double layers are compressed sufficiently close to the particles by a high concentration of counterions in the system, coagulation of the latex will occur. Thus, without satisfactory electrostatic stabilization, the latex will irreversibly coagulate or "preflock" during handling and storage before use. On the other hand, if the concentration of rubber particles is sufficiently great to force their electrostatic double layers close to each other, a secondary electroviscous effect occurs wherein interaction between neighboring particles produces a high viscosity gel. Dilution of the system or compression of the double layers somewhat with additional electrolyte restores such a latex to its original viscosity and behavior characteristics.

Latex solids approaching 70% generally are achieved by blending about 73.5% volume fraction of 200-250 nm diameter particles and about 26.5% of particles one-half or less this size. This ratio allows a maximum solids compaction before the secondary electroviscous effect occurs.

The technology used to produce latexes with specific particle size characteristics for these and other end-use applications is sophisticated and somewhat costly. The particle size obtained in a latex usually ranges from about 40 nm up to about 120 nm. The emulsifier for the latex and the characteristics of the polymerization system govern the specific size achieved. If 200 nm diameter particles are to be produced by agglomeration, several 120 nm particles must coagulate and coalesce together. Sometimes large, relatively monodisperse latexes are produced by using the original latex as seed, adding more monomer, usually in a semicontinuous manner, and polymerizing to the desired size. More emulsifier also is required to maintain colloidal stability of the larger particles, since the total weight of these latex particles is increased considerably. However, care must be exercised that the new emulsifier never reaches a concentration near its critical micelle concentration in the aqueous phase. Else, new particles will be generated with small particle sizes along with the desired 200 nm particles.

## METHODS TO AGGLOMERATE LATEXES

At present there are several methods used to agglomerate commercial latex to the desired particle size and broad size distribution required. One such agglomeration process involves freezing the latex under carefully controlled temperature conditions followed by melting and heat concentration at atmospheric pressure.

Another process for producing synthetic rubber latexes of increased particle size involves mechanical agglomeration under precise conditions of shear followed by atmospheric pressure heat concentration.

A third method employed to agglomerate commercially and control particle size involves agglomerating during emulsion polymerization with only a small amount of emulsifier. When the latex particles have grown until their emulsifier-starved surfaces are colloidally unstable, shear mixing of the latex particles partially agglomerates them. The latex may be kept from totally coagulating by quickly adding some emulsifier. This will again produce a colloidally stable system. Sometimes shear conditions are such that agglomeration will reduce the particle surface area sufficiently that the original emulsifier can saturate the particle surfaces and prevent coagulation. The agglomerated latexes may be then further polymerized to the final product. Since the rate of polymerization is dependent on the number of rubber particles in the system, polymerization is generally much slower after agglomeration.

As can be seen, these commercial methods of latex agglomeration and concentration are complicated and expensive procedures requiring carefully controlled conditions to produce the desired end-products.

The process that I have discovered involves agglomerating the latex to the desired particle size and size distribution by mixing a synthetic rubber latex containing primary or secondary carboxyl groups on the particle surfaces with a controlled amount of potassium salt. If the potassium ion concentration in the serum exceeds the critical coagulation concentration of potassium in this latex and there is adequate agitation at the point of mixing, the particle size of the latex will increase without any coagulum forming. The primary or secondary carboxyl groups can be a part of the emulsifier or polymerized into the rubber backbone as in carboxylated latexes.

After the initial agglomeration, potassium ions interact with carboxyls on the agglomerated rubber to cause the particles to assemble in their most favorable, closely packed structural configuration as a secondary reversible structure of larger size.

The size and size distribution of the agglomerated latex particles depends upon the severity of coagulation conditions before the larger reversible secondary structure immobilizes all particles, preventing further irreversible agglomeration from occurring. Figure 1 shows a plot of latex solids versus the normality of potassium. At potassium ion concentrations from the critical coagulation concentration of 0.3 N to about 1.2 N, the rubber only partially agglomerates before the large secondary structure forms. Therefore, a broad particle size distribution is formed. Between 0.88 N and 1.12 N potassium ion concentration, the best size distribution to allow concentration to nearly 70% solids latex is produced. Above 1.2 N potassium, all the latex particles agglomerate to a uniform particle size before the large secondary structure forms. Some control of this monodisperse particle size is obtained by the degree of agitation. When we carried out our

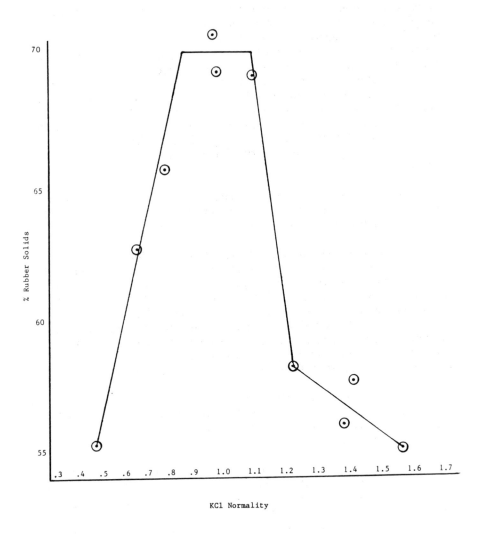

Figure 1.    Centrifuged Rubber Solids vs. KCl Agglomeration Concentration

agitation in a 1/4-inch static mixer, the latex particle size was nearly 340 nm. However, use of similar ingredient flow rates in a 1/2-inch static mixer reduced the particle size to about 240 nm. In other words, the more violent the agitation during ingredient mixing, the larger the latex particle size produced.

The large secondary structure provides another benefit besides stopping agglomeration at a controlled particle size. It makes concentration of the latex easy by centrifugation or, to some extent, even by gravity separation. The rate of gravity separation coupled with the density difference between the rubber particles and serum gives a Stokes law particle size of well over 1 μm for these secondary structure particles.

Gravity separation will increase the solids 10 to 15% leaving a clear aqueous phase below the rubber latex phase. Centrifugation will increase the solids to between 50 and 70% depending on the particle size distribution. With SBR latex, about 1.15 normality in potassium from potassium chloride is required in order to increase the density of the aqueous phase adequately to support spontaneous serum separation. This can be accomplished by adding potassium salt after the initial agglomeration. There is no further increase in particle size after the initial agglomeration, no matter how much additional potassium salt is added to the system. The clear serum containing its proportionate amount of potassium salt that separates from the high solids latex can be cycled into the next batch of latex for agglomeration.

## CONCLUSION

In conclusion, a unique interaction has been found between potassium ion and unhindered carboxyl groups physically or chemically attached on the surface of rubber latex particles. This interaction permits easily controlled agglomeration of the latex to either monodispersed or polydispersed particle sizes. Further, it permits concentration of the latex to high solids by centrifugation.

## ACKNOWLEDGMENT

It was an honor to participate in the seminar marking a milestone in Professor Gilbert's career which led the publication of this book.  Professor Gilbert and I have been friends for 30 years.  The work that is presented here is more related to Dr. Gilbert's activities in industry before his joining the faculty at North Carolina State University than his more recent interests.  It served as the basis for this chapter.

# Chapter 21

# THE USE OF WATER-SOLUBLE POLYACRYLAMIDES IN ENHANCED OIL RECOVERY OPERATIONS

STEPHEN H. HARRIS

*Oil Services Division*
*Allied Colloids, Inc.*
*Suffolk, Virginia*

## INTRODUCTION

Some basic petroleum reservoir engineering must be introduced before explaining how the use of polymers increases the oil recovery from petroleum reservoirs. First of all, oil is trapped inside the pores of rocks in the earth; it is not sitting in large pools or lakes underground. Furthermore, oil, gas and water can occupy the same pore space. The porous rock structures can consist of several layers of rock with different characteristics. These porous rocks are bounded by layers of impermeable rock. Most oil reservoirs are sandstone or carbonate rock.

Oil is trapped under pressure, which can force the oil to flow to the surface when a well is first completed in the reservoir. The oil produced by the natural reservoir energy is defined as the primary recovery. While various forms of energy exist in oil reservoirs (fluids under pressure from overburden rock, expansion of gases, encroachment of aquifers, etc.), this natural pressure is rapidly depleted as the oil is produced. Estimates of primary recovery due to expansion of gases indicate that less than 30% of the oil is produced due to expansion of free gas, and in many cases only 5% of the oil will be produced by expansion of gas coming out of solution.[1]

Far more efficient is the primary recovery due to the encroachment of water. When oil is trapped in strata that is also an artesian water source, the produced oil is replaced by water from the source. The water drive recovers more oil than gas expansion because the pressure in the reservoir is maintained by water influx into the pore space as the oil is produced. This "natural water drive" mechanism can yield 25 to 75% of the oil.[2]

Secondary recovery is defined as the additional oil produced due to the artificial introduction of energy into the reservoir and generally involves the injection of water.[1] Waterflooding the reservoir can theoretically recover

25 to 75% of the oil in the reservoir. However, in practice, less than 50% of the oil is recovered.

Waterflooding recovery efficiencies are affected by several factors but one of the most important is the difference in viscosity between the water and oil.[3] This factor is easy to visualize since the less viscous water will tend to "finger" through the oil rather than push it through the rock to the production well. While the fluid properties and relative amount of the water and oil present in the rock pores are the most important parameters, the rock properties also affect the waterflood performance.

Layers or streaks of rock have different permeabilities (permeability is inversely proportional to the pressure required to flow a liquid through a porous solid per unit time). The water will tend to push oil through the high permeability zones, missing oil in the low permeability zones. Furthermore, the relative permeability to water is a function of the water saturation. This means that as the water moves oil out of the high permeability zone, the problem of poor sweep efficiency only gets worse.[4]

The above problems not only decrease the efficiency of the flood by leaving oil behind in the reservoir, but also result in early breakthrough of water at the production well. This drastically increases the cost to lift the oil to the surface and shortens the economic life of the flood.

### Polymer Flooding

Polymer-augmented waterflooding or polymer flooding is accomplished by dissolving polymers in the water before it is injected into the reservoir. The main purpose of the polymers is to make the water more viscous so that it doesn't finger through the oil.

### Factors Affecting Viscosity

The factors that affect the viscosity achieved with an aqueous polymer solution are shear rate, temperature, molecular weight, polymer type, and chemical make-up of the water the polymers are dissolved in.[5] Acrylamide-acrylic acid copolymers are well suited to polymer flooding due to their solubility in water, high molecular weights, relatively low cost, and nonhazardous nature. They are also highly pseudoplastic in aqueous solutions.

This non-Newtonian, pseudoplastic behavior, decreasing viscosity with increasing shear rate, is used to advantage in the application of polymer floods. At the wellbore where the polymer enters the reservoir, the injection rate and shear rate are quite large, which means that the viscosity is low. This low viscosity at the wellbore means that less energy is required to inject the polymer solution. However, the process must be carefully designed not to exceed a critical injection rate at which shear degradation occurs.[6]

In the radial pattern of the flood, the velocity of the fluid decreases exponentially with distance from the well. As the polymer moves into the reservoir and the flow rate and shear rate diminish, the apparent viscosity of the solution increases. As the shear rate decreases from 500 reciprocal seconds or more at the wellbore to 5 or 10 $s^{-1}$ in the reservoir, the viscosity can increase from just over one centipoise to over 50 cP at fairly low polymer concentrations (1 mg/L or less).

The viscosity of polymer solutions decreases as the reservoir temperature increases. Petroleum reservoir' temperatures range from below 60°F to over 300°F. However, reservoirs with temperatures over 250°F are not suitable for polymer flooding due to the potential for thermal or oxidative degradation of the polymer.[7] Laboratory work must be performed at reservoir temperature when screening polymers for applicability to a particular project.

The temperature and shear rates associated with flow through a specific reservoir rock are mostly out of the control of the engineers and chemists. However, the synthetic copolymers of acrylamide and acrylic acid can be manufactured to obtain a broad range of properties by adjusting the chemical and physical conditions of the polymerization.[8]

The chemical nature of the copolymers can be varied by adjusting the ratio of acrylamide and acrylic acid monomers present in the reaction mixture. Alternatively, the acrylamide homopolymers may be hydrolyzed after polymerization to achieve the same effect. (For this reason, the terms anionic content, degree of hydrolysis and acrylate content are used synonymously.) The resulting copolymers will have different charge densities depending on the percentage of acrylic acid or acrylate salt present in the copolymer. The difference in chemical and physical properties exhibited by acrylamide copolymers with acrylate contents ranging from zero to 40% can be quite dramatic (see Table 1 and Figure 1).

Table 1.

Polyacrylamide Solution Viscosities:
Effect of Molecular Weight and Charge Density

| Polymer | Molecular Weight | Percent Acrylate | Viscosity (cP) |
|---------|------------------|------------------|----------------|
| ONE | 2mm | 25-28 | 29 |
| TWO | 5mm | 24-27 | 31 |
| THREE | 10mm | 22-25 | 36 |
| FOUR | 10mm | 4-5 | 13 |

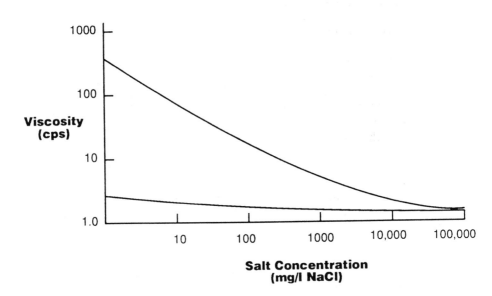

Figure 1. Effect of increasing salt concentration on the apparent viscosity of polyacrylamide solutions. Upper curve is for 70:30 acrylamide-acrylic acid copolymer. Lower curve is for acrylamide homopolymer. Viscosities were measured with a Brookfield LVT viscometer and UL adapter at 24°C and 7.3 s[-1] (6 rpm).

Choosing the proper molecular weight is critical for a successful polymer flood. The viscosity achieved is related to the molecular weight for chemically identical polymers. However, the molecular weight must also be controlled, since the dissolved polymer must be injected into the pores of a rock. Poor injectivity can decrease the effectiveness of a polymer flood, and plugging of the wells can be catastrophic.[9]

In Table 1, the viscosity of various polymer solutions in fresh water (1100 ppm total dissolved solids) is shown. Notice that copolymers One, Two and Three have very similar acrylate content or anionic charge. The viscosity of these solutions increases with increasing molecular weight, although not as much as might be expected. For instance, the five fold increase in measured molecular weight only results in less than 25% increase in apparent viscosity (Brookfield viscometer with UL adapter at 6 rpm).

Far more dramatic is the relationship between polymers Three and Four in Table 1. The polymers have identical molecular weights (within experimental error) but different anionic contents. Clearly the charge density has a far greater effect than molecular weight on the solution viscosity. The hydrodynamic volume of the polymers in water is apparently more sensitive to anionic content and charge repulsion than to the stearic effects from increased molecular weight.

Maximizing the viscosity while minimizing the amount of polymer required is a prime goal in polymer flooding. The charge repulsion effect on viscosity in Table 1 indicates that exceedingly large molecular weights would not be required to achieve acceptable viscosities in fresh water.

However, Figure 1 shows that small amounts of sodium chloride in the water decrease the viscosity dramatically. The sodium cations tend to shield the anionic charges on the polymer, allowing the molecule to shrink in size. The smaller hydrodynamic volume means less interaction between molecules and lower viscosity. This explanation is supported by the minimal effect that the salt has on the nonionic acrylamide homopolymer viscosity shown by the lower curve in Figure 1.

The salt effect is even greater when divalent cations such as calcium or magnesium are present. The sodium salt merely shields the anionic charges from one another, but the divalent salts can actually attract two charges. At very high calcium concentrations, acrylamide copolymers with high percentages of acrylate groups will precipitate due to the anions association with calcium ions.[10]

Figure 2 shows the decreased viscosity resulting from increased calcium concentration in water containing approximately 100 ppm NaCl and 800 ppm Alcoflood 1175L, a high molecular weight 70:30 acrylamide-sodium acrylate copolymer. Notice that 100 ppm calcium carbonate (40 ppm calcium ion) causes a 50% drop in viscosity.

Polymer floods using polyacrylamide copolymers have been performed in waters containing over 100,000 ppm total dissolved salts.[11] However, the use of low salinity water is both technically and economically advantageous. Softened, fresh water offers increased viscosity and long-term stability, both of which are crucial to successful polymer flood projects.[12]

## Screening Polymers

Since waters produced from underground reservoirs (both petroleum reservoirs and aquifers) can vary greatly in terms of salinity and hardness, each water must be tested for compatibility with the polymer. A balance between molecular weight and anionic content to achieve maximum solubility, viscosity and injectivity must be achieved.

Simple viscosity and solution stability tests are generally used for preliminary screening. The initial tests determine the relationship between polymer concentration and viscosity using the water from the particular field under consideration. For these tests, a Brookfield LVT Viscometer fitted with a UL adapter is commonly used to determine the apparent viscosities at the shear rates (5-10 s⁻¹) that are expected in most polymer floods. Although preliminary tests are usually performed at room temperature, data to be used for modeling or designing the flood must be obtained at reservoir temperature.[13]

The long-term stability of the aqueous polymer solutions at reservoir temperature must be determined, since the flood may take several years to complete. The fact that petroleum reservoirs are reducing environments helps the stability at higher temperatures, since oxidative degradation ceases when the oxygen is consumed.[13] However, thermal hydrolysis also occurs at elevated temperatures.[7] As discussed earlier, excessive anionic content in the acrylamide copolymer can lead to precipitation with calcium or other ions.[12]

The stability tests are typically carried out by storing solutions of the polymer in reservoir water (sometimes in the presence of reservoir rock) in sealed ampules in an oven at reservoir temperature.[14] Some of the ampules are removed periodically from the oven and checked visually for solution clarity, and the viscosities measured and compared to the initial viscosity. A cloudy polymer solution would suggest decreased solubility or precipitation of the polymer. Any viscosity loss, whether due to thermal degradation or precipitation of polymer, would be undesirable. The researcher must recommend the polymer that exhibits the best stability as well as maximum viscosity at the minimum dosage.

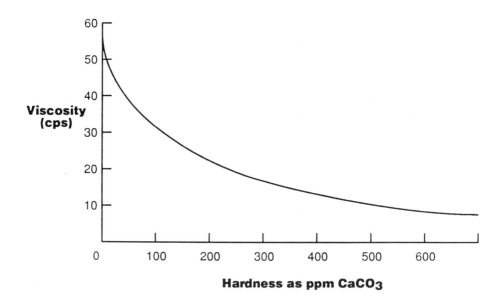

Figure 2. Effect of calcium concentration on the apparent viscosity of solution of acrylamide-acrylic acid copolymer. Viscosity measured with Brookfield LVT viscometer and UL adapter at 50°C and 7.3 s$^{-1}$ (6 rpm).

## Polymer-Rock Interactions

The previous tests have mainly been concerned with the interaction of the polymer and the water in the reservoir, but the reservoir rock also interacts with the polymers in solution. Polyacrylamide copolymers tend to adsorb onto the reservoir rock. The rate and degree of adsorption is a function of the water salinity, type of rock, type and amount of clays present in the pores of the rock, and the anionic content in the acrylamide copolymer.[15] The loss of polymer due to adsorption limits its propagation through the reservoir.

Polymer adsorbed onto the surface of the rock impedes the flow of water or reduces the relative permeability of the rock to water. While this effect can be detrimental to injectivity into the rock and propagation through the reservoir, the property actually improves the sweep efficiency of the polymer augmented waterflood. Like viscosity, the adsorption of polymer increases the resistance to flow through the rock. Therefore, a simple measure of the solution viscosity does not completely describe the behavior of the polymer solution in the reservoir.[15] Furthermore, the advantages and disadvantages of adsorption must be carefully balanced.

After completion of the preliminary screening tests for viscosity and stability, the polymer solution is usually evaluated in rock from the oil reservoir. Core samples can be recovered when wells are drilled into the reservoir. The cores are cut, carefully measured and sealed in containers that will withstand pressure. The polymer solution is injected through these samples with a constant rate pump. By measuring and recording the pressure required to maintain the flow rate, the injectivity of the polymer solution into actual reservoir material is determined.[16] A constant pressure at constant rate is interpreted as good injectivity. A steadily increasing pressure would indicate plugging by the polymer solution. Also, corefloods are the only way to measure the combined effect of viscosity and adsorption that determines the total resistance of the polymer solution to flow through the reservoir rock.[17]

## Flow Through Porous Media

In 1856 Henry Darcy developed a relationship (equation 1) that says the flow rate of a liquid through a porous solid is proportional to the pressure drop across the solid and inversely proportional to the viscosity of the liquid. Darcy's law states that the parameter relating the flow rate, pressure, and viscosity is the permeability. The units of permeability are technically the same as area, reduced from $(cP \cdot cm \cdot cm)/(atm \cdot s)$, but are usually referred to as darcies. In practice, most oil reservoirs have permeabilities much less than one darcy and the permeability is routinely expressed in millidarcies[18]

$$Q/A = k(P_2 - P_1)/\mu L \qquad (1)$$

where

$Q$ = flow rate (cm$^3$/s)
$A$ = area (cm$^2$)
$k$ = permeability (darcies)
$(P_2 - P_1)$ = pressure drop (atm)

$\mu$ = viscosity (cps)
$L$ = length (cm)

Darcy's law is used to experimentally determine the performance of polymer solutions flowing through reservoir rock. By assuming a viscosity of one centipoise (at 30°C) and measuring the pressure drop across a core of known dimensions while holding the flow rate constant, the permeability of the core to water can be calculated. If a polymer solution is then injected at the same rate as the water, the polymer pressure drop can be used with the water permeability to estimate the in situ viscosity of the polymer solution.

However, since adsorption of the polymer decreases the permeability to water, the assumption of constant permeability is not valid. Comparison of the in situ viscosity with the apparent viscosity measured with the Brookfield viscometer gives an indication of the permeability reduction due to polymer adsorption.

Furthermore, polymer flow characteristics are complicated by the non-Newtonian viscosity behavior. In the field, the flow rate decreases with distance from the injection well (assuming radial flow), and therefore the shear rate and viscosity are constantly changing.

The complex relationships between viscosity and permeability are best managed by combining these terms into one variable defined as mobility (permeability divided by the viscosity). The ratio the mobility of the oil to the mobility of the driving fluid was first discussed by Muskat.[19,20] Later authors recognized the mobility ratio as one of the most important factors determining the oil recovery efficiency of a waterflood.[3]

Usually, the mobility is measured at several flow rates to account for the change in viscosity with change in shear rate. Although the adsorption rate is also measured, the mobility is usually determined after the adsorption of the polymer (and subsequent decrease in permeability) has reached equilibrium.

## Polymer Flood Design Considerations

The equilibrium adsorption of polymer onto the rock is used to determine the total quantity of polymer required to maintain the viscosity performance of the polymer solution through the reservoir. The actual concentration of polymer injected is chosen to maximize the injectivity and minimize mobility. The total volume of polymer solution injected is determined by the quantity required and the concentration allowed with a portion allocated to taper the viscosity and concentration back to water at the end of the polymer flood. Engineers typically perform computer simulations to optimize the amount of polymer, concentration of polymer and flow rates for the project.[16]

**Mobility  Control**

The mobility ratio is the mobility of the fluid driving the oil divided by the mobility of the oil (as of 1957, prior to that time there were inconsistencies in the literature).[3] The smaller the mobility ratio, the greater the recovery efficiency of the process. Polymers dissolved in the water used for a flood physically interact with one an other to increase the viscosity of the fluid. The polymers that chemically adsorb onto the rock decrease the permeability to water. Both these actions decrease the mobility of the driving phase and increase the recovery efficiency.

Figure 3 demonstrates the improved sweep efficiency obtained by using polymers. The diagrams represent actual data obtained by flooding square sandpacks from one corner and producing oil from the opposite corner.[21] With a mobility ratio of 71, only 10% of a pore volume (PV) is injected before the water finds a path to the production well. This means that only 10% of the reservoir is swept or ten percent of the oil produced before the water breaks through to the producer. At this point, the water will tend to follow the path that is full of water rather than push additional oil. In contrast, a polymer flood, with a mobility ratio of 2.4, sweeps 60% of the reservoir before water is produced from the test.

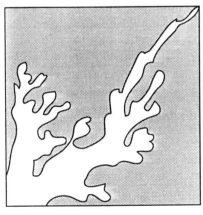

PV = .10
Mobility = 71

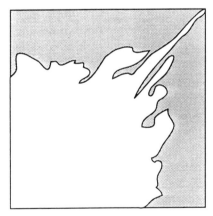

PV = .60
Mobility = 2.4

Figure 3. As the mobility ratio is decreased, a larger pore volume fraction (PV) is injected prior to breakthrough and the sweep efficiency of the flood is increased. (After Habermann,[21] courtesy *Transactions, AIME* 1960.)

Although the experiment illustrated in Figure 3 is an extreme case, improvement in recovery efficiency is obtained whenever an adverse mobility ratio is decreased significantly. Furthermore, the geology and permeability of petroleum reservoirs are never homogeneous and the ability of the polymer to improve the overall sweep efficiency becomes even more important. While the process is difficult to quantify in laboratory experiments and even more difficult to predict with computer simulations, proper application of the water-soluble acrylamide copolymers has resulted in significant production of oil in many full scale field projects.[11,16,22,23,24] Typically, polymer floods produce more of the oil from the reservoir, faster than waterflooding, with less fluid injected and produced. Therefore, polymer flooding saves time, energy, and money while recovering oil that would otherwise be left in the ground and wasted.

## REFERENCES

1.  Clark, N.J.: *Elements of Petroleum Reservoirs,* AIME, Dallas, (1969), 78.
2.  *Drilling and Production Practices,* American Petroleum Institute, Dallas, (1943).
3.  Craig,F.F.: *The Reservoir Engineering Aspects of Waterflooding, Monograph Series,* Society of Petroleum Engineers AIME, Dallas, (1971) **3**.
4.  Dykstra, H. and Parsons, R.L.: The Prediction of Waterflood Performance with Variation in Permeability Profile, *Production Monthly,* (1950) 15, 9-12.
5.  Szabo, M.T.: An Evaluation of Water-Soluble Polymers for Secondary Oil Recovery - Part I, *J. of Petroleum Tech.,* (May 1979), 553-560.
6.  Maerker, J.M.: Shear Degradation of Partially Hydrolyzed Polyacrylamide Solutions, *Society. of Petroleum Eng. J.,* (1975), 311-322.
7.  Ryles, R.G.: Chemical Stability Limits of Water-Soluble Polymers Used in Oil Recovery Processes, *SPE Reservoir Eng.,* ( 1988), 23-34.
8.  Argabright, P.A., Rhudy, J.S., and Phillips, B.L.: Partially Hydrolyzed Polyacrylamides with Superior Flooding and Injection Properties, *Proceedings 57th SPE Technical Conf.,* (1982), SPE 11208.
9.  Ball, J.T. and Pitts, M.J.: Effect of Varying Polyacrylamide Molecular Weight on Tertiary Oil Recovery from Porous Media of Varying Permeability, *Proceedings of 4th SPE/DOE Symposium on Enhanced Oil Recovery,* (1984), SPE/DOE 12650.
10. Muller, G., Laine,J.P., and Fenyo,J.C.: High Molecular Weight Hydrolyzed Polyacrylamides I. Characterization. Effect of Salts on the Conformational Properties, *J.Polymer Sci., Chem.Ed.,* (1979) 17, 659-672.
11. Greaves, B.L., Marshall, R.N., and Thompson, J.H.: Hitts Lake Unit Polymer Project, *Proceedings of 59th SPE Technical Conf.,* (September 1984), SPE 13123.
12. Moradi, A., and Doe, P.H.: Hydrolysis and Precipitation of Polyacrylamides in Hard Brines at Elevated Temperatures, *SPE Reservoir Eng.,* (1987) 2, 189-199.
13. Yang, S.H. and Trieber, L.E.: Chemical Stability of Polyacrylamide Under Simulated Field Conditions, *Proceedings of 60th SPE Technical Conf.,* (1985), SPE 14232.
14. Shupe, R.D.: Chemical Stability of Polyacrylamide Polymers, *J. Petroleum Tech.,* (1981), 1513-1529.
15. Chang, H.L.: Polymer Flooding Technology - Yesterday, Today, and Tomorrow, *J.Petroleum Tech.,* (1978), 1113-1128.
16. Weiss, W.W., and Baldwin, P.W.: Planning and Implementing a Large-Scale Polymer Flood, *J. Petroleum Tech.,* (1985), 720-730.
17. Jennings, R.R., Rogers, J.H., and West, T.J.: Factors Influencing Mobility Control by Polymer Solutions, *J. Petroleum Tech.,* 391-401 (1971).
18. Clark, N.J.: *Elements of Petroleum Reservoirs, AIME,* Dallas, 21-22 (1969).

19. Muskat, M.: *Physical Principles of Oil Production,* McGraw-Hill, New York, 1949.

20. Muskat, M.: *The Flow of Homogeneous Fluids Through Porous Media,* McGraw-Hill ., New York, 1937.

21. Habermann, B.: The Efficiency of Miscible Displacement as a Function of Mobility Ratio, *Trans., AIME, 219,* 264-272 (1960).

22. DeHekker, T.G., Bowzer, J.L., Coleman, R.V., and Bartos, W.B.: A Progress Report on Polymer-Augmented Waterflooding in Wyoming's North Oregon Basin and Byron Fields, *Proceedings of 5th SPE/DOE Symposium on Enhanced Oil Recovery,* (April 1986), SPE/DOE 14953.

23. Hochanadel, S.M., Lunceford, M.L., and Farmer, C.W.: A Comparison of 31 Minnelusa Polymer Floods with 24 Minnelusa Waterfloods, *Proceedings of the 7th SPE/DOE Symposium on Enhanced Oil Recovery,* (April 1990).

24. Maitin, B.K. and Voltz, H.: Performance of Deutche Texaco AG's Oerrel and Hankensbeuttel Polymer Floods, *Proceedings of the 2nd SPE/DOE Symposium on Enhanced Oil Recovery,* (April 1981), SPE/DOE 9794.

# Chapter 22

# OUTDOOR WEATHERING OF POLYSTYRENE FOAM

## JAN E. PEGRAM AND ANTHONY L. ANDRADY

*Research Triangle Institute*
*P. O. Box 12194*
*Research Triangle Park, NC 27709*

## INTRODUCTION

Although weathering and degradation of polystyrene film have been studied by several researchers, little has been done on the weathering of polystyrene foam. This material is widely used by the packaging industry, and these products are therefore an important component of litter and municipal solid waste. Polystyrene foam is also a major component of marine plastics debris,[1,2] a principal source being shipping and fishing vessels. Little or no data on the degradation of polystyrene foam under marine exposure conditions have been reported, but ingestion by birds, turtles and other marine animals has been observed[3-5] with suggested long-term negative effects.[6,7] It is thus timely to study polystyrene foam degradation under marine exposure conditions in view of the current technical, legislative and consumer interest in plastic waste in the environment.

(1)  In-chain peroxides formed during polymerization

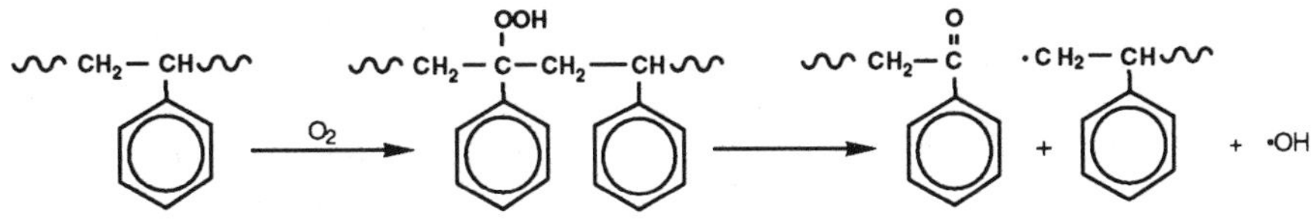

(2)  Terminal phenyl alkyl ketones

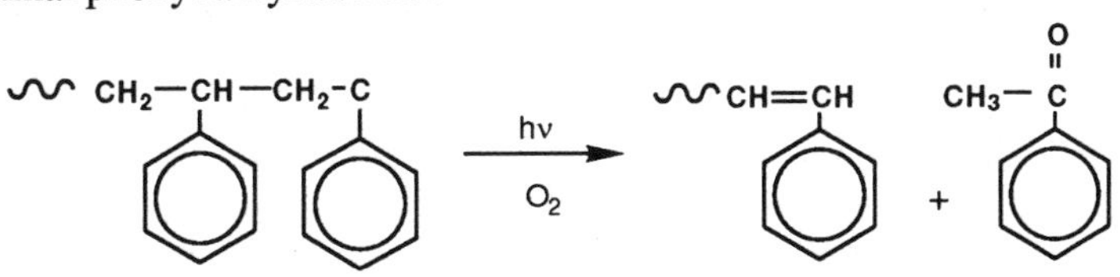

Figure 1.    Proposed degradation mechanisms for polystyrene.

It is not the intent of the present work to study the actual mechanism of degradation. Various sources have shown the reaction to be initiated by ultraviolet light and to occur at UV-absorbing chromophores that are present as impurities in the polymer chain (e.g., in-chain peroxides formed during polymerization[8] or terminal phenyl alkyl ketones[9]). Some proposed reaction mechanisms are shown in Figure 1.

A phenomenon observed with polystyrene degradation is the light-induced yellowing reaction. Several chromophores associated with the yellow color have been proposed,[10-12] but these studies were conducted using wavelengths not typical of sunlight,[13] which limits the applicability of the findings with respect to the outdoor weathering of polystyrene.

The following work discusses polystyrene foam degradation under actual environmental exposure conditions as measured by changes in physical and mechanical properties.

## EXPERIMENTAL

### Samples
Commercially available expanded extruded polystyrene foam trays (used in retail packaging of meats) were used for the study. The samples were stored away from light until the actual exposure was carried out.

### Outdoor Exposure
Polystyrene foam samples were exposed at a coastal location in Beaufort, North Carolina. One set of foam samples was stapled on to a wooden platform and exposed horizontally on the roof of a laboratory building. A second set was exposed floating in sea water. Sea water was continuously pumped through a shallow tank at the beach to maintain a water level of 12 to 18 inches at all times. Floating the samples within the tank minimized accumulation of debris on the samples while providing biologically active fresh sea water. Samples from each set were removed once every two months (dried under ambient conditions in the case of wet samples) and stored in the dark under ambient temperature until tested. The monthly mean temperatures in the Beaufort area for the duration of exposure, as reported by the National Weather Service, are shown in Figure 2. Temperature extremes during the exposure period were 40-87°F in sea water and 19-100°F in air.

### Yellowness Measurements
Yellowness measurements were made using a Macbeth 1500 Colorimeter with an integrating sphere. The specular component of light was excluded, and the ultraviolet light included in the measurements. A white ceramic tile was used as the backing material. The ASTM Yellowness Index was calculated on the basis of the CIE standard illuminant C.

Yellowness Index (ASTM D1925-70) was calculated from the tristimulus values X, Y, and Z using the formula

$$\text{Yellowness Index} = \frac{100(1.28X - 1.06Z)}{Y}$$

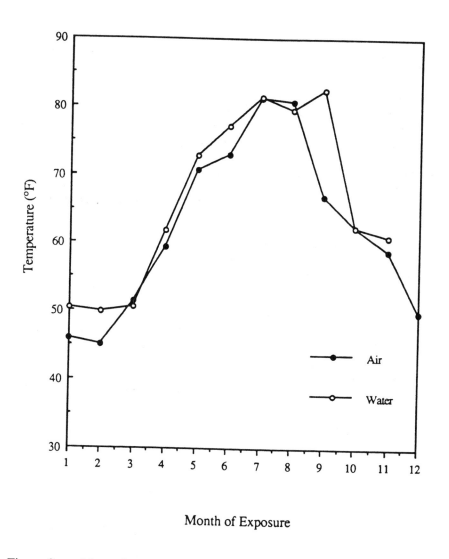

Figure 2.    Mean air and water temperatures for exposure duration.

## Tensile Property Measurements

An Instron tester (Model 1122) was used to determine the tensile properties of the polystyrene foam. A gauge length of 4.0 cm and a crosshead speed of 20 mm/min were used.

## Viscosity Measurements

An Ubbelohde viscometer was used to determine the viscosities of foamed polystyrene samples dissolved in chloroform. The intrinsic viscosity [$\eta$], in dL/g was calculated using the following equation:

$$[\eta] = (1/c)[2(\eta_s - \ln \eta_r)]^{1/2}$$

where $\eta_r$ is the relative viscosity, $\eta_s$ is the specific viscosity and c is the concentration in g/100 mL.

### Gel Permeation Chromatography
A Waters GPC system consisting of a Model M6000A pump, a series of five columns, and a Model R401 differential refractometer detector was used. Columns were Ultrastyragel with a pore size ranging from $10^6$ to $10^2$ Å.

### Fourier Transform Infrared Spectroscopy
Reflectance spectra for polystyrene foam exposed 8 months in air and also unexposed were obtained on an Analect FTIR Spectrometer Model FX-6200.

## RESULTS AND DISCUSSION

### Changes in Yellowness Index
Polystyrene foam samples exposed outdoors in air underwent rapid yellowing on the exposed surface. Figure 3 shows the effect of exposure duration on Yellowness Index for the outdoor samples. The change in sample thickness over time is also depicted. Yellowing was evidently a surface reaction and was accompanied by eventual embrittlement of the affected layer, allowing it to be easily removed by gentle abrasion. As light penetrates the foam, loss of radiation due to scattering occurs at air/polymer interfaces; it is thus reasonable to expect the direct effects of photoreactions to be confined to the surface in the case of polystyrene foam. However, surface studies on photooxidized polystyrene films also show a high degree of surface reaction not typical of the bulk of reacted polymer, suggesting local surface oxidation to be typical of polystyrene film[14] as well as of foam.

Yellowness Index of the exposed surfaces of foam material increased approximately linearly with the duration of exposure for up to 6 months exposure but thereafter decreased. Davis and Sims[15] report a linear increase in Yellowness with the time of exposure for general-purpose polystyrene exposed in Arizona. The present data for polystyrene foam show a higher degree of yellowing for approximately comparable durations of exposure at a North Carolina location receiving less sunlight than Arizona. Furthermore, the present results also show a decrease in yellowing during the 6-12 month period of exposure by as much as 20% of the maximum value. Such a decrease was not obtained in the previously reported study[15] and is likely to be obtained with foams only. The decrease is probably a result of the slow loss of the yellowed brittle layer due to rain and wind rather than to a bleaching reaction at longer exposure times. In fact, thickness measurements of the non embrittled lower layer of exposed samples (Figure 3) support this view and show the thickness of the sample to decrease with time of exposure until it is reduced by nearly 50% at the end of one year.

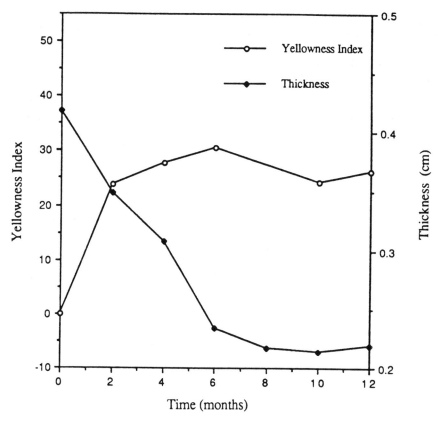

Figure 3.    Yellowness Index and thickness measurements of foam samples exposed outdoors.

Yellowing is accompanied by an increase in ultraviolet absorbance for polystyrene. This effect has been demonstrated for polystyrene film;[16] absorbance spectra for films exposed outdoors showed absorbance at any given wavelength to increase with the duration of exposure. Thus, the outer yellow layer of exposed polystyrene foam could protect the inner layers by filtering out the UV radiation associated with the degradation reaction.

Yellowness Indices of samples exposed floating in seawater could not be conveniently measured due to the extensive surface fouling of the samples. Most surfaces, including plastics, undergo facile fouling[17] when exposed to sea water for even a few weeks. However, it was easy to visually observe yellowing in exposed areas and to remove by gentle abrasion the affected surface layer. The thickness of the remaining nonembrittled layer, presumably unaffected by the light-induced degradation process, decreased more rapidly for samples exposed in seawater than for those exposed in air. The samples exposed in seawater were possibly protected from photodegradation to a lesser extent than those exposed in air, since the yellowed surface layer is washed away by seawater and abraded off gradually, exposing fresh layers of sample to the radiation. Consequently, the sample in water might be expected to undergo faster degradation relative to that in air. This is consistent with the observation that the inner unembrittled layer at a given duration of exposure tends to be thicker in the case of samples exposed in air compared to those exposed in seawater (Figure 4).

### Infrared Spectroscopy

FTIR examination of the yellowed surface layer of foam samples revealed extensive changes in functional groups at the surface during outdoor exposure. The spectrum showed the development of two strong absorption bands, one at 1730 cm$^{-1}$ and one at 1600 cm$^{-1}$, with yellowing. The former was indicative of ketone group formation, which was consistent with general mechanisms of polystyrene oxidation.[18] The band at 1600 cm$^{-1}$ could have been due to a C=C group conjugated with a phenyl group or a β diketone structure. The latter has been proposed as a possible end group formed during polystyrene photooxidation.[12]

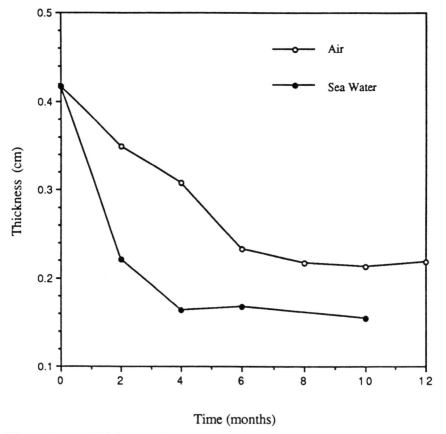

Figure 4.    Thickness of exposed samples after removal of embrittled yellow layer.

**Molecular Weight Changes**

   A consequence of light-induced degradation in polystyrene foam is the reduction of the average molecular weight of the polymer as a result of photolytic chain scission.[19]  Changes in the molecular weight of polystyrene foam as monitored by gel permeation chromatography and solution viscosity are shown in Table 1.  While the yellowing is mostly restricted to the embrittled surface layer of polystyrene foam, the lower seemingly unaffected layer also undergoes some degradation.  The original number average molecular weight of  the outer embrittled layer of polystyrene foam was reduced by 86% on exposure outdoors for a 12 month period. The lower layer lost 45% of the initial number average molecular weight during the same exposure period.   Thus, significant changes in the molecular weight of the lower nonembrittled layer were observed. This is expected in the case of a free radical degradation process where the radicals can easily "move" about the polymer matrix via hydrogen abstraction.

   While the GPC results in Table 1 are based on measurements of the lower, non-embrittled layer, the solution viscosity data are based on the entire cross section of the exposed samples and thus reflect an average molecular weight for each sample.  Some of the yellow layer, however, would have been removed by abrasion, particularly for the samples exposed at sea.

Table 1
Changes in Molecular Weight (GPC and viscosity results) for Polystyrene Foam Exposed Outdoors.

| Duration | $M_n{}^a$ x $10^3$ | | $M_w{}^b$ x $10^3$ | | Intrinsic Viscosity [c] | |
|---|---|---|---|---|---|---|
| (months) | Sea Water | Air | Sea Water | Air | Sea Water | Air |
| 0 | 117.3 | 117.3 | 241.8 | 241.8 | 0.859 | 0.859 |
| 2 | 109.0 | 102.8 | 228.7 | 221.6 | 0.802 | 0.778 |
| 4 | 106.3 | 91.0 | 228.0 | 213.5 | 0.766 | 0.709 |
| 6 | 93.8 | 78.7 | 270.7 | 195.0 | 0.673 | 0.648 |
| 8 | 75.0 | 71.4 | 182.6 | 184.2 | 0.621 | 0.637 |
| 10 | 58.8 | 68.0 | 165.6 | 171.8 | 0.557 | 0.627 |
| 12 | - | 63.9 | - | 160.4 | - | 0.593 |

[a]   $M_n$ = number average molecular weight
[b]   $M_w$ = weight average molecular weight
[c]   Intrinsic viscosity in dL/g. Foam samples were obtained from the entire cross section of the exposed samples.

   In general, for random scission in the absence of concurrent crosslinking, the degree of polymerization changes with time according to the following equation for moderate extents of degradation[20]:

$$1/[DP]_{n,t} - 1/[DP]_{n,o} = kt \qquad (1)$$

where t is time and k is a rate constant. This equation assumes that all main chain links are of equal strength and that the rate of breaking links is proportional to the number of links left at time t. For higher degrees of degradation the above equation becomes

$$\ln (1 - 1/[DP]_{n,o}) - \ln (1 - 1/[DP]_{n,t}) = kt \qquad (2)$$

Assuming photodegradation to be the sole mechanism for molecular weight reduction, the rate constant k is given by

$$k = \Phi I/[n_o] \qquad (3)$$

where $n_o$ is the initial concentration of main chain links, I is the average number of quanta absorbed per unit volumeper second and $\Phi$ is the quantum efficiency of the scission process.

Figure 5 is based on gel permeation chromatographic determination of number average molecular weights for the two sets of samples exposed in air and in seawater. These determinations were carried out on the lower unembrittled layer of samples. Since the rate of change in thickness of the samples is not the same for two conditions of exposure, the comparison implied in the figure is, strictly speaking, not appropriate. Essentially, the number average molecular weight of the embrittled material is regarded as not contributing to the average molecular weight of exposed samples. However, the qualitative features of the analysis are interesting.

The plot of data according to equation (1) above is linear for results pertaining to exposure in air. Those for exposure in seawater are initially linear, with a rate constant lower than that observed with samples exposed in air. This is to be expected in view of the known slower degradation of common thermoplastics when exposed floating in seawater compared to the degradation of those exposed in air.[21,22] At exposure times exceeding about 4 months, the seawater curve deviates from linearity, suggesting a large increase in the k value. This increase is possibly due to the gradual removal of the embrittled yellow layer, which filters out the harmful ultraviolet radiation, continually exposing new material to light. This would correspond to an increase in the parameter I over time in equation (3). The nonlinearity would likely have been observed at an even shorter duration of exposure if the embrittled yellow surface layer had been included in the material for gel permeation chromatography measurements.

## Mechanical Properties

For foamed polystyrene, the changes in mechanical properties of the material with duration of outdoor exposure are also an important consideration. It is generally convenient to monitor environmental deterioration of plastics in terms of tensile properties, particularly ultimate extension. The polystyrene foam, however, has a very low initial (unexposed) extension at break, and the small changes observed on weathering are difficult to interpret.    Changes in tensile strength of the material are, however, of interest.

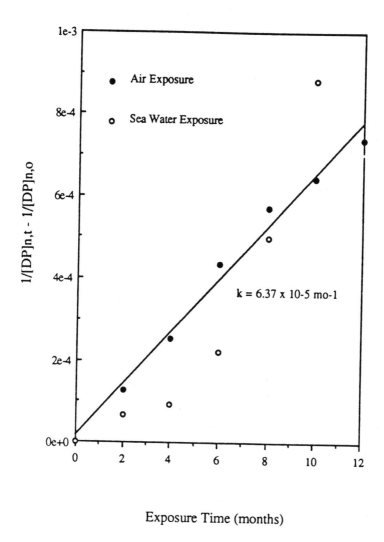

Figure 5.    Molecular weight-time relationship for polystyrene foam exposed in air and at sea.

Table 2 gives values for tensile strength of the complete sample as well as of the lower nonembrittled layer at different durations of exposure. The data again show the degradation process to be faster when exposure is carried out in sea water rather than in air. Tensile strength was calculated both on the basis of the initial, unexposed sample thickness and on the thickness of the residual unembrittled layer. Even when based on the latter thickness, tensile strength decreased over time for the samples exposed at sea, indicating degradation in the inner layers of the foam.

Table 2
Summary of Tensile and Viscosity Data for
Polystyrene Foam Exposed Outdoors.

| Duration (months) | Tensile Strength[a] $(kg/cm^2)$ | | Tensile Strength[b] $(kg/cm^2)$ | |
|---|---|---|---|---|
| | Mean | S. D.[c] | Mean | S. D.[c] |
| (a) Samples Exposed in Air | | | | |
| 0 | 3.89 | 0.50 | 3.89 | 0.50 |
| 2 | 4.31 | 0.34 | 5.16 | 0.40 |
| 4 | 3.46 | 0.59 | 4.70 | 0.80 |
| 6 | 2.45 | 0.27 | 4.37 | 0.48 |
| 8 | 2.39 | 0.27 | 4.60 | 0.52 |
| 10 | 2.61 | 0.14 | 5.09 | 0.27 |
| 12 | 2.37 | 0.27 | 4.53 | 0.51 |
| (b) Samples Exposed in Sea Water | | | | |
| 2 | 2.88 | 0.21 | 5.45 | 0.40 |
| 4 | 1.13 | 0.71 | 5.50 | 1.82 |
| 6 | 1.09 | 0.17 | 3.20 | 0.44 |
| 8 | 1.22 | 0.29 | 2.22 | 0.54 |
| 10 | 0.69 | 0.09 | 2.13 | 0.22 |

a   Tensile strength calculated using the initial area of cross-section.
b   Tensile strength calculated using the area of cross-section based on residual unembrittled layer.
c   Standard deviation based on 4-6 test pieces.

## CONCLUSIONS

Outdoor exposure of expanded, extruded polystyrene foam in air results in rapid discoloration and embrittlement of the surface, increased UV absorbance and the formation of a possibly protective yellow surface layer, the thickness of which varies with time. Yellowing is accompanied by marked decreases in the average molecular weight and tensile strength of the material. The protective yellow surface layer reduces but does not eliminate these degradation effects in the underlying foam layers.

Exposure of the material outdoors while floating in seawater results in similar changes. While the deterioration in air is initially faster, after approximately 4 months of exposure, the rate of deterioration (as measured by changes in molecular weight, solution viscosity, thickness and tensile strength) is found to be greater in seawater than in air. This is attributed at least in part to the continuous loss of yellowed surface layer in water and the consequent reduction in effective screening of sunlight.

## REFERENCES

1.  M. L. Dahlberg and R. H. Day in *Proceedings of the Workshop on the Fate and Impact of Marine Debris*; R. S. Shomura and H. O. Yoshida, Eds., U. S. Department of Commerce, NOAA, Honolulu, Hawaii, 1985, p. 198-212.
2.  A. T. Pruter, *Mar. Pollut. Bull.*, 18(6B), 305-310 (1987).
3.  P. G. Ryan, *Mar. Environ. Res.*, 23, 175-206 (1987).
4.  M. Gochfeld, *Environ. Pollut.*, 4, 1-6 (1973).
5.  G. H. Balazs in *Proceedings of the Workshop on the Fate and Impact of Marine Debris, November 27-29, 1984*; R. S. Shomura and H. O. Yoshida, Eds., U. S. Department of Commerce, Honolulu, Hawaii, 1985, p. 387-429.
6.  R. W. Dickerman and R. G. Goelet, Northern Gannet Starvation After Swallowing Styrofoam, *Mar. Pollut. Bull.*, 14(4), 145-148 (1987).
7.  K. Hjelmeland, G. H. Pedersen and E. M. Nissen, *Mar. Bio.*, 98, 331-335 (1988).
8.  G. A. George and D. K. C. Hodgeman, *J. Polym. Sci., Polym. Symp.*, 55, 195 (1976).
9.  G. A. George, *J. Appl. Polym. Sci.*, 18, 419 (1974).
10. J. B. Lawrence and N. A. Weir, *J. Chem. Soc., Chem. Commun.*, 273 (1966).
11. J. Kubica and B. Waligora, *Europ. Polym. J.*, 13, 325 (1977).
12. O. B. Zapolskij, *Vysokomol. Soedin.*, 7, 615 (1965).
13. N. A. Weir, *Developments in Polymer Degradation, Vol. 4*, N. Grassie, Ed., Applied Science Publishers, London, 1982; p. 189 .
14. J. Peeling and G. T. Clark, *Polym. Degrad. Stab.*, 3, 177 (1981).
15. A. Davis and D. Sims, *Weathering of Polymers*, Applied Science Publishers, 1983; p. 208 .
16. A. L. Andrady and J. E. Pegram, *J. Appl. Polym. Sci., in press.*
17. C. E. Fisher, V. J. Castelli, S. D. Rogers and H. R. Bleile in *Marine Biodeterioration: An Interdisciplinary Study*, J. D. Costlow and R. C. Tipper, Eds., Naval Institute Press, Annapolis, MD, 1984, pp 261-299.
18. J. F. McKellar and N. S. Allen, *Photochemistry of Man-Made Polymers*, Applied Science Publishers, 1979.
19. C. David, D. Baeyens-Volant, G. Delaunois, Q. Lu Vinh, W. Piret and G. Geuskens, *Europ. Polym. J.*, 14, 501 (1978).
20. H. H. G. Jellinek, *Appl. Polym. Symp.*, 4, 41-54 (1967).
21. A. L. Andrady, *Paper Presented at Second International Conference on the Fate and Impact of Marine Debris, Honolulu, Hawaii, April* (1989).
22. J. E. Pegram and A. L. Andrady, *Polym. Degrad. Stabl.*, 26, 333 (1989).

# Chapter 23

# LIQUID BUFFER SYSTEMS IN TEXTILE DYEING

FUAN YANG AND ERNIE DE GUSMAN

*Sybron Chemicals Inc.*
*Wellford, SC*

## INTRODUCTION

For the proper treatment in dyeing of textile materials, it is practice to select a pH range that is best suitable to the particular operation. The pH can fluctuate widely and, if not controlled, can cause erratic results. To control pH fluctuations, chemicals are added to the liquid bath. Among the commonly used materials for buffering pH are monosodium phosphate (MSP), disodium phosphate (DSP), trisodium phosphate (TSP) and soda ash. These chemicals are solids, and users have been faced with difficulties in the measuring, handling and dissolving of these materials.

These difficulties can be eliminated by the use of liquid buffer ingredients. A liquid buffer simplifies life in the dyehouse in a number of ways: it dilutes quickly with cold water in all proportions; eliminates the need to store, transport, handle and dispose of paper bags; does not cause a ring in the mixing tank or scum in the dyeing equipment, will not clog feedlines. It was found that using liquid buffer could improve color yield and provide brilliancy of shade in the reactive dyeing of cellulosic materials.[1]

## EXPERIMENTAL

A few liquids-for-solid products have been manufactured by Sybron Chemical Inc. (Table 1). Their formulations are proprietary. Typical examples of two formulations that have been found suitable for low pH liquid buffer ingredients and for high pH liquid buffer ingredients are tabulated in parts by weight.

A. <u>Low pH Liquid Buffer Ingredients</u>

| | |
|---|---|
| Sodium Hydroxide | 20.6 parts |
| Phosphoric Acid | 52.4 parts |
| Water | 27.0 parts |

B. <u>High pH Buffer Ingredients</u>

| | |
|---|---|
| Potassium Hydroxide | 78.9 parts |
| Phosphoric Acid | 18.0 parts |
| Water | 3.1 parts |

Both formulations can be used alone or used in combination to achieve required buffer zone.

Table 1.        Liquids-for-Solid Buffers

| PRODUCT | Powder Replaced | Buffering Zone | Appication Details |
|---|---|---|---|
| Alkaflo Liquid Alkali | TSP, Soda Ash | 10-11.5 | For reactive dyes; amounts used depend on pH required |
| Tanatex Buffer-in H | TSP | 6-7.5 10.5-12 | Tanatex Buffer-in H and L are used in combination to |
| Tanatex Buffer-in L | MSP | 2-3 6-7.5 | achieve required buffer zone. Especially useful in carpet dyeing. |
| Tanatex Buffer-in RTP | TSPP* | 10-11 | Used in bleaching to replace TSPP. |
| Tanatex Buffer-in SAH | Soda Ash | 9-11 | Replaces soda ash. |

\* - Tetrasodium pyrophosphate

## RESULTS AND DISCUSSION

The titration curve of 0.43 g of formulation A in 100 mL of water with 0.1 N KOH has been plotted in Figure 1. It shows buffering in the pH range of about 5 to 7.5. The low pH liquid buffer ingredient performs in the range in which solid MSP is used (Figure 2).

The titration curves of 0.57 g of formulation B in 100 mL of water with 0.1 N HCl has been plotted in Figure 3. It shows pH buffering in the range of 10.5 to 12. The high pH liquid buffer ingredient performs in the range in which solid TSP is used (Figure 4).

When formulation A is mixed with formulation B, the free phosphoric acid from the low pH liquid buffer ingredients and free potassium hydroxide from the high pH liquid buffer ingredients will react to form a potassium phosphate that would be a buffer formed in situ. The pH values of 1% solution from a mixture of formulations A and B are 2.6 (100/0, w/w), 3.1 (90/10), 5.6 (80/20), 6.4 (70/30), 6.8 (60/40), 7.7 (50/50), 11.1 (40/60), 11.4 (30/70), 11.7 (20/80), 12.0 (10/90) and 12.1 (0/100).

The combination of both liquid buffers provides a pH setting and/or buffering system that is beneficial to the treatment of textile materials. Titration curves of 0.43 g/0.57 g of the blend formulation A/formulation B in 100mL of water with 0.1 N KOH and 0.1 N HCL are shown in Figures 5 and 6, respectively. They demonstrate the buffer properties of MSP and TSP.

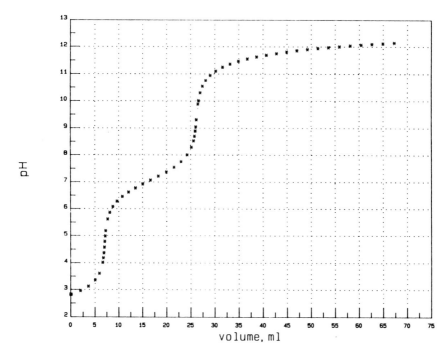

Figure 1.  Formulation A:  0.43 g in 100 mL of water titrated with 0.1N KOH

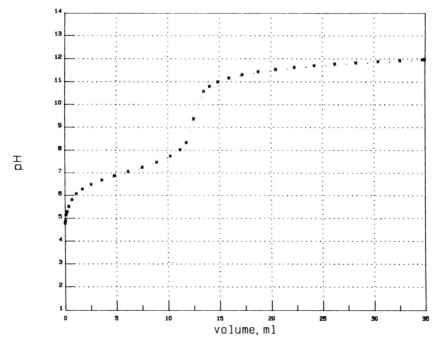

Figure 2.    MSP:  0.1727g in 100 mL of water titrated with 0.1 N HCl

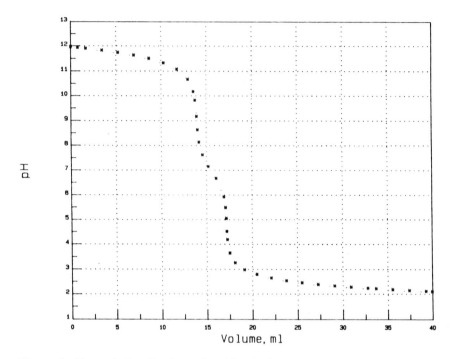

Figure 3.  Formulation B:  0.57g in 100 ml of water titrated with 0.1 N HCl

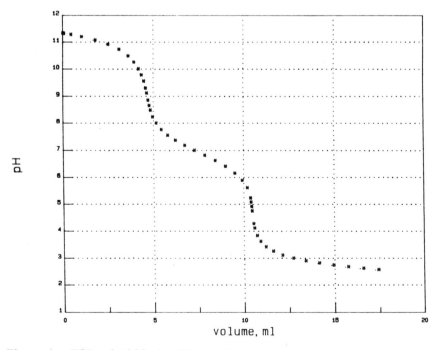

Figure 4.   TSP:  0.1089g in 100 mL of water titrated with 0.1 N KOH

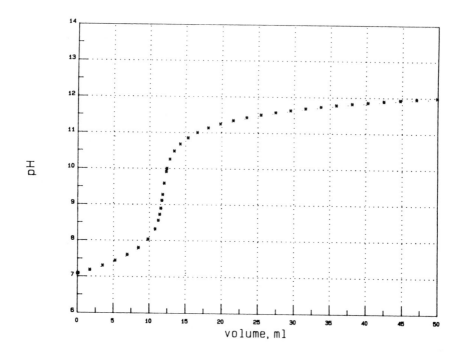

Figure 5        Formulation A/B: 0.43 g/0.57 g in 100 mL of water titrated
with 0.1 N KOH

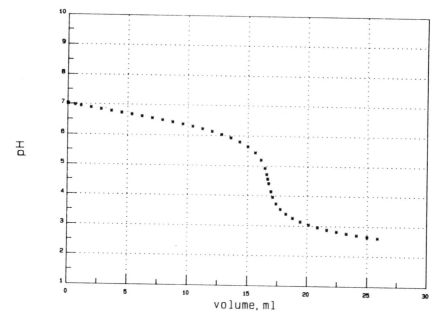

Figure 6.        Formulation A/B: 0.43 g/0.57 g in 100 mL of water
titrated with 0.1 N HCl

In addition to providing a pH setting and/or buffering, a low pH liquid buffer may be used alone to set the final pH of a bleaching bath from about 6.0 to 8.0. The following example illustrates the operation. [2]

Into a suitable dyeing beaker containing an agitator, 10 g of cotton knit are placed in a bath of 200 ml of water. Next are added 0.2g of a sequesting agent (Plexene 280) (Sybron Chemical Inc.) and 1.2 gm of 35% hydrogen peroxide. The bath is heated to boil at 212°F. and held for one hour. The bath is drained and the cotton washed and 200 mL of water and 0.2 g of low pH liquid buffer ingredient A. Then the cotton knit is washed again in 200 mL of water. Neutralizations carried out in this manner with low pH liquid buffer ingredient A produce results and fabric comparable to those of neutralizations done with solid phosphate pH buffer agents.

The high pH liquid buffer can be used as a replacement for the commonly solid alkaline materials (including trisodium phosphate (TSP), sodium silicate and sodium carbonate) in the reactive dyeing of cellulosic textile materials (e.g., rayon, cotton, flax). An alkali is needed to provide conditions that promote formation of a chemical bond between the reactive dye and the cellulosic material.[3] Formulation B performs comparably with the solid alkaline materials in creating the necessary reaction conditions and in producing level full-shade dyeings. The following example illustrates an application of the use of a Formulation B in a reactive dyeing operations.[2]

Into a suitable dyeing beaker containing an agitator, 5 grams of bleached 100% cotton fabric, is placed in a bath of 125 mL of water, 6.25 g of common salt, and 0.2 g of Remazol Red 3FB dye (American Hoechst Company). The bath is stirred for 15 minutes, warmed to 104°F and held for 15 minutes. Then 1.25 g of high pH liquid buffer ingredient B is added to the bath. The bath is heated to 140°F and held for one hour and then allowed to cool to room temperature. The cotton is removed from the bath and washed thoroughly. The use of 1.25 g of liquid buffer B in this procedure results in a dyeing of equal shade depth and fastness properties as results using 2.50 g of TSP as the high pH buffer in the same procedure.

## CONCLUSIONS

Liquid buffers are easily diluted with water in all proportions. They offer ease of handling over the use of solid products such as MSP and TSP. Bag disposal is eliminated also, as is the problem of torn bags, with resultant dusting and loss of contents. The pH buffering is achieved using formulations A and B from a low of pH 3 to a high of pH 10.

## REFERENCES

1.    A. De Maria, *American Dyestuff Report,* 75(10), 22 (1985).
2.    J. C. Moran, U.S. Patent 4,555,348.
3.    E.R. Trotman, *Dye & Chem.Tech..of Text. Fibers*, London, 1963, 492

# Chapter 24

# FRICTIONAL BEHAVIOR OF FIBROUS MATERIALS

B. S. GUPTA

*Fiber and Polymer Science Program*
*College of Textiles*
*North Carolina State University*
*Raleigh, North Carolina 27695-8301, U.S.A.*

## INTRODUCTION

Fiber to fiber and fiber to other surface friction has been the subject of numerous investigations during the past four decades. Considered in these studies have been the behaviors of different types of fiber and the effects of several procedural and environmental factors. In some, the focus has been on the extent to which the actual frictional behavior of a fiber material deviated from that given by the classical laws. In the absence of a standard test method, the primary goal in some of the others has been the development of suitable test devices for measuring friction.

Friction plays an important role in governing the behavior of fibers during processing and the properties of the final structure. Most processes for optimum performance require friction between fibers and between fibers and other surfaces to lie in certain ranges. Too high or too low values have generally proved detrimental to the process efficiency and the quality of the resulting product. The need for knowledge concerning the frictional behavior of fibrous materials becomes quite critical if new fibers and processes must be developed. However, although this property has been studied over the past several decades, its nature in fibers is still largely not understood. A review of the existing literature shows that whereas a comparatively adequate unified concept of friction exists for materials that deform plastically, no such concept or theory exists for materials that deform differently (i.e., elastically or viscoelastically). At the present time, the frictional phenomena in these materials, which include polymers and fibers and structures based on them, are described purely in empirical terms.

Presented in this chapter is a summary of the major activities of the writer in the area of friction. His earliest work was focused on evaluating the frictional behavior of surgical sutures and how it was affected by the factors used in tying knots and how it correlated with knot security. Based on tests on a number of sutures and under a variety of conditions, conditions leading to a secure knot were identified and a model characterizing it was proposed. In a second project, the effects of the exposure of human hair to chlorine in swimming pool water and to popular cosmetic treatments on characteristics that were important in hair (e.g. appearance, combability and manageability) were studied. The results showed that most of the effects could be understood in light of the changes that took place in hair friction and the correlation of these changes with those in surface morphology.

In an investigation recently completed, the departure of the frictional behavior of fibers from that given by the classical laws was examined. The work led to the development of a general structural model that provided an insight into the nature of friction in fibrous materials. It gave a clearer meaning to the classical parameter the coefficient of friction $\mu$ and gave structures to the constants a and n of the empirical equation $F = aN^n$. The model brought to light the factors, structural as well as procedural, that affected the values of the three parameters. The theoretical work was followed by an experimental investigation in which the effects of a number of structural variables [molecular orientation, setting and annealing, and cross-sectional shape] and procedural variables [testing environment (dry, wet) mode of contact (line, point) and contact pressure] on friction in acrylic and polypropylene yarns were studied and rationalized in light of the proposed model.

In a study presently being conducted, the frictional behavior of textile fabrics is being examined in detail. The objectives of this work are to understand the effects the fabric structural factors have on fabric to fabric and fabric to other surface friction, and to determine and control the effect fabric friction has on processes such as automated handling, transport and joining associated with apparel manufacturing.

Selected results from each of these investigations are presented below. Also given at the end is a brief discussion of the gaps that still exist in our understanding of friction and the studies that may be undertaken to close some of them.

## THEORETICAL TREATMENT: STRUCTURAL MODEL

Amonton as early as 1650 proposed that the frictional force (F) was directly proportional to the normal force (N) and the proportionality constant was a property of the material called the coefficient of friction (Equation 1).

$$F = \mu N \tag{1}$$

The meaning to the parameter $\mu$ was given by Bowden and Tabor[1] in 1950, who proposed their well-known Adhesion-Shearing theory. According to these authors, adhesion developed at points of contact between two bodies

(Figure 1), which must be sheared in order for sliding to occur. Frictional force was thus proposed to be given mostly by the product of the specific shear strength of the junctions (S) and the true area of contact (A):

$$F = SA + P \cong SA \tag{2}$$

The term (P) existing in this equation was the additional force needed for asperities of the harder body to plough through the surface layers of the softer body. This term in rigid solids with relatively smooth surfaces was assumed to be small and thus generally dropped. The authors showed that Amonton's law was valid only on materials that deformed plastically as given by the pressure area curve in Figure 2. When two bodies are brought into contact, the total load N is distributed over points of contact. Initially the area is small and thus the pressure high; the latter, being above the yield pressure, leads to plastic flow, giving an increase in the area and, therefore, a decrease in the pressure, as depicted by the isoload curves in Figure 2. This continues untill the pressure has dropped to the yield pressure and the load can be expected to be supported elastically at each asperity. Thus,

$$W_i = \text{constant} = P A = P_y A_i \tag{3}$$

$$N = \Sigma W_i = P_y \Sigma A_i = P_y A$$

$$F = S A = \left(\frac{S}{P_y}\right) N = \mu N \tag{4}$$

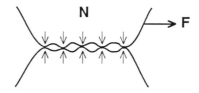

Figure 1. Distribution of load over points of contact.

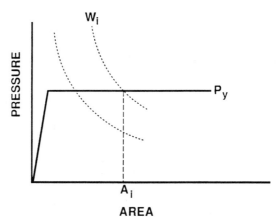

Figure 2. Pressure - area curve applicable to materials deforming plastically.

The coefficient of friction is thus given by the ratio of the two material properties, the strength of the junctions formed between the two bodies and the yield pressure of the asperities. According to this equation, the frictional force is linearly proportional to the normal force and the coefficient of friction is a material property, independent of the normal force.

In fibers and polymers, however, the frictional force is found to be nonlinearly related with the normal force as shown by the curve in Figure 3. Consequently, the coefficient of friction in these materials varies with the normal force as well as the apparent area of contact. Several authors have studied the observed dependence of $F$ on $N$[2-11] and suggested equations relating them. The simplest and the most widely accepted empirical equation fitting the experimental data is as given by equation (5).

$$F = a N^n \qquad (5)$$

Defining the coefficient of friction as the ratio of $F$ to $N$ (equation (4)) and substituting for $F$ from (5), one gets the following for $\mu$:

$$\mu = a N^{n-1} \qquad (6)$$

The values of the constants $a$ and $n$ have been found to vary with the fiber type; however, it is not known what the natures of these constants are and how their values might be affected by the factors in fiber structure. The focus of this study, the details of which can be found in a thesis,[12] and in a paper submitted recently for publication,[13] thus, has been on developing a structural model that characterized friction in fibrous materials and gave theoretical meanings to these constants.

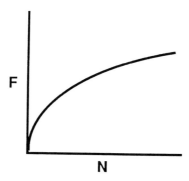

Figure 3. Frictional force versus normal force in fibers in general

In developing the model[12,13] it was assumed that the mechanism governing friction in fibers was the adhesion-shearing of Bowden and Tabor. There are some limitations associated with this assumption and these will be pointed out later. From the physical standpoint, the differences among plastic, elastic and viscoelastic materials lie in the shapes of the force-deformation curves obtained. In frictional studies, the concern is with these properties in compression. To account for different behaviors, a general pressure-area curve is assumed and given by equation (7).

$$P = K A^{\alpha} \tag{7}$$

In this equation, $K$ and $\alpha$ are constants whose values differentiate between different materials. Behaviors covering a broad range can be obtained, as seen in Figure 4. When two bodies are pressed together, the asperities supporting load deform according to equation (3) until the isoload curves intersect with the pressure-area curve given by equation (7) (Figure 5).

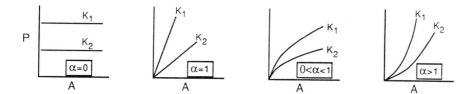

Figure 4. Pressure-area curves showing a broad range of behaviors.

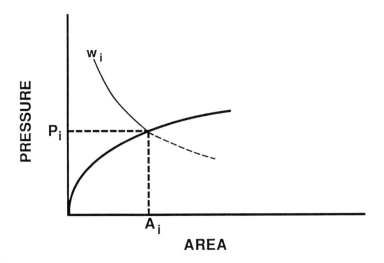

Figure 5. Interaction of an isoload curve with the pressure-area curve
given by equation (7).

At that point, the area of contact at each point of contact is defined. These
areas are then summed up to obtain the total true area of contact, which can
then be substituted in equation (2) to obtain the value of F.

$$w_i = P_i A_i = K A_i^{\alpha+1}$$

$$A_i = K^{-\gamma} w_i^{\gamma}$$

(where, $\gamma = (\alpha + 1)^{-1}$ )

$$\text{or} \quad A = K^{-\gamma} \Sigma^m w_i^{\gamma} \tag{8}$$

In this, m is the number of asperities or the points of contact over which the
load is distributed. The solution for Equation 8 was obtained by assuming a
number of stress distributions, namely uniform or rectangular, spherical, and
conical (Figure 6), and generalizing the result. While the detailed treatments of
these can be seen elsewhere,[12,13] the result obtained is as given by equation
(9) in which C is the model constant.

$$A = C K^{-\gamma} m^{1-\gamma} N^{\gamma} \tag{9}$$

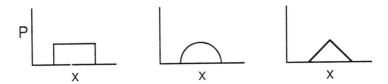

Figure 6. Stress distributions considered for theoretical treatment.[12,13]

Computations show that the value of C is relatively constant and very nearly 1[12]. Plugging A from (9) into (2) gives the following equation for F.

$$F = S C K^{-\gamma} m^{1-\gamma} N^{\gamma}$$

If the above equation of F is compared with (5), structures of the constants a and n become immediately obvious:

$$n = \gamma = (\alpha + 1)^{-1} \tag{10}$$

$$a = S C K^{-n} m^{1-n} \tag{11}$$

By combining (11) and (6), one gets the following equation for the coefficient of friction $\mu$:

$$\mu = S C K^{-n} m^{1-n} N^{n-1} \tag{12}$$

Equation 10 shows that the parameter n is a material constant, dependent only on the shape of the pressure-area curve (Figure 4). Its value is 1 for materials that deform plastically, 0.5 for materials that deform linearly or elastically (in compression) and a value other than these for materials which deform nonlinearly or viscoelastically. In most experimental investigations, the value of n has been found to lie above 0.5, most frequently between 0.7 and 0.9. The value of the parameter a (equation 11) is governed by the values of S, C, K, n and m. Of these, S, K and n can be considered as the constants of the material. C is generally a constant as mentioned earlier. The factor that affects the value of a and is not a material constant but dependent upon the procedure used is m, the number of points of contact. The latter could be expected to vary with the apparent area of contact used during testing. For example, the line contact between two fibers should lead to a greater value than the point contact. The value of $\mu$ (equation 12) is affected by all the factors associated with a and additionally by the normal force N. Thus, both the parameters a and $\mu$ are affected by the apparent area of contact during testing; however, the former, being independent of the effect of the normal force, could be considered to be more of a material constant than the latter. In the special case of the materials deforming plastically, i.e. n = 1 (or $\alpha$ = 0), both a and $\mu$ become equivalent and reduce to the dimensionless quantity S/K.

## GENERAL METHODOLOGY

Since standard test methods for studying frictional behavior of fibers, yarns and fabrics did not exist, suitable devices and procedures were developed to characterize friction in these materials. The three devices constructed measured friction by the point contact (filaments and yarns),[12] line contact (filaments and yarns)[12,14,15] and areal contact (fabrics).[16,17] The measurements by the first two methods could be carried out in both the standard atmospheric conditions and the fluid environments. Schematics of the devices are shown in Figures 7, 8 and 9. In the point contact method (Figure 7), a modification of the classical Capstan system is used. A fiber is held taught in a bow attached to the crosshead of the Instron and another held in the load cell and passed over it at right angles. With both the angle of wrap $\theta$ and the diameter of the fiber kept small, the area of contact could be restricted to a small region. With $T_0$ as the fixed tension and T as the tension required to slide one fiber over the other, the coefficient of friction $\mu$ is given by:

$$\mu = \frac{1}{\theta} \ln \frac{T}{T_0} \tag{13}$$

The above equation is based on the assumption that the frictional force and the normal force were linearly related with each other (equation 1). However, if the latter were related by Equation 5, the work of Howell[18,19] provided the following approximate relationship between $\mu$, a and n in which r was the radius of the fiber or the yarn serving as the fixed element.

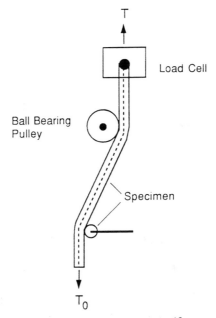

Figure 7. Schematic of the point contact device.[12]

$$\mu = a\,(\,T_0\,/\,r\,)^{n-1} \qquad\qquad (14)$$

Thus from the known value of $r$ and the values of $\mu$ at several values of $T_0$, one could determine the values of the constants $a$ and $n$.

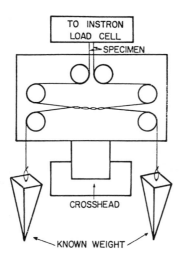

Figure 8. Schematic of the line contact device.[14]

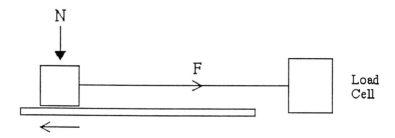

Figure 9.  Schematic of the areal contact device.[16,17]

For the line contact tests (Figure 8), the twist method proposed by Lindberg and Gralen[3] was used.  The method involves twisting two strands a given number of turns and applying different loads to the opposite ends. Gradually increasing the difference in tensions results in slippage.  At this point the values of the two tensions are noted and used in the following equation, based on the classical law (equation 1) for calculating the value of the coefficient of friction.

$$\mu = \left( \frac{1}{\pi t \beta} \right) (\ln T/T_0) \qquad (15)$$

In this equation, $T$, $T_0$, $t$ and $\beta$ are the tension activating slippage, the initial tension, the number of turns of twist and the twist angle, respectively. El-Mogahzy[12] related the value of $\mu$ obtained by the above equation with the parameters a, and n, as follows:

$$\mu = a (T_0 \beta^2 / 4r)^{n-1} \qquad (16)$$

Thus from the values of  r, the radius of the strand,  $\beta$, the twist angle, and $\mu$, the coefficient of friction, obtained at different levels of the fixed tension $T_0$, the values of the parameters a and n could be determined.

The frictional behavior of fabrics was characterized by the method illustrated in Figure 9.  A piece of fabric is laid flat and held securely in place on the traversing platform.  A square block also holding a piece of fabric rides over the platform.  The block is connected to a load cell and carries on its back a predetermined load.  The frictional force  F  is measured at several different levels of the normal force  N  and by plugging these in equation 1 the values of $\mu(N)$ or $\mu(N/A_0)$ could be obtained.  Also, by using the following modification of Equation 5, the values of a' and n' could be determined.

$$F / A_O = a' (N / A_O)^{n'} \qquad (17)$$

In this, $A_0$ is the apparent area of contact between the fabrics and is given by the area of the bottom face of the top block.  This empirical equation has been found in the literature to effectively characterize the frictional behavior of a fabric.[20-23] The relationship between $\mu$, a' and n'  applicable to a fabric is as follows:

$$\mu = \frac{F}{N} = \frac{F/A_O}{N/A_O} = a' (N/A_O)^{n'-1} = a' (N')^{n'-1} \qquad (18)$$

In the above equation,  N' is the normal pressure.

## FRICTIONAL STUDIES IN SURGICAL SUTURES

### Introduction
Sutures are used in surgery for reapproximating and repairing cut vessels and tissues, transplanting body organs and implanting prosthetic devices.  Among many criteria used for the selection of sutures, two are the ease of the suturability of the body tissues or the prosthetic materials, and the security of the knot placed in the suture. Different sutures perform differently in these regards, and a surgeon must select the right suture and procedure for each application.  A physical characteristic that directly affects the performance of a suture in these respects is the surface friction.  An understanding of the frictional behavior of sutures and of the effect the factors used in surgical procedures have on suture friction are important.  Such understanding could not only be expected to provide an insight into the differences noted in the performance of the various sutures in clinical practice but also provide the necessary criteria for selecting the right suture and the procedure for each surgery.

### Materials and Methods
In this study, the frictional behavior of eight different sutures (Table 1) was evaluated using the twist method with t equaling three turns of twist (Figure 8, Equation 15).  The sutures used, all of size 00, covered five different materials (polyester, polyglycolic acid, polypropylene, nylon and silk), two different constructions (monofilament and braided multifilament), and several different types of coatings.  The tests were conducted at several different tensions.[15,24] From the stick-slip profiles, the values of $\mu_s$ and

$\mu_k$ were evaluated.

Table 1.
Sutures Used in the Frictional Studies[15]

| Suture | Manufacturer | Description |
|--------|--------------|-------------|
| Ethilon® | Ethicon | Monofilament nylon |
| Prolene® | Ethicon | Monofilament polypropylene |
| Silk | Ethicon | Braided |
| Mersilene® | Ethicon | Braided polyester |
| Tevdek II® | Deknated | Braided, Teflon® impregnated polyester |
| TiCron® | Davis + Geck | Braided, silicone-treated polyester |
| Dexon® | Davis + Geck | Braided polyglycolic acid |
| Surgilon® | Davis + Geck | Braided, silicone-treated nylon |

## Results

Statistical analysis of the values of $\mu$ showed that both the suture material and the applied tension had significant effects on the values of $\mu$. The trends obtained with $\mu_s$ and $\mu_k$ were simila[24] and those pertaining to the former are illustrated in Figure 10.[15] Some interesting observations can be made. The two monofilament sutures started out having the highest $\mu$ value at 0.125 pound tension, but ended up having the lowest values of all sutures at higher tensions. The values of Dexon®, Surgilon®, Mersilene® and silk are generally close. Although the materials of these sutures are different from each other, they all have the same assembly structure, namely the braided. The effect of surface finish can be seen by comparing the curves of the three polyester sutures, all braided but finished differently, namely, Mersilene® (uncoated), TiCron® (silicon-treated) and Tevdek® (Teflon®-

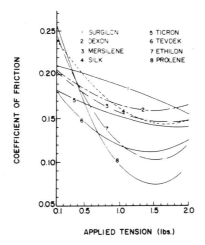

Figure 10. Frictional behavior of surgical sutures.[15]

impregnated). The latter two, having special finishes, generally have lower frictional values.

Although the causes for stick-slip are not yet fully understood, the phenomena is speculatively attributed to the deformational and the visco-elastic characteristics of the material. The stick-slip profiles of these materials, shown in Figure 11,[15] are as different from each other as are their frictional values. They differ widely in terms of amplitude and period of vibration as well as in the general character of the profile. While most of these seem to show only a primary cycle, at least one (Prolene®) shows a composite curve consisting of a secondary cycle superimposed over a primary cycle, each with a characteristic frequency and amplitude. The above characteristics of the stick-slip profiles varied with the number of turns of twist (t) and the applied tension.[24]

## Discussion

Under given conditions of test, friction is found to be different in different suture materials. This shows that all factors in the structure, namely the constituent material (the fiber), the assembly structure (braided, monofilament), and the surface finish, had significant effects on the coefficient of friction. On a given suture, applied tension and the number of turns of twist affected the value.[24] Although at the time this work was done, the values of a and n were not computed, it is clear from the shapes of the curves in Figure 10 that they are different for different sutures. The differences thus could be expected to result from the differences in the four important parameters of the model, namely S, K, n and m. The value of S is expected to vary with the chemical nature of the fiber, those of K and n with the viscoelastic behavior of the material in compression and that of m with the surface morphology and the finish. Measurements by Wolf[24] with different values of t showed an increase in $\mu$ with the latter. This can be

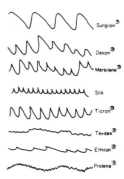

Figure 11.    Stick-slip profiles of various suture materials (tension 1 pound).[15]

expected to be due to an increase in the length and thus in the number of points of contact with an increase in  t.

In many applications of sutures, friction plays a dual role.  Low friction is needed for pulling a suture thread through the tissues and for transporting a throw by pressure from a surgeon's finger from above the surface to the wound in the body cavity.  To assist in this process, some suture surfaces are coated with special finishes.  However, high friction is desirable to provide security to a knot.  The latter consists of the mechanical interlacing of the ends of a thread.  Knot holding force, the force at which a knot slips, is given by the magnitude of the total frictional force generated within the structure of the knot.  If the force generated is high, the knot may hold firmly; if not, the components of it may slide and result in failure.  Thus a higher coefficient of friction is desirable in sutures from the standpoint of knot security.  This fact has been confirmed by the results of another investigation in which, for a given size of suture and number of throws, the knot holding force—the force at which a knot slips—was found to be low in sutures with Teflon® or silicone coatings and in sutures in in vitro  and in vivo  fluid environments.[25,26]  One or more additional throws were required in these cases to develop a secure knot.  Based upon these and other similar results, a friction based hypothesis of a secure knot could be proposed.[26]

## FRICTIONAL BEHAVIOR OF HUMAN HAIR

### Introduction

Some of the important characteristics of human hair are manageability, appearance and hand.  During normal use, hair may be exposed to a wide variety of conditions and treatments.  Normal hair maintenance involves cleansing, drying and styling.  The effects of these procedures can be complex due to the variety of cleansing and conditioning emulsions available and by the popularity of blow-dryers and curlers.  Additionally, hair is subjected to other treatments that may be voluntary or involuntary in nature.

The former include the cosmetic treatments of coloring, bleaching or permanent waving, while the latter includes the exposure of hair to sun light, environmental pollutants, and chlorine, the latter as found in swimming pools. These factors affect hair behavior primarily by changing surface morphology and frictional characteristics. This is evidenced by the fact that the exposed hair becomes increasingly reflective and more difficult to untangle and comb. This project was undertaken to examine and understand the nature of the changes that take place in surface friction and morphology when hair was given some of the above treatments. Full details of the study can be found in a thesis[27] and a number of publication.[14,28-30] Selected treatments (primarily chlorination) and results obtained with them are, however, reviewed here.

Materials and Methods
   The fibers used were natural blond and dark brown Caucasian hair. The variables of the study discussed here are the chlorine concentration (parts per million, ppm), the number of one-hour cycles of chlorination and the pH of the solution. Hair was extracted in a 50:50 chloroform/methanol mixture and dried prior to any treatment. The chlorine solution was prepared by dilution of a sodium hypochlorite solution and the concentration was monitored and controlled by an iodometric analysis. The pH of the solution was varied over a broad range and controlled with HCl and NaOH. Each cycle of chlorination consisted of soaking hair for one hour in a chlorine solution, rinsing in deionized water and drying for 15 minutes in an air circulating oven at 40-50°C. The frictional tests were conducted by the twist method (Figure 8, Equation 15). Two hair fibers were twisted together by a given number of turns of twist (t = 2) and subjected to low initial forces ($T_0$ = 3 gf), the forces as may be encountered during handling and combing. The hair fibers having a directional character, tests were conducted in both the "with" and the "against" scale directions. For surface morphology, fibers were chosen for examination under SEM from each friction test direction and treatment. The regions of the fibers examined were those where the fibers had actually rubbed during friction tests. This procedure provided an excellent tool for understanding the state of the surface and the causes for the changes in the frictional behavior.

**Results**
   The blond and the brown hair gave similar trends. Also, the static and the kinetic friction values varied with the variables of the study in a similar way. Some of the typical results are shown in Figures 12 and 13[14] Incremental increase in chlorine concentration from 0 to 10 ppm gave increases in $\mu$ (Figure 12A) that were significant at the 99% confidence level. Further increases to 50 ppm did not produce additional significant change. On the effect of the number of cycles of treatment also it is seen that the greatest increase in $\mu$ had occurred by 20 cycles and further increase in cycles produced little additional change. These results thus indicate that a mild chlorination treatment, as might be experienced in swimming, could produce a marked effect on frictional properties of hair. The effect of cycles

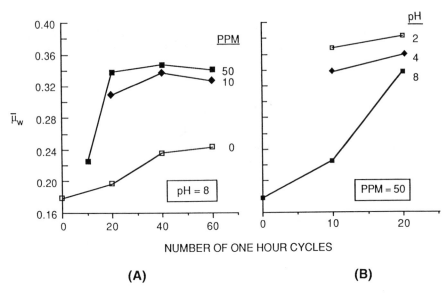

12. Frictional behavior of dark brown human hair. The values plotted are the average of the static and the kinetic obtained in the "with" scale direction.[14]

and concentration on DFE ($\mu_a$ - $\mu_w$) is illustrated in Figure 13A. It is seen that DFE decreases significantly with increase in concentration, i.e. in transition from 0 ppm (control) to 10 ppm concentration (treatment with water void of chlorine). The DFE value of control doesn't change with cycles. Also the DFE corresponding to 10 ppm and 20 cycles treatment being already so low doesn't undergo a significant further change with either the cycles or the concentration.[14]

The results in Figure 12B frictional properties show a strong pH dependence. In all comparisons, a decrease in pH from 8 to 2 (highly acidic) gave an increase in $\mu$, with the greatest change occurring at lower number of cycles[10], and a much greater change in transition from 8 pH to 4 than from 4 to 2. The results shown in Figure 13B indicate a marked decrease in DFE with pH at 10 cycles. At higher cycles, the DFE being already low at neutral pH, an increase in acidity has little further effect.

### Discussion

A friction trace of fiber is frequently stick-slip in character. This is usually attributed to the viscoelastic nature of fibers. The effect of a treatment may reflect as much as or more in a change in the nature of stick-slip profile as in a change in the average value of $\mu$. Usually, a hard elastic surface has been seen to give a straight line profile whereas a soft viscoelastic surface a stick-slip profile. Typical profiles found in this work are included in Figure 14.[27] They clearly indicate a marked alteration in the properties of the

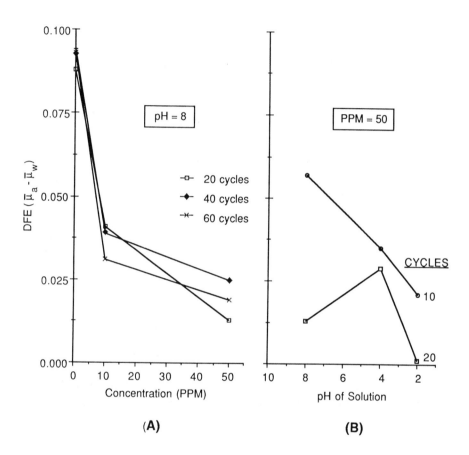

Figure 13. The DFE behavior of dark brown human hair.[14]

fiber surface with increase in cycles of chlorine treatment (Figure 14B) or increase in acidity (Figure 14A). Comparison of these profiles clearly indicates a transition from a hard and elastic surface of the control and less severely chlorinated specimens to a relatively soft and plastic surface of the more severely chlorinated specimens.

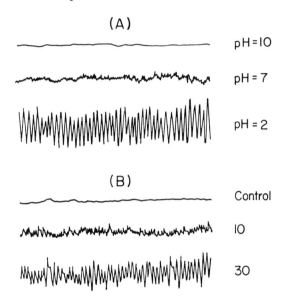

Figure 14. Stick-slip profiles obtained under different conditions of tests: (A) 10 cycles and 10 ppm dark brown hair, and (B) 7 pH and 10 ppm blond hair.[27]

Transition from hard elastic to soft plastic surface is verified by examination of the fiber under SEM. The particular area of hair examined was that where the rubbing of fibers had actually taken place during friction tests. The control samples and those mildly chlorinated in neutral solutions showed a definite scale structure and damage due to friction tests minimum (Figure 15).[14] With increase in the number of cycles of treatment, the fiber lost its scale structure (Figure 16)[14]. With increase in acidity of the chlorine solution, the surface became soft and the low forces used in friction rubbing were enough to plough the surface by causing bulk deformation (Figure 17).[27]

Examination of the literature on wool (a hair-like fiber) indicates a number of degradative reaction possibilities,[27,28] including cystine oxidation and breakage of disulfide bonds, tyrosine degradation, and peptide cleavage. It is suggested that the rate of reaction is least under mild and neutral chlorination conditions but extensive under concentrated and acidic chlorination conditions. The results found indicate that during chlorination of hair under acidic conditions, significant peptide bond cleavage and disulfide bond oxidation may have taken place. The increase in stick-slip character noted with chlorination is most likely due to the surface behaving increasing viscoelastic or plastic under low rubbing forces. Reduction in DFE is no doubt due to the destruction or dissolution of cuticular scales. And an

Figure 15. SEM photograph of dark brown human hair given 60 cycles of treatment with deionized water ( control). Friction test made in the against scale direction.[14]

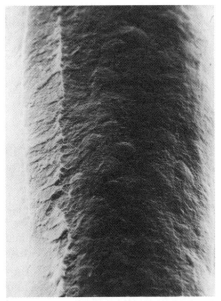

Figure 16. SEM photograph of dark brown human hair given 20 cycles of 50 ppm chlorine treatment under 8 pH conditions. Friction test made in the against scale direction.[14]

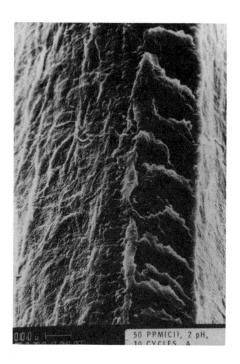

Figure 17. SEM photograph of dark brown human hair given 10 cycles of 50 ppm chlorine treatment under 2 pH conditions. Friction test made in the against scale direction.[27]

increase in the coefficient of friction observed is expected to have resulted from changes in the values of the model parameters m, K and n (equation 12), most particularly from an increase in m (the area of contact) and a decrease in K (the hardness).

## FRICTIONAL STUDIES WITH VARIABLES IN FIBER STRUCTURE

### Introduction

As discussed earlier, the fibers are viscoelastic in nature, their frictional behavior is not supported by the classical laws. An empirical model that fits the data on these materials involves the constants a and n (equation 5). The theoretical work discussed earlier showed the structure of these constants, and of the classical parameter µ. Little information is available in the literature on the effect the factors in fiber structure may have on friction. The availability of the structural model provided an opportunity to not only examine but also to understand the effects of such as well as other variables of interest. Investigated in this project, therefore, were the effects of several of these variables, namely, the molecular orientation, setting and annealing, cross-sectional shape, mode of contact and the testing environment on the parameters a, n and μ. The results obtained were examined in light of the structural model presented. Since full details of the study are being submitted for publication elsewhere,[32] only brief comments are included here.

## Materials and Methods

The materials used in this study were the polypropylene and the acrylic continuous filament yarns specially produced and supplied by Hercules Incorporated and the Monsanto Company, respectively. The polypropylene yarns were obtained in two sets: in one, draw ratio was the variable; and in the other (monofilament yarns), cross-sectional shape was the variable. The acrylic yarns also consisted of two sets, one annealed and the other unannealed. Each of these included molecular orientation as the variable. The friction between the yarns was measured by both the point contact (Figure 7) and the line contact (Figure 8) methods and, in selected cases, in both the standard atmospheric and the aqueous environments. The tests were conducted at several different levels of the initial tension $T_0$ (equations 13–17) and the values of the parameters n, a, and $\mu(T_0)$ were computed.[12]

## Results

The results showed that the value of n did not vary with a change in any of the factors except the molecular orientation. Thus the main effect of these factors was on the value of the parameter a. With n remaining constant, an increase or a decrease in a with a variable should cause a similar change in $\mu(T_0)$ (equation 6). Thus annealing of acrylic yarns led to an increase in the value of a, and also in the value of $\mu(T_0)$. Similarly, a change in the cross-sectional shape from the trilobal to the circular, in the mode of contact from the point to the line, or in the testing environment from the dry to the wet, caused an increase in the value of a and, therefore, an increase in the value of $\mu(T_0)$. In the case when the molecular orientation was changed, the values of both a and n were affected. An increase in orientation led to a decrease in a and an increase in n. The net effect of these was an increase in the value of $\mu(T_0)$ (Table 2).[13]

## Discussion

Using the experimentally found values of a and n and assuming certain values of the factors S, C and m, one could for qualitative comparisons estimate a value of K from equation (11). The value so calculated could then be plugged into equation (7) to predict the pressure-area curve. Such curves obtained with different levels of a given variable could then be compared with each other to obtain a clue as to the effect the variable might have on the overall frictional response of the material[12] A qualitative summary for four of the variables is given in Figure 18. The latter shows that the level Y, as compared to X, of the variables led to a higher value of the true area of contact A and through it to a higher value of $\mu(T_0)$.

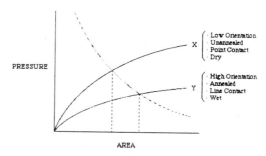

Figure 18.      Pressure - area curves showing qualitative effects of the variables of the study.

The results showed that the value of  a  was affected by all the factors studied, the structural as well as the procedural. The value of  n, however, was constant with respect to most variables but affected significantly by the molecular orientation. These results, thus, indicated that molecular orientation was an important structural factor that affected fiber frictional behavior.

Table 2

Effect of Orientation on $\bar{\mu}$ in Acrylic Yarns
(Line contact; $T_0 = 11$ gf)[12,31]

| Draw Ratio | Orientation Factor * | $\bar{\mu}$ |
|------------|----------------------|-------------|
| 2x | 0.695 | 0.186 |
| 4x | 0.756 | 0.230 |
| 6x | 0.785 | 0.238 |

* Sonic modulus orientation factor.

## FRICTIONAL BEHAVIOR OF TEXTILE FABRICS
### Introduction

A study of the frictional behavior of a fabric is important from the standpoints that (1) friction affects some of the important aesthetic and mechanical properties of a fabric and (2) surface forces associated with fabrics greatly affect automated handling, transport and other processes used in converting a fabric into final products. Greatly increased emphasis on high-speed processing and automation in apparel manufacturing in recent years has made it necessary that fabric to fabric and fabric to other surfaces frictional responses be fully understood. Relatively little is known about a fabric's frictional properties and how they may vary with the fiber and yarn frictional properties, fabric structure, and processing variables. This study was, therefore, recently undertaken to develop such an understanding.

## Materials and Methods

Twenty-six different commercially available woven fabrics varying broadly in terms of material and construction parameters were used.[16,17] The weave was mostly plain, but a limited number of others, e.g. twill and sateen, were also included. The fiber types used were cotton, rayon, polyester, silk, linen and wool, and blends of polyester and cotton. The yarns used in the fabrics were both staple and continuous filament and ranged significantly in the value of the tex (5–450). The fabrics varied in terms of the finish given and the openness of the structure; the latter ranged from a very open weave to tightly woven construction (cover factor[33] ranged from 16 to 27).

The method used for characterizing friction has been described earlier (Figure 9). The fabric to fabric friction was measured using the same fabric but with the contact between the back (the top block) and the face (the bottom plate) of the fabric. The apparent area of contact ($A_0$) was maintained constant at 6.45 cm$^2$. The values of $\mu(N/A_0)$ or $\mu(N')$, and of a' and n' (equation 18) were determined.

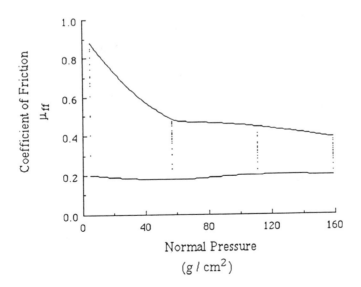

Figure 19. The range of the $\mu(N')$ values found among the fabrics tested.[16]

## Results

The value of the coefficient of inter-fabric friction varied with the normal pressure in conformation with equation 18 in which the exponent n' was less than 1. The value of $\mu(N')$ varied greatly among the fabrics and covered a broad range (0.1–0.9).[16] The upper and lower bounds of this range are shown in Figure 19. Finished fabrics gave higher $\mu(N')$ values than the unfinished (greige) fabrics and the continuous filament yarn fabrics gave $\mu(N')$ values that were very low—of the order of 0.1. These lay along the lower bound of the range shown in Figure 19. An interesting difference between the frictional behaviors of continuous filament yarn fabrics and the spun yarn fabrics is seen in Figure 20 in which the values of a' are plotted against n'. A distinct grouping is seen between the spun and the filament yarn fabrics. The fabrics from spun yarns had low n' and high a' values whereas those from filament yarns had high n' and low a' values. Another interesting result, also noted in the literature,[11] is that the values of a' and n' are intimately related. The results presented here refer to the weft-weft friction. Similar results were found with the friction measured in the warp-warp direction.[17]

## Discussion

The frictional results on fabrics followed approximately the same general trends as found on fibers and yarns, i.e. the value of $\mu$ decreased exponentially with an increase in N. An interesting but expected result noted on fabrics (Figure 19) is that at low normal pressures the fabrics differed greatly from each other in their frictional response, whereas at high normal pressures the differences were relatively much less. This result can be expected to have important practical consequences in operations such as apparel manufacturing in which a broad range of pressures are encountered. Thus, at low pressures such as found in ply separation and fabric transport, different fabrics in terms of their frictional behavior differ greatly from each other whereas at high pressures such as encountered in sewing, the differences are not perceived to be as great.

The relationship between a' and n' noted in Figure 20 could be generally as expected if the model developed on fibers (Equation 11) also applied to fabrics and the values of the parameters S, K and m either did not vary much from one fabric to another or, if they did vary significantly, they changed in a manner that their combined effects were minimal.

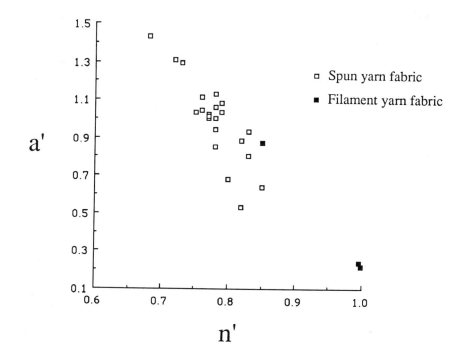

Figure 20. Plot of a' against n' for the fabrics tested.[16]

## CONCLUDING REMARKS

Friction is an important characteristic of fibrous materials, and an understanding of this phenomenon is essential for both developing optimum processing conditions for a product and predicting and controlling product behavior. Its role in governing the mechanical properties of yarns and fabrics, the behavior of assemblies during processing and the fiber-machine interaction during manufacturing is well known. Additionally, its effect on performance of sutures and of knots tied in them in surgery and its relationship with appearance and manageability of human hair have been illustrated. These examples, involving both traditional and nontraditional applications of textiles and polymers, indicate that any physical phenomenon in which a fibrous material comes in contact with another such material or with other surfaces involves friction as an important factor. Thus a need has existed and continues to exist for a thorough understanding of the frictional phenomena in these materials.

The theoretical work recently completed and reviewed in this chapter has shed some light on the nature of friction in fibers by giving meaning to the parameters a and n , and by showing the structure of the classical parameter μ. The availability of this information has served a useful purpose in interpreting and understanding the effects produced by some of the variables of the study. However, understanding is still lacking in a number of areas. Both Bowden and Tabor's classical theory and the model from the work of the writer and El-Mogahzy are based on the assumption that adhesion develops at junctions and these must be sheared in order for sliding to occur. It is not clear what other mechanisms, if any, also play a role in governing friction in fibrous materials. Is deformation of a surface that absorbs or stores work during sliding a significant factor? In fabrics, where yarns or fibers are held strictly by frictional forces, sliding of one surface over another is likely to lead to some permanent displacement of constituent elements. It would be important to know if the mechanism governing friction in assemblies is different from the one governing friction in fibers and films and, if so, what are these differences.

Another area in which understanding is lacking is the mechanism that gives rise to stick-slip in friction tests. As seen from the results on sutures and hair, different materials and treatments lead to different profiles. In many cases, a treatment has led to a relatively greater change in the nature of the friction trace than in the average value of the coefficient of friction. A thorough understanding of the causes responsible for this phenomenon is expected to significantly advance our knowledge of the field of friction.

From the experimental standpoint, some areas that are worth pursuing and the progress made in which could be expected to close some of the gaps alluded to above are as follows: developing a sensitive technique for determining the pressure-area curves in fibrous materials; studying the effect of temperature on frictional behavior of fibers; and further exploring the effects of structural factors on friction in fibers, yarns and fabrics.

## ACKNOWLEDGMENTS

The work reviewed in this chapter was supported by funds from a number of sources, including the Johnson and Johnson Baby Products Company, the North Carolina State University's Biomedical Research Support Grant, the organized research funds of the College of Textiles and the NCSU Center for Research on Apparel, Fiber and Textile Manufacturing (CRAFTM). The graduate students and the faculty colleagues participating in this work were Dr. Yehia El-Mogahzy, Dr. Nancy Fair, Mr. Neelesh Timble, Dr. Kay Wolf, Dr. Tim Clapp and Dr. R. W. Postlethwait.

# REFERENCES

1.  Bowden, F. P., and Tabor, D., *The Friction and Lubrication of Solids, Oxford University Press*, London (1950).
2.  Bowden, F. P., and Tabor, D., *The Friction and Lubrication of Solids, Part II*, Oxford, London (1964).
3.  Lindberg, J. and Gralen, N., Measurement of Friction between Single Fibers. II. Frictional Properties of Wool Fibers Measured by the Fiber-Twist Method, *Textile Res. J., 18*, 287-301 (1948).
4.  Bowden, F. P. and Young, J. E., Friction of Diamond, Graphite and Carbon and the Influence of Surface Film, *Proc. Roy. Soc., A208*, 444 (1951).
5.  Lincoln, B., Frictional and Elastic Properties of High Polymeric Materials, Brit. *J. Appl. Phys., 3*, 260 (1952).
6.  Gralen, N., Olofsson, B., and Lindberg, J., Measurement of Friction between Single Fibers. Part VII. Physicochemical Views of Inter-fiber Friction, *Textile Res. J., 23*, 623-629 (1953).
7.  Howell, H. G., The Laws of Static Friction, *Textile Res. J., 23, (8)*, 589-591 (1953).
8.  Howell, H. G., and Mazur, J., Amonton's Law and Fiber Friction, *J. Tex. Inst., 44*, T59-69 (1953).
9.  Lodge, A. S., and Howell, H. G., Friction of Elastic Solids, *Proc. Phys. Soc. London, B67*, 89 (1954).
10. Mazur, J., Friction Between Dissimilar Fibers, *J. Tex. Inst., 46*, T712-714 (1955).
11. Viswanathan, A., Frictional Forces in Cotton and Regenerated Cellulosic Fibers, *J. Tex. Inst., 57*, T30-41 (1966).
12. El-Mogahzy, Y., A Study of the Nature of Friction in Fibrous Materials, Ph.D. dissertation, North Carolina State University, Raleigh, North Carolina, 1987.
13. Gupta, B. S., and El-Mogahzy, Y. E., A Study of Friction in Fibrous Materials. Part I. Structural Model, *Text. Res. J.*, In Review (1991).
14. Fair, N., and Gupta, B. S., Effects of Chlorine on Friction and Morphology of Human Hair, *J. Soc. Cosmet. Chem., 33*, 229-242 (1982).
15. Gupta, B. S., Wolf, K. W., and Postlethwait, R. W., *Effect of Suture Material and Construction on Frictional Properties of Sutures, Surg. Gynecol. Obstet., 161*, 12-16 (1985).
16. Timble, N. B., Structural Factors Affecting Interfacial Forces Between Fabrics. Ph.D. dissertation research in progress, North Carolina State University, Raleigh, North Carolina.
17. Clapp, T. G., Timble, N. B., and Gupta, B. S., The Frictional Behavior of Textile Fabrics. Fiber Science and Technology, Applied Polymer Symposium #47, *J. Appl. Poly. Sci.*, John Wiley and Sons, New York (1991).
18. Howell, H. G., The General Case of Friction of a String Round a Cylinder, *J. Tex. Inst., 44*, T359-362 (1953).
19. Howell, H. G., The Friction of a Fiber Round a Cylinder and Its Dependence upon Cylinder Radius, *J. Tex. Inst., 45*, T575-579 (1954).

20.    Wilson, D., Study of Fabric-on-Fabric Dynamic Friction, *J. Textile Inst., 54, T143* (1963).

21.    Ohsawa, M., and Namiki, S., Anisotropy of the Static Friction of Plain-woven Filament Fabrics, *J. Text. Machinery Soc. of Japan, Vol. 12, No. 5* (1966).

22.    Zurek, W., Jankowiak, D., and Frydrych, I., Surface Frictional Resistance of Fabrics Woven from Filament Yarns, *Text. Res. J., 55*, 113 (1985).

23.    Carr, W. W., Posey, J. E., and Tincher, W. C., Frictional Characteristics of Apparel Fabrics, *Text. Res. J., 58*, 129 (1988).

24.    Wolf, K. W., An Experimental Study of Interfiber Friction in Surgical Sutures by the Twist Method. Master of Science hesis, North Carolina State University, Raleigh, North Carolina, 1979.

25.    Good, E. D., Mechanical Analysis of Knot Strength and Security in Polyester and Silk Surgical Sutures. Master of Science thesis, North Carolina State University, Raleigh, North Carolina, 1978.

26.    Gupta, B. S., and Postlethwait, R. W., An Analysis of Surgical Knot Security in Sutures. In: *Biomaterials* 1980, G. Winter, D. Gibbons and H. Plenk, Jr., Eds., pp. 661-668, Wiley-Interscience, New York, 1982.

27.    Fair, N. B., Frictional Studies in Human Hair, Ph.D. dissertation, North Carolina State University, Raleigh, North Carolina, 1984.

28.    Fair, N. B., and Gupta, B. S., The Chlorine-Hair Interaction. I. Review of Mechanisms and Changes in Properties of Keratin Fibers, J. Soc. Cosmet. Chem., *J. Soc. Cosmet. Chem., 38*, 359-370 (1987).

29.    Fair, N. B., and Gupta, B. S., The Chlorine-Hair Interaction. II. Effect of Chlorination at Varied pH Levels on Hair Properties, *J. Soc. Cosmet. Chem.*, 38, 371-384 (1987).

30.    Fair, N. B. , and Gupta, B. S., The Chlorine-Hair Interaction. III. Effect of Combining Chlorination with Cosmetic Treatments on Hair Properties, *J. Soc. Cosmet. Chem., 39*, 93-105 (1988).

31.    Gupta, B. S., El-Mogahzy, Y. E., and Selivansky, D., The Effect of Hot-Wet Draw Ratio on the Coefficient of Friction of Wet-Spun Acrylic Yarns, *J. App. Poly. Sci.*, 38, 899-905 (1989).

32.    Gupta, B. S., and El-Mogahzy, Y. E., A Study of Friction in Fibrous Materials. Part II. *Text. Res. J.*, to be submitted.

33.    Booth, J. E., *Principles of Textile Testing*. Chemical Publishing Company, Inc., New York (1969)

# Chapter 25

# FLUID FLOW THROUGH NEEDLE-PUNCHED GEOTEXTILE FABRICS

V. CHAHAL, D. R. BUCHANAN AND M. H. MOHAMED

*Department of Textile Engineering, Chemistry, and Science*
*College of Textiles*
*North Carolina State University*
*Raleigh, NC 27695-8301*

## INTRODUCTION

Flow rate and pressure drop in geotextiles are generally characterized by the properties of permittivity and transmissivity (often also referred to as permittivity) in transplanar and in-plane flow respectively. These properties, however, are dependent upon the fabric thickness and therefore do not truly characterize the flow behavior. Fabric thickness measurement under confined flow conditions poses great difficulties and hence accuracy in determining permeability coefficients is lost.[4] In the present study, accurate thickness measurement was possible with the use of precision thickness transducers and the overall accuracy of the experimental data was enhanced by the use of automated data acquisition and analysis.

The internal fabric structure is characterized by an equivalent mean pore size and the void volume fraction. The pore size distribution is normally determined from the sieve analysis for geotextiles and various methods are available for determining the void volume fraction (porosity). In this study the pore size distribution was obtained from the desorption curves obtained using water as the wetting fluid and successively applying pressure to drain fluid from the presoaked geotextile fabric.[1,5] The equivalent pore size based on the hydraulic radius theory was expressed as a function of the fabric thickness and other fabric parameters. The changes that occur in the mean pore size due to the normal pressures existing on the fabric have been considered in the pore size analysis.

Needle-punched nonwoven fabrics made from crimped polyester staple fibers were used in this study. All the fabrics were tested under high normal pressures (100 - 500 kPa), both in the transplanar and the in-plane flow permeameters. The same pressures were used for each fabric, thereby facilitating comparison between the permeability coefficients.

## EXPERIMENTAL TECHNIQUES

### Transplanar Permeability Measurements
Darcy's Law states that the equation for a one-dimensional flow in the transplanar permeability test is given by:

$$k_t(m/s) = \frac{Q(m^3/s)\ h(m)}{A(m)\ \Delta P(m)} \tag{1}$$

The flow rate Q is measured experimentally, h is the fabric thickness in the flow direction, and A is the cross-sectional area of the fabric. $\Delta P$ is the observed pressure drop resulting from the applied pressure head on the fabric through the water column. A material independent permeability coefficient having units of $(length^2)$ is expressed in darcy units (1 darcy = $9.87 \times 10^{-9}$ $cm^2$).[6]

### Test Apparatus
The test apparatus is a hybrid design incorporating the features of permeameters given by the ASTM committee[7] and Koerner et al.[8] (Figure 1). A continuous flow of filtered water is maintained to a constant head reservoir with an overflow, and the water temperature is monitored constantly. The elevation of the reservoir can be changed, thereby changing the applied pressure head. The test specimen is held between a pair of porous stones and distributor plates. The distributor plates distribute water evenly across the fabric and also provide a water seal at the sides. The highly porous stones disperse the flow from the distributor plate into the sample. The pressure head at the entrance and the exit surfaces of the sample is monitored with four piezometers, two at either surface, which are attached to the distributor plates. A piston, driven by an aircylinder, compresses the test sample through the distributor plates. The sample thickness is measured with a linearly variable displacement transducer (LVDT). The drain is located at the top end of the tube as four equally space holes. Water is collected in an outer cylinder and transferred to a graduated container.

### Test Procedure
Before conducting the permeability tests, the test specimens and the porous stones were kept underwater for a period of 24 hours. This removes the entrapped air inside the sample and therefore eliminates the possibility of two-phase flow.

The bottom distribution plate ($D_B$) and the porous stone ($P_B$) are inserted inside the sample tube (T). These are kept under water and remain inside the sample tube until the completion of the experiment. The sample (S), or a number of samples, is then inserted inside the tube from the top. Finally the top porous stone ($P_T$) is placed into the tube (T) from the top. The top distributor plate ($D_T$) is fixed to the air-cylinder piston, and it moves in and out of the sample tube (T) when the air pressure is applied.

The air pressure is kept at the minimum initially and increased gradually in steps during the subsequent tests. Water is allowed to flow through the samples while a constant head is maintained. The entrapped air in the system is expelled through a purge valve (V). A periodic expulsion of air eliminates the accumulation of air bubbles in the flow path.

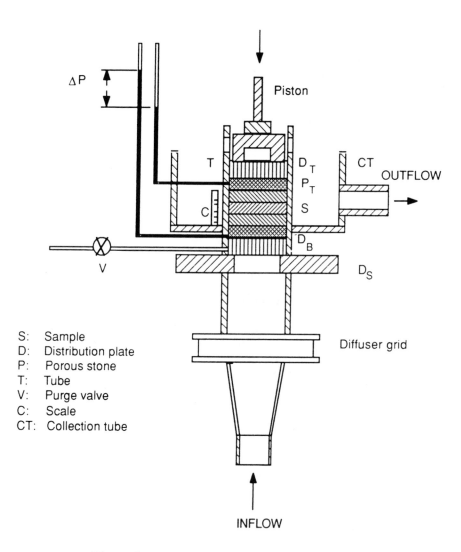

S:   Sample
D:   Distribution plate
P:   Porous stone
T:   Tube
V:   Purge valve
C:   Scale
CT:  Collection tube

Figure 1.        Transplanar flow permeameter.

Flow is maintained for about 5 minutes before recording any data. The sample thickness h, the piezometric pressure $\Delta P$, the volumetric flow rate Q and the compressive loads are recorded. Subsequent tests can be conducted at higher applied pressure heads by raising the elevation of the reservoir. At a height where the flow rate violates Darcy's law, an upper limit is established. The next set of tests can be performed from the lowest elevation of the reservoir, but at higher compressive loads.

## Permeability Correction

The porous stones were calibrated to account for pressure head losses. Even though the permeability of these stones was 20 to 30 times higher than that of the fabrics, a permeability correction was applied for all of the fabric samples.

For the low range of flow rates, the flow behavior of the stones is characterized by a set of linear equations relating the pressure drop and the flow rate:

$$\Delta P_{stone,\sigma} = -C_{1,\sigma} + C_{2,\sigma}\, Q \tag{2}$$

This correction procedure is incorporated into the data acquisition program as:

$$K_t^c = Q\,h/\,A\,\Delta P_c \tag{3}$$

where:          $\Delta P_c = \Delta P_0 - \Delta P_{stone} \tag{4}$

$\Delta P_c$ = corrected pressure drop

$K^t_c$ = corrected permeability

The corrected pressure drop is obtained from the pressure drop vs. flow rate relationship given by Equation (4) at the experimentally observed flow rates in the sample. The corrected permeability values were within 1-2% of the observed permeability values for the finer, 6 and 9 denier fabrics. However for the coarser 45 denier fabrics, the correction increased the permeability by 30-50%.

## Thickness Correction

The fabric thickness (h) is measured in terms of the displacement of the air-cylinder piston. This piston movement also reflects the small deformation that occurs in the porous stones at the applied loads. Therefore the thickness change at various stress levels in the porous stones ($\Delta t$) must be taken into consideration and incorporated in the equation for the fabric thickness. This fabric thickness correction is given by Equation (6) for $\sigma_0 = 100$ kPa:

$$\Delta t = t_\sigma - t_{\sigma,0} = \{\sigma - \sigma_0\} \, 1.04625 \times 10^{-6} \tag{5}$$

$$h_c = h_o + \Delta t \tag{6}$$

**Water Temperature Correction**
    The flow rate (Q) is observed at an experimental water temperature and is converted to the standard temperature of 20° C. The equation used for the correction [7] is:

$$Q_c = 1.76 \, Q \, [1 + 0.0337 \, T + 0.00022 \, T^2 \,]^{-1} \tag{7}$$

where T is the water temperature in degrees Celsius and $Q_c$ is the corrected flow rate. This correction was applied for the in-plane permeability measurements as well.

**In-Plane Permeability Measurements**
    In-plane permeability measurements were conducted with a radial flow permeameter, whose design is based on the work of Raumann[9] and Koerner et al.[8]. A circular disk-shaped sample was preferred over a rectangular shape to eliminate end edge effects. The effective in-plane permeability determined with this apparatus is:

$$k_p(m/s) = \frac{Q(m^3/s)}{2\pi h(m) \, \Delta P(m)} \, \ln \, (r_0/r_i) \tag{8}$$

where: $k_p = (K_{xx} \, K_{yy})^{1/2}$

    The flow rate Q, the thickness h, and the pressure drop $\Delta P$ are determined experimentally. The effective in-plane permeability $k_p$ is defined as the square root of the product of the in-plane permeability coefficients along the two principal directions of the fabric.

**Test Apparatus**
    The test specimen is held between two steel disks that form two impermeable bounding surfaces, such that the flow occurs radially outward in the fabric plane (Figure 2). An inverted cup-shaped chamber is mounted rigidly on the top disc, which is maintained under a constant load through the piston of an air-cylinder. The thickness of the compressed fabric is monitored constantly. The pressure head of water on the fabric is maintained through a reservoir with an overflow. The water is supplied to the fabric through the inverted chamber. A bleed valve is used to expel the entrapped air inside this chamber. A flow channel, at the outer periphery of the bottom disk accumulates the drained water, and this water is transferred to a graduated container to measure the flow rate. The pressure drop across the fabric is

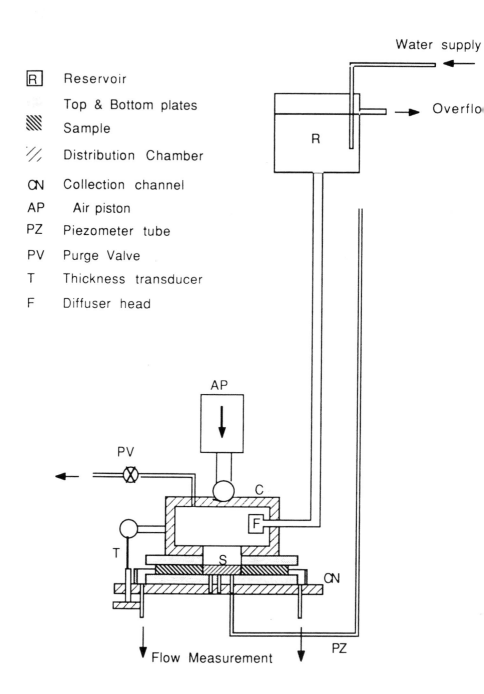

Figure 2.        In-plane flow permeameter.

measured using four piezometric capillaries with their sensing ends located inside the annular sample.

## Test Procedure

The test samples are kept under water for 24 hours before the measurement. The bleed valve (PV) must be in the open position. The test specimen is placed on the bottom disk and an external load is applied through the piston (AP). The supply air pressure is increased in uniform steps from a predetermined minimum value after successive tests. A continuous water supply to the reservoir (R) must be maintained and the overflow must be in operation. Water is then allowed to run through the fabric for some time until it is visible in the bleed tube. The bleed valve (PV) is closed as soon as no air bubbles are visible in the bleed tube. The flow of water is maintained for 3 to 5 minutes to ensure steady state conditions before data are recorded. The sample thickness h, the piezometric pressure $\Delta P$, the volumetric flow rate Q and the compressive loads are recorded.

An experimental procedure similar to that discussed for transplanar permeability measurements is followed to determine permeability coefficients for subsequent compressive loads and pressure gradients.

## Data Acquisition

The experimental variables were measured using precision transducers for the following physical properties: fabric thickness, flow rate, and normal pressure. The data acquisition was accomplished with a Hewlett-Packard Multiprogrammer[TM] unit. A computer program was written for a Hewlett-Packard 9817 computer to establish communication between devices, record data for the experimental variables, scan data, average and otherwise process the data and output results.

The thickness transducer was calibrated at the beginning of each experimental session with a plate of known thickness (8 mm). The slope of the calibration curve and the subsequent thickness measurements were calculated by the computer.

The flow rate was derived indirectly from time-based gravimetric measurements made with a Mettler PC-440 top-loading digital balance. Times were measured by the computer internal clock. Flow rate data were averaged for at least two measurements.

The normal pressure on the fabric was determined with a pressure transducer installed in the compressed air supply line. The transducer signal was conditioned by a Measurements Group signal conditioner, sampled repeatedly and averaged. The transducer voltage measurements were converted into the normal pressure using the transducer calibration curve and the air cylinder geometry.

The flow chart in Figure 3 illustrates the flow of data from the transducer to the computer.

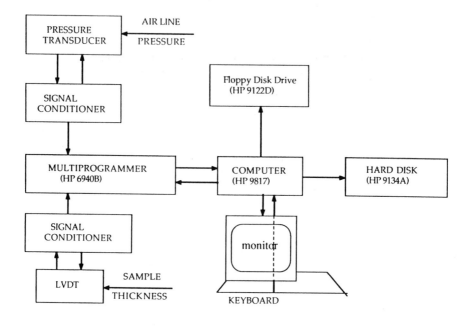

Figure 3.    Schematic of interfacing between instruments and data
             aquisition hardware.

### Averaging of Permeability Data
Measurements were conducted on five different samples taken from different locations in each fabric roll. Each of the fabric samples was tested under five different stress levels (100, 200, 300, 400, and 500 kPa). The permeability was calculated for each of the fabric samples, and the average of five measurements over different stress levels was used to determine fabric thickness and permeability.

## DARCY'S LAW VALIDITY

### Transplanar Permeability
Both equations (1) and (8) require that the pressure drop be a linear function of the volumetric flow rate (Darcy's law). To verify this linearity, fabrics were tested under different applied stresses. The results for 9 denier heavyweight fabric (4409, 400 npi) are shown in Figure 4 for stress levels of 100, 200 and 500 kPa. Each of the flow rate vs pressure head curves is linear, verifying the applicability of Darcy's law, and the slope (permeability) decreases with increasing applied stress. Darcy's law validity was assumed for the remaining fabric samples.

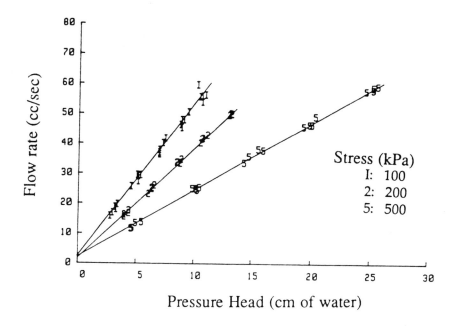

Figure 4.     Flow rate versus pressure head in transplanar flow permeameter at different stess levels in heavyweight fabric (denier: 9, needling: 400)

### In-Plane Permeability

A similar behavior for in-plane permeability measurements was observed for 45 denier heavyweight fabric (4445, 400 npi). These data also obey Darcy'sLaw, as is shown in Figure 5 for stress levels of 100, 200, 300, 400, and 500 kPa.

## NONWOVEN FABRIC CHARACTERIZATION

### Fabric Variables

Needle punched fabrics designed for geotextile applications were selected for this study. The fabric characteristics and processing parameters that contribute most to fluid permeability are:

(a)     fiber diameter ($d_f$, m), derived from the fiber linear density ($\rho_{f,l}$, denier),

(b)     fiber mass density ($\rho_{f,m}$, kg/m$^3$),

(c)     fabric areal density ($\rho_{F,a}$, g/m$^2$),

(d)     fabric mass density ($\rho_{F,m}$, kg/m$^3$),

(e)     needling density (N, punches/in$^2$), and

(f)     thickness at a standard pressure (h, mm @ 0.1 psi).

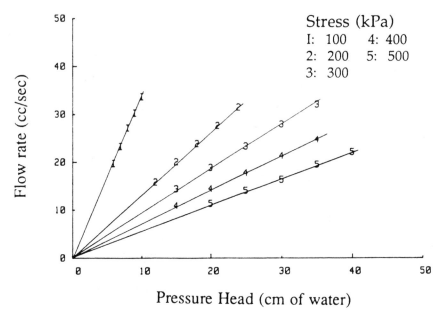

Figure 5.    Flow rate versus pressure head in in-plane flow permeameter at different stess levels in heavyweight fabric (denier: 45, needling: 400)

The experimental set of twenty-one fabrics is classified in Table 1.

**Permeability Variables**
        Past phenomenological studies have led to a definition of permeability of porous media in terms of the following parameters:

(a)    porosity of the medium ($\varepsilon$),
(b)    mean pore size ($D_p$, m),
(c)    pore size distribution, and
(d)    fluid Reynolds number ($N_{Re}$)

**Porosity of a Nonwoven Fabric**
        The porosity $\varepsilon$ of a nonwoven fabric is defined as:

$$\varepsilon = 1 - s \ ; \qquad s = 1 - \varepsilon = \text{solid fraction}$$

$$s = \frac{\text{Volume of fiber}}{\text{Volume of fabric}} = \rho_{F,m}/(\ \rho_{f,m})$$

Since,    $\rho_{F,m} = \dfrac{\text{Fabric Areal Density}}{\text{Fabric Thickness}} = \dfrac{\rho_{F,a}}{h}$

then    $\varepsilon = 1 - \rho_{F,a}/(\ h\ \rho_{f,m})$    (9)

Table 1  Classification and Variables of Needle Punched Geotextile Fabrics.

| Fabric # | $\rho_{f,l}$ | N | T | $\rho_{F,a}$ | $\rho_{F,m}$ | Porosity [$\varepsilon$] Calc. | Expt. |
|---|---|---|---|---|---|---|---|
| 2106 | 6 | 2 | 4.65 | 352.8 | 75.9 | 0.946 | 0.933 |
| 2109 | 9 | 2 | 6.45 | 469.4 | 72.8 | 0.945 | 0.944 |
| 2145 | 45 | 2 | 6.50 | 375.9 | 57.8 | 0.958 | 0.942 |
| 4106 | 6 | 4 | 4.30 | 323.8 | 75.3 | 0.946 | 0.925 |
| 4109 | 9 | 4 | 4.78 | 455.3 | 95.3 | 0.928 | 0.905 |
| 4145 | 45 | 4 | 6.00 | 426.7 | 71.1 | 0.948 | 0.929 |
| 6106 | 6 | 6 | 4.05 | 326.4 | 80.6 | 0.942 | 0.919 |
| 6109 | 9 | 6 | 4.25 | 466.3 | 109.7 | 0.917 | 0.890 |
| 6145 | 45 | 6 | 5.50 | 438.8 | 79.8 | 0.942 | 0.920 |
| 4206 | 6 | 4 | 6.90 | 746.9 | 108.3 | 0.922 | 0.892 |
| 4209 | 9 | 4 | 7.86 | 1039.0 | 132.2 | 0.900 | 0.868 |
| 4245 | 45 | 4 | 8.60 | 816.0 | 94.9 | 0.931 | 0.905 |
| 6206 | 6 | 6 | 6.17 | 729.9 | 118.3 | 0.915 | 0.881 |
| 6209 | 9 | 6 | 6.50 | 909.5 | 139.9 | 0.894 | 0.860 |
| 6245 | 45 | 6 | 7.41 | 806.9 | 108.9 | 0.921 | 0.891 |
| 4406 | 6 | 4 | 10.39 | 1300.6 | 125.2 | 0.910 | 0.875 |
| 4409 | 9 | 4 | 12.14 | 1845.5 | 152.0 | 0.885 | 0.848 |
| 4445 | 45 | 4 | 13.78 | 1612.6 | 117.0 | 0.915 | 0.883 |
| 6406 | 6 | 6 | 8.92 | 1252.7 | 140.4 | 0.899 | 0.860 |
| 6409 | 9 | 6 | 10.42 | 1912.7 | 183.6 | 0.861 | 0.816 |
| 6445 | 45 | 6 | 13.31 | 1926.6 | 144.8 | 0.895 | 0.855 |

$\rho_{f,l}$ = Fiber Denier  = # g/9000m

N = Needling Density in 100 punches/in$^2$

T = Fabric Thickness @ 0.1PSI in mm

$\rho_{F,a}$ = Fabric Areal Density (g/m$^2$)

$\rho_{F,m}$ = Fabric Mass Density (Kg/m$^3$)

The fabric areal density ($\rho_{F,a}$), fabric thickness (h) and the fiber mass density ($\rho_{f,m}$) can be determined experimentally. The solid fraction s of these fabrics was determined experimentally at 700 Pa and is shown as a function of the fabric mass density in Figure 6. There is generally a linear relationship between solid fraction and fabric mass density, except possibly for the finer fibers at lower fabric densities.

## Mean Pore Size in a Nonwoven Fabric

The internal pore structure and specifically the mean pore size is the most important parameter governing the fluid flow from the Darcian to the inertial flow regime. This parameter can be defined in various ways, but it must be truly representative of the pore size and the pore geometry of the porous medium, and it also must be definable in terms of the porous media construction parameters.

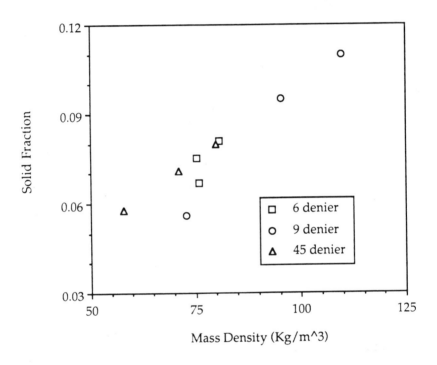

Figure 6. Mass density of nonwoven fabrics as a function of solid fraction.

The simplest approach in the case of fibrous assemblies using fibers with circular cross section would be to idealize the internal structure using the concepts of the cell model techniques proposed by Happel and Kuwabara.[10,11] The cell dimension in this case would be governed by the medium porosity ($\varepsilon$) and the fiber diameter ($d_f$). The hydraulic diameter $D_H$, expressed in terms of the cell dimension b, and the fiber diameter $d_f$ is:

$$D_H = (b^2 - d_f^2)/d_f \tag{10}$$

Since          $d_f^2/b^2 = 1 - \varepsilon$, then

$$D_H = d_f \, \varepsilon/(1-\varepsilon) \tag{11}$$

Johnston[12] proposed a mean pore diameter $(D_p)$ as a function of the fiber diameter $(d_f)$ and a porosity function $[f(\varepsilon)]$ from Davies'[13] analysis. The reduced equation is:

$$D_p = d_f /( \varepsilon \, [2 \, f(\varepsilon)]^{1/2}) \tag{12}$$

where: $\qquad f(\varepsilon) = (1 - \varepsilon)^{1.5} \, [1 + 56 \, (1 - \varepsilon)^3]$

Mlyarek[14] used a modified form of Equation (11) with a correction factor that is correlated with porosity:

$$D_p = 3.3 \, d_f \, \varepsilon \tag{13}$$

A more complete analysis of the equivalent mean pore size in a nonwoven fabric requires analysis on the basis of material properties and the external load, which is written as:

$$D_p = f(\varepsilon, \, s, \, d_f, \, h, \, \rho_{f,l}, \, \rho_{f,m}, \, \rho_{F,m}, \, \sigma) \tag{14}$$

where:
$\varepsilon =$ porosity
$s =$ solid fraction
$d_f =$ fiber diameter
$h =$ fabric thickness
$\rho_{f,l} =$ fiber linear density
$\rho_{f,m} =$ fiber mass density
$\rho_{F,m} =$ fabric mass density
$\sigma =$ normal stress.

From the channel theory, the hydraulic diameter $D_H$ is:

$$
\begin{aligned}
D_H &= 4 \, (\text{void volume/unit volume})/\text{pore surface area/unit volume} \\
&= 4\varepsilon/sS_0 \tag{15}
\end{aligned}
$$

where $S_0$ is the specific surface of the porous medium per unit of solid volume. From Equation (9), expressions for $\varepsilon$ and $s$ lead to the following equation for the mean hydraulic diameter for a fibrous material with fibers of circular cross section:

$$D_H = (4 h \rho_{f,m})/(S_0 \rho_{F,a}) - 1 \qquad (16)$$

$$= \frac{4 \times 10^{-6}}{9\pi} \{(h \rho_{f,m}/\rho_{F,a}) - 1\}\{\rho_{f,m} d_f/\rho_{f,l}\} \qquad (17)$$

Since the fiber diameter $d_f$ is difficult to measure directly, it must be expressed in terms of the measurable quantities: fiber linear density $\rho_{f,l}$ and fiber mass density $\rho_{f,m}$. Then equation (17) becomes:

$$D_H = \frac{[2 \times 10^{-3}]}{3 \pi^{1/2}} \cdot [\{h \rho_{f,m}\}/\rho_{F,a} - 1]\rho_{f,l} /\rho_{f,m} \qquad (18)$$

In Equation (18), only the fabric thickness h varies with the external applied stress $\sigma$.

The mean pore sizes calculated through equations (11), (12), (13), and (18) are compared to the experimentally determined pore sizes for light, medium, and heavy fabrics in Figure 7. The best agreement for lightweight fabrics is given by Equation (11) (Channel theory), while Equations (12) (Johnston's model) and (18) (Model) are more suitable for medium and heavyweight fabrics. Equation (13) (Mlynarek's model) tends to be useful only in the mid-range of applied normal pressures (300 kPa); above this pressure it overpredicts and below it underpredicts the mean pore size.

### Pore Size Distributions

The pore size distribution curve for a fabric gives a measure of the variation of its internal pore structure; it also highlights the mean, the most probable, and the maximum pore radii. These distribution histograms have been obtained experimentally from the "pore-size distribution experiment".[1,5] The pore size histograms for two fabrics of 400 needling density (4206 and 4209) from 6 denier fibers under 700 Pa compressive stress are shown in Figures 8 and 9 for a medium weight (747 g/m$^2$) and a heavy weight (1301 g/m$^2$) fabric, respectively. The broken lines indicate cumulative drainage curves, which show the cumulative volume of the displaced water; the histograms were obtained by observing the volume displaced under successive pressures. In general, fabrics made from 6 and 9 denier fibers have internal pore sizes in the range of 50 to 200 μm, while those made from 45 denier fibers have internal pore sizes in the range of 100 to 400 μm.[1]

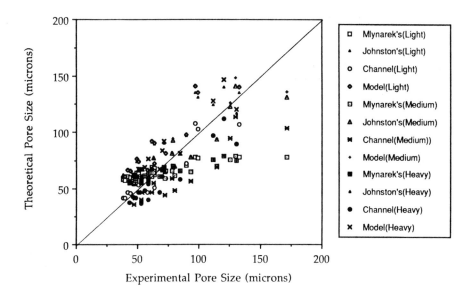

Figure 7.    Pore size analysis:  mean pore size in nonwoven fabrics using theoretical models.

## Reynolds Number

The Reynolds number is the ratio of the inertial to the viscous losses in any flow regime.  In the Darcian flow regime the flow is highly viscous, and therefore Reynolds numbers are extremely small.  If U is the macroscopic fluid velocity, $\rho$ is the fluid density and $\mu$ is the dynamic fluid viscosity, the Reynolds number is defined as:

$$N_{Re} = \rho\ U\ d_f/\mu$$

But for fibrous structures a modified Reynolds number, obtained by taking into account the solid volume fraction (s), is used:

$$N'_{Re} = \rho\ U\ d_f/\mu s \qquad (19)$$

These modified Reynolds numbers are large because of the small solid volume fraction in nonwoven fabrics.  For 6 and 9 denier fabrics, $N'_{Re}$ is between 0.01 and 1.0; for 45 denier fabrics, $N'_{Re}$ is between 0.1 and 5.0.[1]

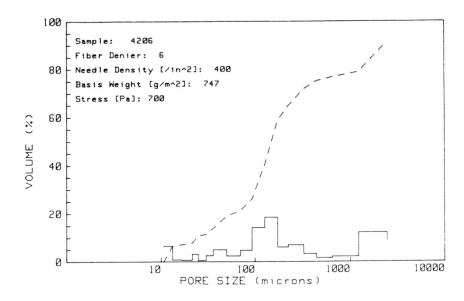

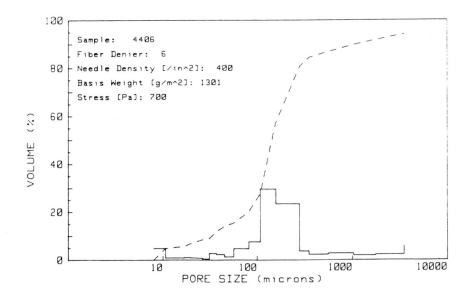

Figures 8 and 9. Pore size histograms: medium weight fabric (denier: 6, needling: 400) and heavyweight fabric (denier: 6, needling: 400)

### Intrinsic Permeability Function

Intrinsic permeability functions developed for fibrous assemblies have been compared with the experimental permeability data. In the past permeability coefficients have been developed on the basis of analytical and empirical work by Davies,[13] Ingmanson,[15] Emersleben,[16] Iberall,[17] Chen[18] and Raumann[9] as functions of fiber diameter ($d_f$), solid fraction (s), porosity ($\varepsilon$), Reynolds number ($N_{Re}$), fiber linear density ($\rho_{f,l}$), and/or fiber mass density ($\rho_{f,m}$). The applicable equations are:

$$k = \quad d_f^2/\{64\ s^{3/2}\ [1 + 56\ s^3]\} \qquad \text{(Davies[13])} \qquad (20)$$

$$k = \quad d_f^2/\{56 s^{3/2}\ [1 + 57\ s^3]\} \qquad \text{(Ingmanson[15])} \qquad (21)$$

$$k = \quad \pi d_f^2\ \ln\ [0.64\ s^{-1/2}]/24.4 \qquad \text{(Chen[18])} \qquad (22)$$

$$k = d_f^2\ \varepsilon\ (4 - \ln N_{Re})/\{9.4\ s\ (2.4 - \ln N_{Re})\} \qquad \text{(Iberall[17])} \qquad (23)$$

$$k = 3\ d_f^2\ \varepsilon\ (2 - \ln N_{Re})\{16s\ (4 - \ln N_{Re})\} \qquad \text{(Emersleben[16])} \qquad (24)$$

$$k = \rho_{f,l}\ \varepsilon^3\{\ 24\ \pi\rho_{f,m}\ s^2\} \qquad \text{(Raumann[9])} \qquad (25)$$

The permeability coefficients from equations (20) through (25) are compared with those obtained experimentally in Figure 10. It appears that the Iberall and Emersleben equations (equations (23) and (24)), which take explicit account of the Reynolds number involved in the flow process, yield the closest approximation to the experimental results; even so, they are consistently overpredictive for fabrics from 45 denier fibers. The results from all the other treatments are consistently underpredictive, even for 6 and 9 denier fibers, and appear not to be suitable for characterizing these nonwoven fabrics.

### Fabric Load-Deformation Behavior

A knowledge of the constitutive behavior under compressive loads is necessary to relate fabric permeability characteristics to practical applications. The fabric load-deformation behavior was analyzed from permeability measurements under compressive stress. A power law relationship (equation (26)) between the fabric thickness (h) and the compressive stress ($\sigma$) was observed for these fabrics, consistent with the relationship reported by Peterson[19] in his study of fluid flow in a system of loose nylon fibers.

$$h = C_1\ \sigma^{-m} \qquad (26)$$

A higher value of the exponent m indicates a fabric with low bulk modulus. The same fabric will have different values of $C_1$ and m in the two permeability experiments because, for the in-plane measurement, the lateral deformation is unrestrained, while in the transplanar measurement, the lateral deformation is restricted by the flow tube.  This variation is observed to a greater extent in thick fabrics than in thin fabrics.

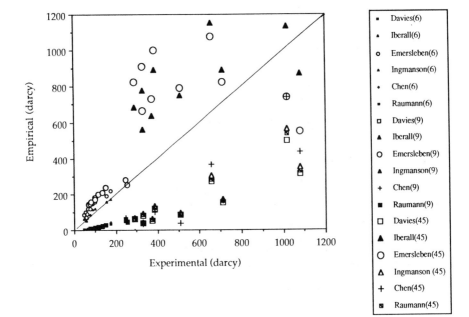

Figure 10.  Permeability analysis: experimental versus empirical for different fiber linear densities.

Thickness vs. normal stress results for three of the fabrics (light, medium and heavy weight) from 9 denier fiber at 400 needling density are shown in Figure 11.  Increasing basis weight alone does not change the bulk modulus, although increasing needling density will produce a significant change in bulk modulus.

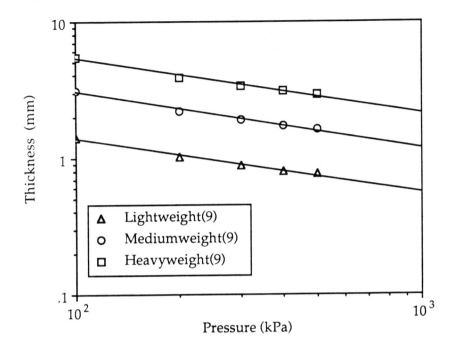

Figure 11. Log thickness versus normal pressure in nonwoven fabric
(denier: 9, needling: 400).

## STATISTICAL ANALYSIS

### Model Parameters

Permeability is a material-dependent property and it depends upon the
fabric parameters as well as the following experimental variables:

1. Applied normal stress on the fabric ($\sigma$, kPa),
2. Fabric areal density ($\rho_{F,a}$, g/m$^2$),
3. Fiber diameter ($d_f$, $\mu$m),
4. Fabric needling density (N, punches/in$^2$).

The fabric variables 2, 3, and 4 affect the pore size and the internal
pore geometry of the fabric and its structural stability under externally applied
loads (experimental variable 1). The fabric geometry under a specific load can
be defined in terms of its thickness (h):

$$h = h\ (\sigma, d_f, \rho_{F,a}, N) \tag{27}$$

Permeability is derived from the pressure drop $\Delta P$ and the flow rate (Q) measurements observed in a porous medium (equation 1). Hence the flow rate can be expressed as a function of the following variables:

$$Q = Q (\Delta P, h) \tag{28}$$

or    $$Q = Q (\Delta P, \sigma, d_f, \rho_{F,a}, N) \tag{29}$$

The variables appearing on the right side of equation 29 are treated as the independent variables, and the permeability, which is a linear function of flow rate) is treated as the dependent variable.

In the Darcian flow regime, the flow rate varies as a linear function of pressure drop with a zero intercept; therefore the permeability remains constant in the Darcian flow regime. Hence the variable pressure drop ($\Delta P$) in equation (29) can be eliminated as an independent variable, and the permeability model becomes:

$$k = k (h) \tag{30}$$

or    $$k = k (\sigma, d_f, \rho_{F,a}, N) \tag{31}$$

**Model Analysis**

The model described by equation (31) was analyzed statistically with a univariate analysis of covariance. Permeability and its logarithmic transformation were tested with the General Linear Model, a procedure developed by SAS[20] for analysis of variance of an unbalanced data set. Scaling factors of $10^3$ for permeability and $10^4$ for log permeability were used. Fiber diameter $d_f$ was studied indirectly for its effects with another fiber variable -- fiber linear density $\rho_{f,l}$, since the two are related by:

$$\rho_{f,l} = \rho_{f,l}(d_f^2, \rho_{f,m}) \tag{32}$$

where $\rho_{f,m}$ is the mass density of the fiber. Hence equation (31) can be modified to:

$$k = k (\sigma, d_f^2, \rho_{F,a}, N) \tag{33}$$

The fiber mass density was uniform (only polyester fibers were used in the fabrics) and therefore was not included in the analysis.

Each fabric parameter was studied at three treatment levels, and the stress was varied at five equally spaced intervals, as follows:

1. Stress ($\sigma$)                 100, 200, 300, 400, 500 (kPa)
2. Fiber linear density ($\rho_{f,l}$)  6, 9, 45 (denier)
3. Fabric areal density ($\rho_{F,a}$)     400, 800, 1600 (g/m$^2$)
4. Needling density (N)          200, 400, 600 (punches/in$^2$)

# RESULTS

## Permeability and Stress

The external loads under which geotextiles are tested is a major contributing factor toward the variation observed in the resulting permeability. This variability accounts for about 20% or more of the total variability in all cases, and the linear component of the stress effect is highly significant. High stresses resulted in reduced permeability in all of the tested fabrics. The nonlinear contribution due to stress is also statistically significant, and contributes 2 - 3% to the total variability.

## Permeability and Fiber Linear Density

The fiber linear density is the single most significant factor, contributing from 59 to 75% of the total variability in permeability. As the fiber linear density is also a measure of fiber diameter, this basic factor is the most significant for a permeability model. Coarse fibers form a more permeable needled fabric than fine fibers, as also reported by Lunenschloss[21].

The combined effects of fiber linear density and stress on permeability also were significant, accounting for about 12% of the total variability of the model. This effect is observed by the increased linearity in the log-permeability vs. log-stress plots shown in Figure 12.

## Permeability and Needling Density

The needling density is a fabric parameter that contributes towards the mechanical bonding of the fibrous web. High needling density results in a more closely packed structure, and therefore a fabric of lower porosity (see Table 1). Hence, high needle density fabrics should have a lower permeability, which was found by Mohamed and Afify[22]. This does not always hold, however, since the increase in needling density also affects the constitutive behavior of the fabric by compacting it. Due to these competing effects, therefore, the effect of needling on permeability is not statistically significant. Similar results have been reported by Kothari and Newton[23].

## Permeability and Fabric Areal Density

Fabric areal density can be defined as the product of the fabric mass density and its thickness. Due to the planar geometry of fabrics and their low thicknesses, the fabric mass per unit area is used and expressed as an areal density. For the transplanar permeability model, fabric areal density is not statistically significant; however it is statistically significant for the in-plane permeability model, providing a variability contribution of about 0.25%.

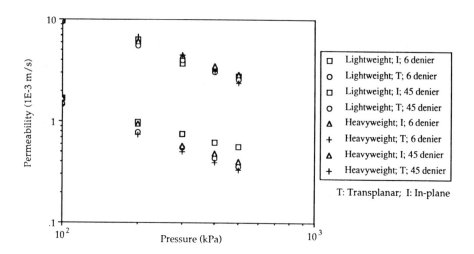

Figure 12. In-plane and transplanar permeability versus pressure in the needle-punched nonwoven fabrics (needling: 600).

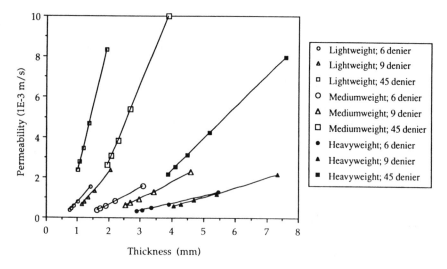

Figure 13. Transplanar permeability versus fabric thickness in the needle-punched nonwoven fabrics (needling: 400).

## Permeability and Fabric Thickness

A permeability-thickness model based on equation (30) was analyzed with a one way correlation analysis using "PROC CORR" from the SAS library[20] Data for each fabric type were tested independently for correlation; regression coefficients were between 0.98 and 0.99. These correlations are demonstrated in Figure 13, in which there is a highly linear relationship between permeability and thickness for each fabric type.

## Prediction of Permeability from Load-Deformation Behavior

Substituting the fabric thickness-pressure relationship given by equation (26) into the permeability equation (equation (1)) shows that fabric permeability can be predicted form the fabric thickness observed during the permeability test. In Figures 14 through 16, experimental and predicted permeabilities are plotted against pressure (and therefore thickness).

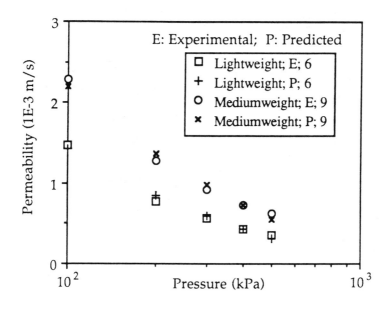

Figure 14. Experimental and predicted permeability versus normal pressure in the needle-punched nonwoven fabrics (6 and 9 denier).

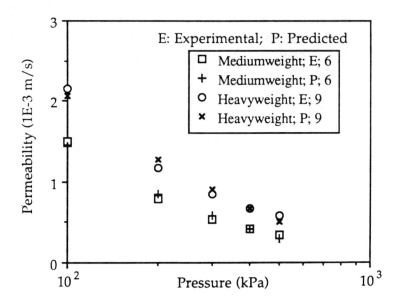

Figure 15.  Experimental and predicted permeability versus normal pressure in the needle-punched nonwoven fabrics (6 and 9 denier).

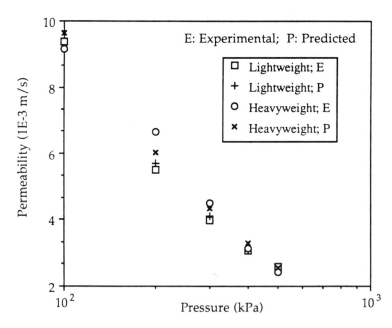

Figure 16.  Experimental and predicted permeability versus normal pressure in the needle-punched nonwoven fabrics (45 denier).

## CONCLUSIONS

Fiber diameter is the most important fabric parameter to be considered when designing needle-punched nonwovens for fluid transmission under Darcian flow conditions. This is due to the fact that the internal pore structure is directly dependent on the fiber diameter. As expected, a strong correlation was observed between fiber linear density and fabric permeability.

The second most important factor is the normal pressure under which the fabrics are confined during the flow. This reflects the structural changes that occur on external loading and that alter the internal pore size distribution, and hence the permeability coefficient.

The fabric areal density and the needling density of these fabrics showed statistically small or insignificant effects on in-plane and transplanar permeability. However, a significant difference was observed between fabrics made of fine fibers (6 and 9 denier) and those made of coarse fibers (45 denier).

A strong linear correlation was observed between fabric permeability and thickness under the experimental stresses. For this reason, the constitutive material behavior of the fabric under load can be a good indicator of its behavior toward fluid flow.

## REFERENCES

1.  Chahal, V., Fluid Flow Through Needle Punched Nonwoven Fabrics, Ph.D. dissertation, North Carolina State University, Raleigh, NC (1989).
2.  Chahal, V., D. R. Buchanan, and M. H. Mohamed, "Fluid Flow Through Needlepunched Geotextile Fabrics", *INDA Book of Papers* (1988).
3.  Chahal, V., D. R. Buchanan, and M. H. Mohamed, "Fluid Flow Through Needlepunched Geotextile Fabrics", *INDA-TEC 89 Book of Papers* (1989).
4.  Sluys, L. van der, and W. Dierickx, *Geotext. Geomembr. 5*, 283 (1987).
5.  Miller, Bernard and Ilya Tyomkin, *Text. Res. J. 56*, 35 (1986).
6.  Muskat, M., and R. D. Wyckoff, *The Flow of Homogeneous Fluids Through Porous Media*, McGraw-Hill (1946).
7.  *ASTM d-35.03.85.04*, "Proposed Standard Test Method for the Water Permeability of Geotextiles Under Load by the Permittivity Method" (1986).
8.  Koerner, R. M., J. A. Bove, and J. P. Martin, *Geotext Geomembr. 1*, 57 (1984).
9.  Raumann, G., "Inplane Permeability of Compressed Geotextiles", *Second International Conference on Geotextiles 3A*, 55 (1982).
10. Happel, J., *AIChE J. 5*, 174 (1959).

11. Kuwabara, S., J. Phys. *Soc. Japan 14*, 527 (1959).
12. Johnston, W. E. and J. N. Breston, *Producers Monthly 15*, 10 (1951).
13. Davies, C. N., *Proc.    Inst. Mech. Eng. 1B*, 185 (1952).
14. Mlynarek, J., *Geotext. Geomembr. 2*, 65 (1985).
15. Ingmanson, W. L., B. D. Andrews, and R. C. Johnson, *TAPPI J. 42*, 840 (1959).
16. Emersleben, O., *Phys. Z. 26*, 601 (1925).
17. Iberall, A. S., *J. Res. Natl. Bur. Stds. 45*, 398 (1950).
18. Chen, C. Y., *Chem. Rev. 55*, 595 (1955).
19. Peterson, R. M., *TAPPI J. 53*, 71 (1970).
20. *SAS User's Guide: Statistics, Version 5 Edition*, SAS Institute Inc., Cary, NC, (1985).
21. Lunenschloss, J. and W. Albrecht (editors), *Non-woven Bonded Fabrics*, Ellis Harwood Ltd., London (1985).
22. Mohamed, M. H., and E. Afify, "Efficient Use of Fibrous Structures in Filtration", *Environmental Protection Technology Series, EPA-600/2-76-204* (1976).
23. Kothari, V. K. and A. Newton, *J. Text. Inst. 61*, 525 (1974).

# Chapter 26

# DELIGNIFICATION OF HARDWOODS DURING ALKALINE PULPING: REACTIONS, MECHANISMS AND CHARACTERISTICS OF DISSOLVED LIGNINS DURING SODA-AQUEOUS PULPING OF POPLAR

ALBERT D. VENICA, CHEN-LOUNG CHEN AND JOSEF S. GRATZL

*Department of Wood and Paper Science*
*North Carolina State University*
*Raleigh, North Carolina 27695-8005*

## INTRODUCTION

The observation that addition of a small amount of AQ in alkaline pulping of woody tissues enhances removal of lignin[1] led to extensive investigations of the prevailing delignification reactions during the soda-AQ pulping processes. Model compounds have been used extensively to study the reaction mechanisms involved in the delignification during soda-AQ pulping of woods. As shown in Figure 1, phenolic $\beta$-aryl ether ($\beta$-$\underline{O}$-4) substructures were found to undergo nucleophilic attack on C-$\alpha$ by AHQ$^-$ via the corresponding quinonemethide intermediates to give adducts that are susceptible to Grob-type heterolytic fragmentation under the conditions of alkaline pulping.[2-4] The fragmentation results in cleavage of the $\beta$-aryl ether bond with concomitant regeneration of AQ via a redox $\beta$-elimination.[2,3] In addition to this method of reductive $\beta$-aryl ether cleavage, other mechanisms have been proposed to account for the high efficiency of quinone additives, such as oxidation of carbinol groups in lignin side chains.[5-9]

By introducing carbonyl groups, lignin side chains become prone to base-catalyzed C-C bond cleavages such as reverse aldol additions and $\beta$-eliminations provided that hydroxyl groups and ether linkages are present in proper positions, as illustrated in Figure 2. It should be noted that these cleavages could occur in both phenolic and nonphenolic structures. Furthermore, the cleavage of $C_\alpha$-$C_\beta$ bonds by reverse aldol additions is not affected by the type of interunit linkage at the C-$\beta$. This suggests that the degradation of lignins should be greatly enhanced not only by promoting the splitting of the very stable nonphenolic $\beta$-aryl ether linkages in $\alpha$-$\underline{O}$-4 substructures but also by the cleavage of $C_\alpha$-$C_\beta$ in side chains.

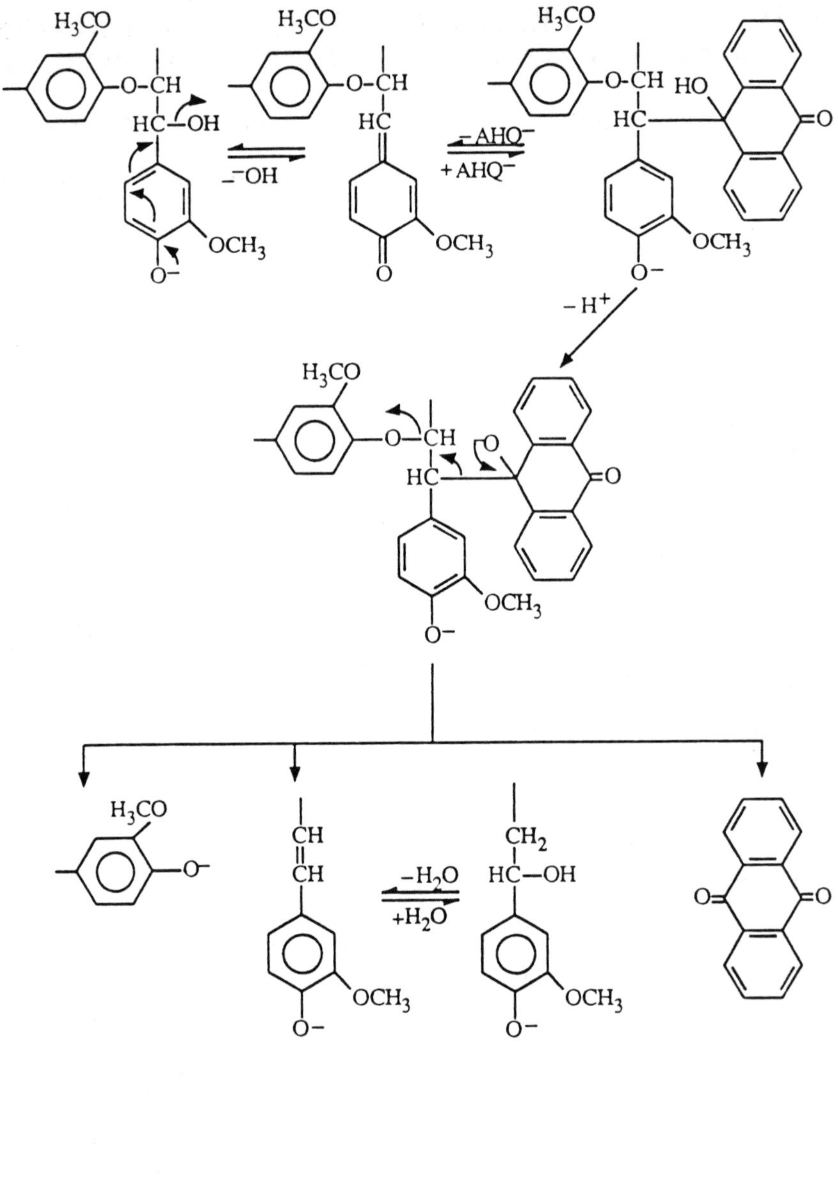

Figure 1. Reaction mechanism for inductive cleavage of β-aryl ether structures by AHQ.[2]

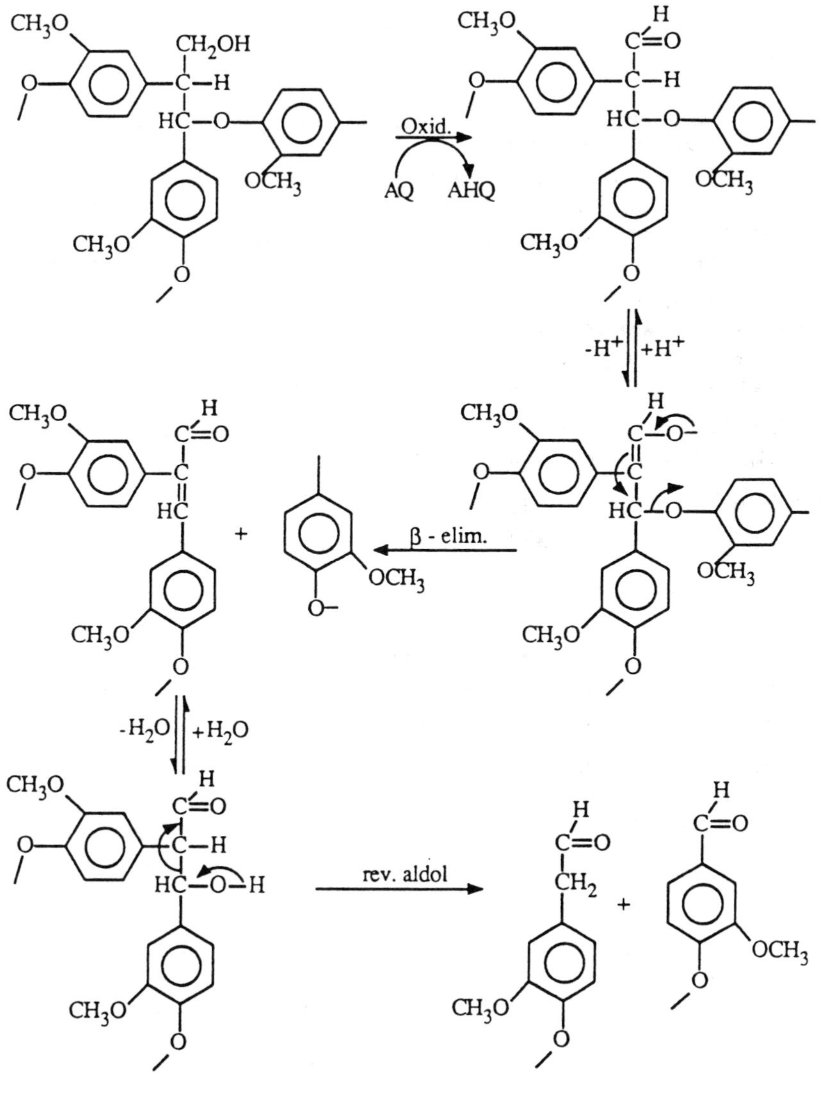

Figure 2.    Postulated mechanism for the reaction of nonphenolic α-aryl ether structures in lignins with AHQ.[2]

Hardwoods respond to soda-AQ pulping much better than softwoods. Moreover, soda-AQ pulping of softwoods yields, in general, pulps with somewhat inferior strength properties as compared to the corresponding kraft pulps.[10] In contrast, hardwoods including aspen (_Populus tremuloides_) can be pulped by soda-AQ process to give pulps that are comparable to the corresponding kraft pulps in terms of yield, strength properties and bleachability[11-13].

The chemical structure of hardwood lignins differs from that of softwoods. Hardwood lignins comprise a considerable amount of substructures containing syringylpropane units in addition to those consisting of guaiacylpropane units, which are the major, primary substructures of softwood lignins. Because syringylpropane unit has an additional methoxyl group at C-5 as compared to the guaiacylpropane unit, hardwood lignins are less condensed and contain more aryl ether linkages, in particular those in β-$O$-4 type aryl ether substructures, than softwood lignins do. Consequently, the average $M_w$ of hardwood lignins is somewhat lower than that of softwood lignins. In addition, the lignins of _Populus_ species were found to contain $p$-hydroxybenzoic acid units.[14,15] These units are present at the periphery of the lignin and link C-β mostly in the form of benzoic acid ester and, to a lesser extent, in the form of α-aryl ether linkages to the lignin core, which is predominantly composed of guaiacyl- and syringylpropane units.[16]

Our understanding of the major and important chemical reactions occurring during alkaline pulping is based on results of experiments conducted under the pulping conditions using model compounds featuring the most abundant and reactive substructures in lignins. Considerable attention has been given to the reactivity of interunit ether linkages, in particular to β-aryl ether bonds in β-$O$-4 type substructures.[2-4,17-23] On account of the importance of softwoods as raw material for production of chemical pulp, the model compound experiments were primarily focused on guaiacyl type model compounds, whereas only very few studies were undertaken with syringyl type models.[22,23]

The present study was initiated in view of the deficiencies in our understanding of the fundamental knowledge in the chemistry of delignification during alkaline pulping of hardwoods. The objective is to investigate the degradation of hardwood lignins by monitoring the formation of simple phenols as well as of higher molecular weight fragments.

The experiments were conducted in a flow-through reactor (FTR) in order to diminish the chance for occurrence of secondary reactions, since the liquor samples are exposed only a few minutes to high temperature. The products thus obtained are assumed to originate mostly from primary degradation reactions and are expected to shed some light on structural changes of hardwood lignins and on the reaction mechanisms by which the lignins are degraded during the soda and soda-AQ pulping processes.

## RESULTS AND DISCUSSION

The cooks in the FTR were carried out in a 45 ml stainless steel bomb with poplar chips, which were extracted for 36 hours with benzene/ethanol (2:1, *v/v*) and cut to approximately matchstick size. After evacuation, the bomb was placed in a oil bath and the pulping liquor was introduced at a flow rate of approximately 15 mL/min. The bomb was heated according to a pre-determined heating schedule. The temperature was increased at a rate of $4^\circ$C/min. The cooks were divided in three stages in which the concentrations of sodium hydroxide solutions were 30, 15 and 4.7 g/L, respectively. These concentrations were selected to simulate the alkaline profile of batch pulping process with 14% active alkali as shown in Figure 3. The residence time of the liquor in the bomb was only about 2.5 minutes. Anthraquinone (0.25 g/L) was added as AHQ, which was prepared by reducing AQ with glucose at $80^\circ$C in a nitrogen atmosphere.

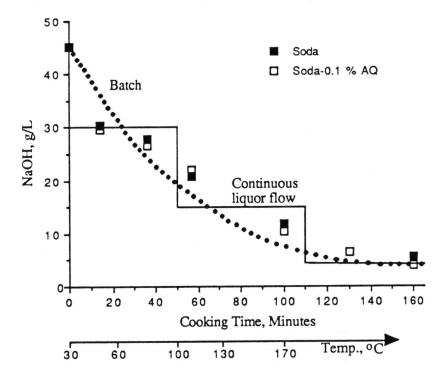

Figure 3. Alkali profile of batch and flow-through reactor (FTR) cooks.

Samples were taken during the heating-up period at 110, 120, 130, 140, 150$^\circ$C and when the maximum temperature of 170$^\circ$C was reached, and at 30 and 60 minutes thereafter. Each sample was neutralized with 4 N hydrochloric acid and was extracted with 1,2-dichloroethane for 24 hours. The extracts were then analyzed by GC and GC-MS. In all cases, concentration of a degradation product is the amount of the compound present in the liquor during 2 minutes of cooking at a specific temperature and expressed in ppm scale based on the total amount of lignin dissolved at the end of cook.

As shown in Figure 4, the formation of coniferyl alcohol (**1**; CA) and sinapyl alcohol (**2**; SA) during the heating period of soda-cook increased abruptly when compared to their formation in soda pulping. This clearly shows the effectiveness of AQ-AHQ system in alkaline delignification. The formation of a quinonemethide intermediate containing a β-aryl ether bond can initiate a type of "peeling" reaction in the presence of AQ. The reductive cleavage of this bond by AQ/AHQ redox system results in the formation of new phenolic structures. This is analogous to the role hydrosulfide ions play in kraft pulping.[20]

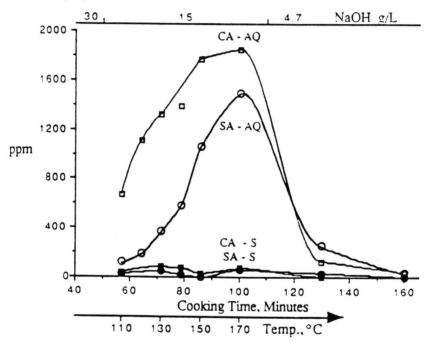

Figure 4.  Formation of coniferyl alcohol (CA) and sinapyl alcohol (SA) during soda (-S) and soda-AQ (-AQ) pulping of poplar wood (*Populus deltoides*) using flow-through reactor.

When the CA formation was studied under batch conditions,[8,9,24,25] it was found that the maximum concentration of CA in the cooking liquor occurs at 140°C, and at this temperature the rate of CA degradation rapidly becomes significant.[24] The disappearance of CA usually follows two major pathways, i.e., (a) condensation with dissolved lignins[25] or (b) degradation to vinylguaiacol (**7**), apocynol (**9**) and vanillin (**3**).[8]

In the FTR experiments both pathways are restricted by the short residence time. The highest CA and SA concentrations were found at 170°C. This suggest that under the reaction conditions, CA and SA are generated without being subjected to condensation and/or degradation as observed in the batch cooking. Conceivably, this type of "peeling" reaction continues until a linkage in lignin is encountered that is resistant to further degradation and/or

CA and SA precursors (i.e., uncondensed phenolic β-$\underline{O}$-4 type substructures) rapidly undergo degradation after the maximum temperature has been reached. As a result, the concentration of CA and SA in the cooking liquor decreases sharply, as observed in Figure 4.

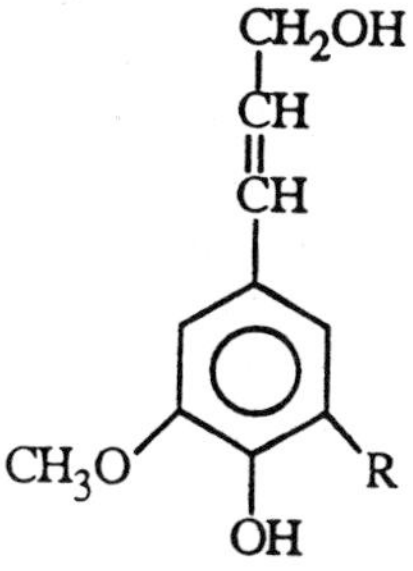

1, R = H
2, R = OCH₃

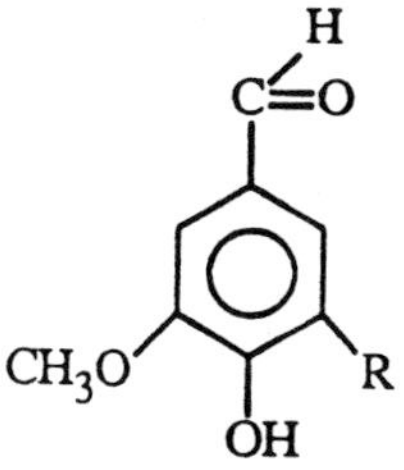

3, R = H
4, R = OCH₃

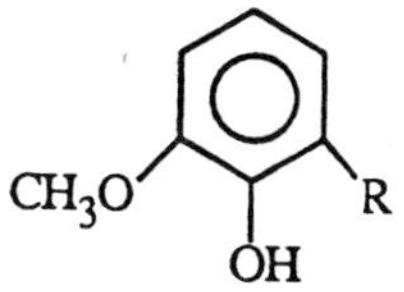

5, R = H
6, R = OCH₃

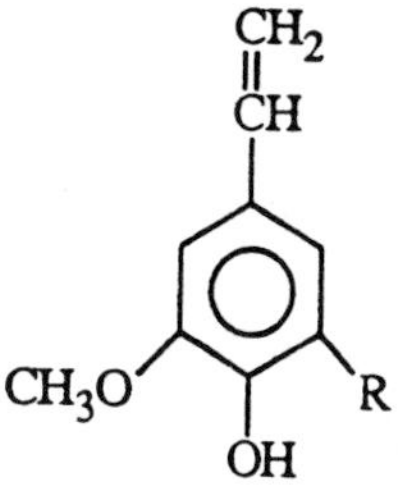

7, R = H
8, R = OCH₃

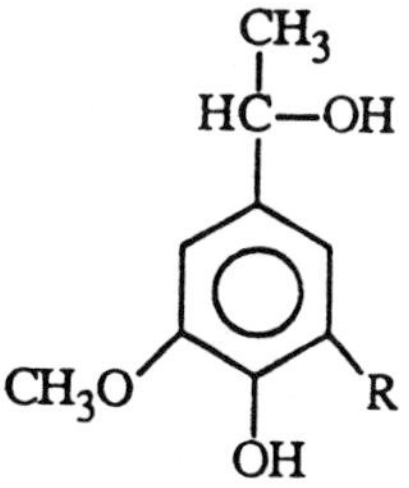

9, R = H

Unexpectedly, the concentration of SA in the cooking liquor up to reaction time of 120 minutes was lower than that of CA (Figure 4). Generally, β-aryl ether bonds in syringyl-syringyl- and guaiacyl-syringyl-type β-_O_-4 model compounds are more susceptible to base-catalyzed cleavage than those in the corresponding guaiacyl-guaiacyl-type model compounds.[23,24] When the β-substituent has a guaiacyl nucleus, the β-aryl ether bond is more resistance to based-catalyzed hydrolysis involving an internal $SN_2$ reaction with neighboring group participation, i.e., hydroxyl group at C-α.[22] In the absence of strong nucleophilic species, however, the reaction rate for the base-catalyzed hydrolysis of β-aryl ether bond is very slow because of the rather high pKa value for the hydroxyl group at C-α, and the major reaction is then reverse aldol addition by way of the corresponding α-carbonium ion (benzylic cation) intermediate, resulting in formation of an arylvinyl ether structure with elimination of formaldehyde. Evidently, the latter reaction does not contribute to fragmentation of lignins significantly, since the reaction involves only cleavage terminal $C_β$-$C_γ$ bond. No data are available so far about the behavior of the aforementioned β-_O_-4 type model compounds. However, the results from nitrobenzene oxidation of the poplar wood and two lignin fractions from the soda-AQ cook as shown in Table 1 may explain the lower SA concentration. The first lignin fraction (LF-1) was obtained from the cooking liquor during the heating-up period, i.e., the cooking temperature ranging from 110 to 170°C. The second lignin fraction (LF-2) was collected after the maximum cooking temperature (170°C) had been reached.

As given in Table 1, the results of nitrobenzene oxidation show that the molar ratio of syringaldehyde (**4**) to vanillin (**3**) (S/V molar ratio) for LF-1 is greater than that of the lignin in the original poplar wood. This implies that the number of uncondensed syringylpropane units relative to the number of uncondensed guaiacylpropane units in the lignin solubilized during the heating-up period increases as compared to those in the lignin present in the original poplar wood. Thus, it is evident that uncondensed guaiacylpropane units are more susceptible to AHQ-catalyzed, reductive cleavage of β-aryl ether bonds in the β-_O_-4 type substructures during the heating-up period and/or to base-catalyzed condensation than uncondensed syringylpropane units. The fact that the S/V molar ratio for LF-2 is similar to that of the lignin in the original poplar wood further demonstrates that the increase in the S/V molar ratio for LF-1 is not due to a selective solubilization of syringyl-rich lignin from the original lignin during the pulping process. Similar results were observed during kraft pulping of birch wood.[26]

Figure 5 shows [13]C NMR spectra of lignins LF-1 and LF-2, while Table 2 summarizes the assignment of signals in the spectra of LF-1 and LF-2. The spectra are in good agreement with the results of nitrobenzene oxidation, discussed previously. In the aromatic region, the spectrum of LF-

1 exhibits a very strong signal 2 at δ 152.2 (ppm) corresponding to C-3/C-5 of the etherified, uncondensed syringyl group. The rather weak signal 3 at δ 149.3 corresponds C-3 of the etherified, uncondensed guaiacyl group, while the moderately intensive signal 4 at δ ≅ 147.9 is an overlapped signal corresponding to C-4 in the etherified, uncondensed guaiacyl group, C-3 in nonetherified guaiacyl group and C-3/C-5 in the nonetherified, uncondensed syringyl group. The moderate signal 6 at δ 134.6 is also an overlapped signal corresponding to C-1 in etherified guaiacyl group and C-4 in the nonetherified syringyl group. Moderate to weak signals 10, 11 and 12 at δ 119.2, 115.5 and 111.2 correspond to C-6, C-5 and C-2 in etherified and nonetherified

Table 1.    Yield of vanillin (V) and syringaldehyde (S) obtained by nitrobenzene oxidation of poplar wood (*Populus deltoides*) and lignin preparations obtained from soda-AQ pulping the wood.

| Lignin Preparation | Yield, μmole/100 mgLignin | | S/V Molar Ratio |
|---|---|---|---|
| | S | V | |
| Poplar Wood[a] (*Populus deltoides*) | 154 | 116 | 1.33 |
| LF-1[b] | 124 | 80 | 1.55 |
| LF-2[c] | 96 | 72 | 1.33 |

[a] Pre-extracted wood meal passed through a 60 mesh screen.
[b] Lignin recovered from cooking liquor during heating-up period (100-170°C).
[c] Lignin recovered form cooking liquor after maximum temperature (60 min at 170°C).

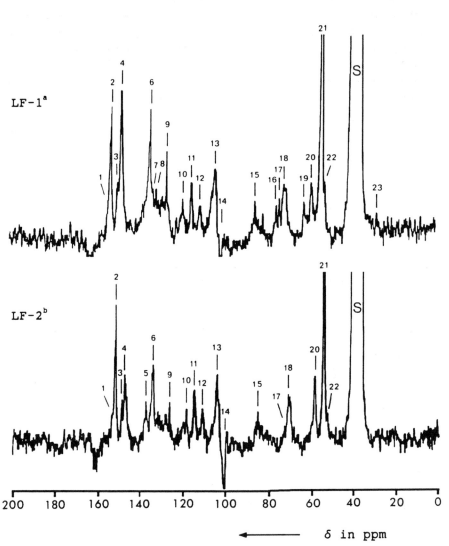

a LF-1 = Lignin recovered from cooking liquor during
    heating-up period (100-170°C).

[a] LF-1 = Lignin recovered from cooking liquor during
    heating-up period (100-170°C).
[b] LF-2 = Lignin recovered from cooking liquor after
    maximum temperature (60 min at 170°C).

Figure 5. $^{13}C$ NMR spectra of LF-1 and LF-2 fractions obtained from soda-AQ pulping of poplar wood (_Populus deltoides_) using flow-throughreactor.

guaiacyl group, respectively. The moderate signal 13 corresponds to C-2/C-6 of both the etherified and nonetherified syringyl groups. Thus, the $^{13}$C-NMR spectrum of LF-1 reveals that the lignin contains an appreciable amount of etherified syringyl group, probably in the form of β-aryl ethers, and the molar ratio of syringyl to guaiacyl groups is roughly in close proximity of 1, although the spectrum is not a quantitative one. In addition, the spectrum also indicates that guaiacyl groups are present in the lignin mostly not in the form of β-aryl ether linkage, as evidenced by the very low intensity of signal 3.

The aromatic region of the $^{13}$C NMR spectrum of LF-2 is similar to that of LF-1, except for the intensity of some signals. For example, the intensity of signals 4 and 6 in the former is much stronger than the corresponding signals in the latter. The intensity of other signals in both spectra are by and large similar in the magnitude. It must be noted that signal 4 at $\delta \cong 147.9$ is a overlapped signal corresponding to C-4 in etherified, uncondensed guaiacyl group, C-3 in nonetherified guaiacyl group and C-3/C-5 in nonetherified, uncondensed syringyl group, while signal 6 at $\delta \cong 134.6$ is also a overlapped signal corresponding to C-1 in etherified guaiacyl group and C-4 in nonetherified syringyl group. Since LF-1 does not contain appreciable amount of etherified guaiacyl groups as discussed previously, the increase in the intensity of signals 4 and 6 in the spectrum of LF-2 as compared to that of the corresponding signals in the spectrum of LF-1 must be caused by formation of terminal nonetherified syringyl groups at the final stage of the soda-AQ pulping, i.e., during the period after the maximum cooking temperature has been reached. Thus, both nitrobenzene oxidation and $^{13}$C NMR spectra data of solubilized lignins indicate that (a) uncondensed guaiacylpropane units are more susceptible to AHQ-catalyzed cleavage of β-aryl ether bonds during the heating-up period, resulting in the higher concentration of CA than that of SA in the cooking liquor, and (b) the abrupt decrease in concentration of CA and SA after the maximum cooking temperature is caused by the presence of linkages that are resistant to AHQ-catalyzed reductive cleavage and/or base-catalyzed hydrolysis under the reaction conditions.

When coniferyl alcohol was treated with an aqueous alkaline solution containing anthraquinone monosulfonate (AMS), vanillin was produced as the major product. This finding strongly suggests that coniferyl alcohol is first oxidized to coniferyl aldehyde, which in turn undergoes reverse aldol condensation, resulting in cleavage of the $C_\alpha$-$C_\beta$ bond with concomitant formation of vanillin. In the absence of AMS, only a small amount of vanillin was produced, but vinylguaiacol (**7**) and apocynol (**9**) were formed as the major products. Moreover, only 25 % of the starting material was recovered as simple phenols.[8] As shown in Figure 6, the small differences between the concentrations of syringaldehyde (**4**; S) and vanillin (**3**; V) in soda and soda-AQ cooking liquors suggest that these compounds are primary degradation products of certain lignin substructures. The slight difference could be

Table 2.    Assignment of signals in $^{13}C$ NMR spectra of lignin preparations LF-1 and LF-2[a] obtained from soda-AQ pulping of poplar wood (_Populus deltoides_).

| Signal | Chemical Shift ($\delta$ in ppm) | Intensity[b] | | Assignment of Signals[c] |
|---|---|---|---|---|
| 1 | 153.4 | w | w | C-3/C-3' in etherified 5-5 |
| 2 | 152.3 | s | s | C-3/C-5 in etherified S $\beta$-_O_-4 |
| 3 | 149.3 | w | w | C-3 in etherified G $\beta$-_O_-4 |
| 4 | 147.9 | m | s | C-4 in etherified G $\beta$-_O_-4<br>C-3 in nonetherified G<br>C-3/C-5 in nonetherified S |
| 5 | 138.1 | w | - | C-1 in nonetherified S |
| 6 | 134.6 | m | s | C-1 in etherified G<br>C-4 in nonetherified S |
| 7 | 133.0 | - | w | C-1 in nonetherified G and S |
| 8 | 131.4 | - | w | C-1 in nonetherified G and S |
| 9 | 126.8 | w | m | C-5/C-5' in nonetherified 5-5 |
| 10 | 119.2 | w | w | C-6 in etherified and nonetherified G |
| 11 | 115.5 | m | m | C-5 in etherified and nonetherified G |
| 12 | 111.2 | w | w | C-2 in etherified and nonetherified G |
| 13 | 104.4 | m | s | C-2/C-6 in etherified and nonetherified S without $\alpha$-C=O |
| 14 | 100.7 | vw | vw | C-1 in xylan |
| 15 | 86.0 | w | w | C-$\beta$ in G and S $\beta$-_O_-4 |
| 16 | 75.5 | - | m | C-4 in xylan |
| 17 | 74.2 | w | m | C-3 in Xylan |
| 18 | 72.3 | m | m | C-$\alpha$ in G and S $\beta$-_O_-4 |
| 19 | 63.1 | - | m | C-$\beta$ in $\beta$-5 and G and S $\beta$-_O_-4 with $\alpha$-C=O |
| 20 | 60.2 | m | m | C-$\gamma$ in G and S $\beta$-_O_-4 |
| 21 | 55.9 | vs | vs | O$\underline{C}H_3$ in Ar-OCH$_3$ |
| 22 | 53.7 | w | m | C-$\beta$ in $\beta$-$\beta$ and $\beta$-5 |
| 23 | 28.8 | - | vw | CH$_2$ in Ar-CH$_2$-Ar' |

[a] See Table 1 for origin of lignin preparations LF-1 and LF-2.
[b] vs = very strong; s = strong; m = moderate; w = weak; vw = very weak.
[c] G = Guaiacylpropane; S = Syringylpropane.

attributable to reverse aldol fragmentation initiated by oxidation of γ- or α-carbinols by AQ.[8,9]

Guaiacol (**5**; GL) and syringol (**6**; SL) are produced by side chain elimination from units with an α-hydroxyl group via base-catalyzed reverse aldol addition.[27]  As shown in Figure 7, the concentration profiles for the formation of guaiacol and syringol in soda and soda-AQ pulping reveal that these compounds are formed in higher amounts when the quinone additive was present.

Vinylguaiacol (**7**) and vinylsyringol (**8**) were detected only when AQ was present.  This suggests that these compounds are not secondary products formed by way of CA and SA, respectively, but are products resulting from reductive cleavage of vinyl ether structures in the presence of AHQ.[18]  The vinyl ether structures are produced from phenolic β-aryl ether (β-$\underline{O}$-4) substructures by base-catalyzed elimination of formaldehyde via the corresponding quinonemethide intermediates.  The compounds are quite stable toward alkali.

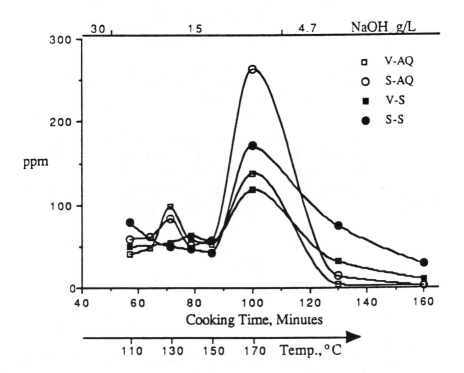

Figure 6.  Formation of vanillin (V) and syringaldehyde (S) during soda (-S) and soda-AQ (-AQ) pulping of poplar wood (*Populus deltoides*) using flow-through reactor.

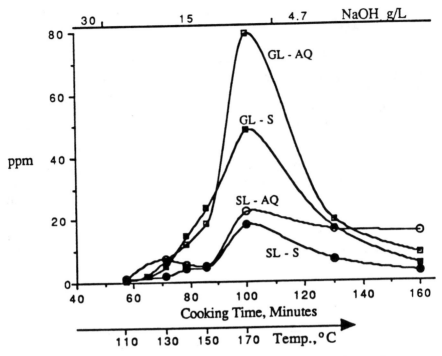

Figure 7. Formation of guaiacol (GL) and syringol (SL) during soda
(-S) and soda-AQ (-AQ) pulping of poplar wood (*Populus
deltoides*) using flow-through reactor.

Table 3.    The effect of changes in the composition of third stage liquor
on yield and kappa number during soda-AQ pulping of poplar
wood (*Populus deltoides*) using flow-through reactor.

| Cook | Composition of 3$^{rd}$ Stage Cooking Liquor | Kappa Number | Yield (%) |
|------|---------------------------------------------|--------------|-----------|
| AQ-1 | 4.7 g/L NaOH + 0.3 g/L AQ | 23.0 | 53.8 |
| AQ-2 | 15 g/L NaOH + 0.3 g/L AQ | 12.0 | 50.7 |
| AQ-3 | 4.7 g/L NaOH | 21.9 | 52.2 |

The common characteristic of the concentration profiles for each of the
compounds is the increase in their concentration until 170°C is reached. It is
likely that the precursors for the compounds are consumed shortly after the
maximum cooking temperature is reached. The phenolic precursors would be
then generated, e.g., the β-aryl ether linkages in etherified units are cleaved

by internal nucleophilic attack of neighboring hydroxyl groups. This postulation is in good agreement with the results of an experiment, in which the last stage in a FTR soda-AQ cook (see Figure 3) was conducted with sodium hydroxide solution in the absence of AQ. As given in Table 3, the elimination of AQ in the third stage cook did not affect the pulp yield and the degree of delignification. This finding strongly suggests that the rate and reaction mechanism of lignin degradation in the final stage of alkaline pulping are independent of the quinone additive. Increasing the alkali charge in the final stage from 4.7 g/L to 15 g/L (Table 3, Cook AQ-2) resulted in a substantially better degree of delignification, an improvement of approximately 45%. This can interpreted that in the final phase the removal of lignin depends exclusively on the hydrolytic cleavage of nonphenolic β-aryl ether substructures with hydroxyl groups in neighboring positions at C-α or C-γ via epoxide intermediates. The rate of this internal $SN_2$ reaction depends on the concentration of base.

## CONCLUSIONS

The formation of low molecular weight compounds during alkaline pulping increases greatly when AQ is added to the system. The ability of AQ/AHQ system to attack and degrade lignins is evident by the high concentration of coniferyl alcohol (**1**) and sinapyl alcohol (**2**) found in the cooking liquors withdrawn at different stages of the cook, using a flow-through reactor. In addition to coniferyl alcohol and sinapyl alcohol, several low molecular weight lignin degradation products are also formed. Evidently, these compounds are generated by primary degradation of lignins. During the heating-up period (maximum cooking temperature, 170°C), the lignins were mainly degraded by reductive β-aryl ether cleavages, in which the cleavage of arylglycerol-β-guaiacyl ether units is predominant over that of arylglycerol-β-syringyl ether units. Moreover, residual lignins remaining in the fibers seem to contain an appreciable amount of non-phenolic β-syringyl ether units. These β-aryl ether linkages are cleaved at high temperature (150-170°C) via base-catalyzed internal nucleophilic substitution. The results also show that extended delignification can be achieved by maintaining the concentration of sodium hydroxide solution higher than that found in the normal cooks during the last phase of delignification.

## ACKNOWLEDGEMENTS

The authors are grateful to the Organization of American States and Centro de Investigaciones en Celulosa y Papel (CICELPA-INTI), Buenos Aires, Argentina, for providing a scholarship for ADV.

## REFERENCES

1.    H. H. Holton, *Pulp and Paper Can.*, 78(10), T 218 (1977).

2.   J. Gierer, O.Lindeberg and I.Noren, *Holzforschung*, 33(6),213 (1979).
3.   L. L. Landucci, *Tappi*, 63(7), 95 (1980).
4.   H. Aminoff, G. Brunow, G. E. Miksche, and K. Poppius, *Paperi ja Puu*, 61, 441 (1979).
5.   D. H. Hawes, M. C. Schroeter, C.-L. Chen, and J. S. Gratzl, Paper presented at the Cellulose, Paper and Textile Division, ACS Spring Meeting, Appleton, Wisconsin, USA, May 1978.
6.   H. Araki, D. H. Hawes, M. C. Schroeter, C.-L. Chen, and J. S.Gratzl, Proc. 1979 Canadian Wood Chemistry Symposium, p. 117, Harrison Hot Spring, British Columbia, Canada, Sept. 19-21, 1979.
7.   M. C. Schroeter, "Hydrolysis of Models Featuring β-ether Structures of Lignin in Alkali and Sodium Anthraquinone Monosulfonate", Ph. D. Thesis, North Carolina State University, Raleigh, NC, USA (1980).
8.   R. G. Hise, C.-L. Chen, and J. S. Gratzl, Proc. 2nd International Symposium on Wood and Pulping Chemistry, Vol.2, p. 166-177, Tsukuba Science City, Japan, May 23-27, 1983.
9.   R. G. Hise, D. K. Seyler, C.-L. Chen, and J. S. Gratzl, Proc. 4th International Symposium on Wood and Pulping Chemistry, pp. 391-397, Paris, France, April 27-30, 1987.
10.  N.H. Shin, "A Modified Soda-AQ/Oxygen Pulping Process as an Alternative to Kraft Process for the Production of Bleached Softwood Pulp", Ph.D. thesis, North Carolina State Uni., Raleigh, NC, (1988).
11.  T. J. Blain, *Trans. Tech. Sect. CPPA.*, 5(1), TR3 (1979).
12.  K. L. Ghosh, V. Venkatesh, W. J. Chin, and J. S. Gratzl,*Tappi*, 60(11), 127 (1977).
13.  J. M. MacLeod, and N. Cyr, *Pulp & Paper Can.*, 84(4), T81 (1983).
14.  I. A. Pearl, D. L. Beyer, B. Johnson, and S. Wilkinson, *Tappi*, 40(5), 374 (1957).
15.  C. J. Venverloo, *Holzforschung*, 25(1), 18 (1971).
16.  K. Kratzl, and J. Okabe, *In* "Biosynthesis of Aromatic Compounds", p. 67, Pergamon Press, London (1966).
17.  G. Miksche, *Acta Chem. Scand.*, 26, 4137 (1972).
18.  J. Gierer, and S. Ljunggren, *Svensk Papperstidn.*, 82(3), 71 (1979).
19.  J. Gierer, and S. Ljunggren, *Svensk Papperstidn.*, 82(17), 503 (1979).
20.  S. Ljunggren, *Svensk Papperstidn.*, 83(13), 363 (1980).
21.  J. R. Obst, *Holzforschung*, 37(1), 23 (1983).
22.  G. E. Miksche, *Acta Chem. Scand.*, 27, 1355 (1973).
23.  R. Kondo,Y.Tsutsumiand, and H.Imamura, *Holzforschung*, 41(2), 83 (1987).
24.  R. D. Mortimer, Proc. 1979 Canadian Wood Chemistry Symposium, p. 166, Niagara Falls, Ontario, Canada, Sept. 1979.
25.  R. Kondo, and K.V. Sarkanen, Proc. 2nd Intern. Symp. on Wood and Pulping Chem., V.2, p.155, Tsukuba Science City, May 23-27, 1983.
26.  G. Gellerstedt, K. Gustafson, and R. A. Northey, *Nordic Pulp and Paper Research J.*, 3(2), 93 (1988).
27.  R. G. Hise, "Delignification Mechanism in Kraft-AQ and Soda-AQ Pulping: The Role of Oxidative-Hydrolytic Processes in the Cleavage of Carbon-Carbon Bonds", Ph.D. Thesis, North Carolina State University, Raleigh, NC, USA (1984).

# Chapter 27

# THE EVOLUTION OF EPM/EPDM CATALYST SYSTEMS DURING R. D. GILBERT's PROFESSIONAL CAREER

E. K. EASTERBROOK AND
E. G. KONTOS

*Elastomers R & D*
*UNIROYAL CHEMICAL COMPANY*
*Nagautuck, Connecticut 06770*

## INTRODUCTION

There are two types of ethylene-propylene rubber, EPM and EPDM[1]. The designation EPM applies to the simple copolymer of ethylene and propylene ("E" for ethylene, "P" for propylene and "M" for the polymethylene type backbone). In the case of EPDM, the "D" designates a third comonomer, a diene, which introduces unsaturation into the chain (Figure 1).

**EPM:**

EPM represents a copolymer of monomer units E

(ethylene) and P (propylene).

Ethylene $\quad$ $CH_2 = CH_2$

Propylene $\quad$ $CH_2 = CH - CH_3$

M refers to a polymethylene chain

with no unsaturation in "backbone".

$$- CH_2 - CH_2 - CH_2 - CH_2 - CH_2 - \overset{\overset{\displaystyle CH_3}{|}}{CH} - CH_2 - CH_2 - CH_2 - \overset{\overset{\displaystyle CH_3}{|}}{CH}$$

**EPDM:**

EPDM represents an EPM containing a Diene (D)

monomer unit to introduce unsaturation

for curing purposes.

Figure 1. $\quad$ What is EPM/EPDM?

Three dienes are employed to introduce the unsaturation (Figures 2 and 3).  They are dicyclopentadiene, ethylidene norbornene and 1,4-hexadiene.  The dienes are so structured that only one of the double bonds will polymerize and the unreacted double bond acts as a site for sulfur crosslinking.  In addition the unreacted double bond does not become part of the polymer backbone and therefore the excellent aging properties of the saturated backbone are maintained.

♦    UNSATURATION REQUIRED FOR SULFUR VULCANIZATION

♦    INCORPORATION OF A THIRD MONOMER, A NON-CONJUGATED DIENE, INTRODUCES
PENDANT UNSATURATION

EXAMPLES:

DICYCLOPENTADIENE

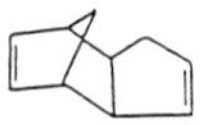

ETHYLIDENE NORBORNENE

$= CH - CH_3$

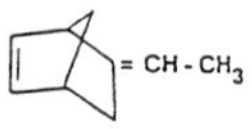

1,4-HEXADIENE

$CH_2 = CH - CH_2 - CH = CH - CH_3$

Figure 2.      Examples of EPDM terpolymers. (Unsaturation required for sulfur vulcanization; incorporation of a third monomer, a nonconjugated diene, introduces pendant saturation).

These polymers are synthesized via Ziegler-Natta catalysis (Figures 4 and 5).  Ziegler made the original discovery of this catalyst formed by reacting a metal alkyl or hydride of a Group I to III metal with a salt of a Group IV to VIII transition metal under an inert atmosphere.  Ziegler primarily applied this to polyethylene whereas, Natta's contribution was in the field of polypropylene and elastomeric copolymers of ethylene and propylene.

The formation of the active catalyst site results from the reduction of the transition metal halide to a lower valence state (Figure 6).  The reduction occurs through an alkylation of the transition metal halide via the aluminum alkyl.  The active species are considered to be the alkylated state of a lower valence state.  In the case of titanium and vanadium, it is often times the +3 state.

COPOLYMER

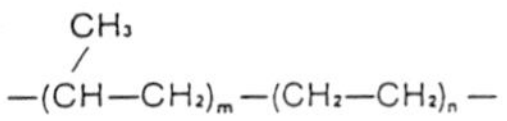

$$-(\overset{\displaystyle CH_3}{\overset{\displaystyle /}{CH}}-CH_2)_m-(CH_2-CH_2)_n-$$

DCPD TERPOLYMER

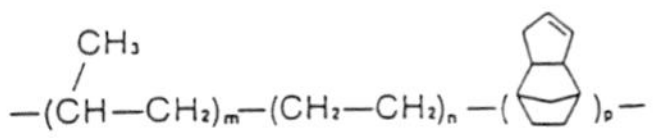

$$-(\overset{\displaystyle CH_3}{\overset{\displaystyle /}{CH}}-CH_2)_m-(CH_2-CH_2)_n-(\ \ )_o-$$

ENB TERPOLYMER

$$-(\overset{\displaystyle CH_3}{\overset{\displaystyle /}{CH}}-CH_2)_m-(CH_2-CH_2)_n-(\ \ )_o-$$  CH—CH₃

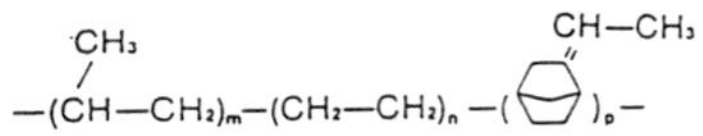

Figure 3.    Molecular structures.

<u>ZIEGLER</u>    -    HIGH DENSITY POLYETHYLENE USING TiCl$_4$ AND AL(Et)$_3$

<u>NATTA</u>    -    ISOTACTIC POLYPROPYLENE USING TiCl$_3$ AND AL(Et)$_3$

GENERIC

CATALYSTS FORMED BY REACTING A METAL ALKYL OR HYDRIDE OF A GROUP I TO III METAL WITH A SALT OF A GROUP IV TO VIII TRANSITION METAL UNDER AN INERT ATMOSPHERE.

MOST COMMON

FORMED BY REACTING ALUMINUM ALKYLS WITH SALTS OF TITANIUM OR VANADIUM.

Figure 4.    Ziegler-Natta catalysis of polymerization.

- ♦ CAN POLYMERIZE OLEFINS AT RELATIVELY LOW PRESSURES AND TEMPERATURES

- ♦ CAN POLYMERIZE ALPHA-OLEFINS TO VERY HIGH MOLECULAR WEIGHT

- ♦ CAN POLYMERIZE OLEFINS AND DIENES TO DIFFERENT STEREOCHEMICAL STRUCTURES

- ♦ POLYMERIZATION CAN BE DONE IN SOLUTION, SLURRY OR VAPOR PHASE WITH SOLUBLE, COLLOIDAL OR HETEROGENEOUS CATALYSTS

Figure 5.    Uniqueness and versatility of Ziegler-Natta catalysis.

## POSSIBLE MECHANISM FOR
## ZIEGLER/NATTA POLYMERIZATION

### Formation of Active Site:

$$AlEt_3 + VCL_4 \longrightarrow AlEt_2Cl + VCl_3Et$$

$$VCl_3Et \longrightarrow VCl_3 + Et$$

$$VCl_3 + AlEt_3 \longrightarrow \underline{VCl_2Et + Et_2AlCl}$$
Active Site

### Polymerization:

$$Cl_2VEt + CH_2 = CH_2 \longrightarrow Cl_2V-CH_2-CH_2- Et$$

$$Cl_2V-CH_2-CH_2-Et \xrightarrow{\overset{H}{\underset{CH_3}{CH_2=C-CH_3}}} Cl_2V-CH_2-\overset{H}{\underset{CH_3}{C}}-CH_2-CH_2-Et$$

Figure 6.    Possible mechanism for Ziegler-Natta polymerization

This chapter deals with the modification and improvements primarily in catalyst efficiencies that have taken place with catalyst systems used to synthesize EPM/EPDM during the time period of R. D. Gilbert's career.

## Highly Dispersed Catalyst System

Natta and his coworkers at the Milan Polytechnic Institute of Industrial Chemistry pioneered the preparation of polypropylene and ethylene/propylene copolymers. For the preparation of the various stereospecific polypropylenes, Natta employed the so-called heterogeneous, insoluble type of catalyst systems. These systems were prepared by reacting various aluminum alkyls with insoluble transition metal halides like titanium and vanadium trichlorides (Figure 7). For stereospecific control, a surface is usually required.

| Cocatalyst | Catalyst * |
|------------|-----------|
| $R_n Al X_{3-n}$ | $TiCl_3$ |
| | $VCl_3$ |

Figure 7.    Ziegler-Natta Polyoelfin catalyst systems (heterogeneous).
             * Insoluble in hydrocarbon solvent

Attempts at producing copolymers of ethylene and propylene with the above described catalysts led to a mixture of homopolymers and copolymers with variable composition including block copolymers. Natta discovered[2] that if soluble vanadium halides were employed then much more homogeneous copolymers were obtained (Figure 8). This was confirmed by extraction with various solvents, and x-ray analysis as well as determination of composition by infrared spectroscopy.

| Cocatalyst | Catalyst ** |
|------------|------------|
| $Al R_3$ | $VOCl_3$ |
| | $VCl_4$ |
| $LiAlR_4$ *** | $TiCl_4$ |

* Catalyst efficiencies of 100 to 200 gms. polymer/gm. transition metal halide.

** Soluble in hydrocarbon solvent.

*** Al/Ti ratio must be close to 1/1

Figure 8.    Highly dispersed Ziegler-Natta catalysts systems for EPM*

Using this information as a basis, efforts were made to develop a copolymer catalyst system based on titanium tetrachloride also soluble in hydrocarbon solvents. This was highly desirable, since the compound was readily available, cheap and not a pro-oxidant of the polymer like vanadium. Eventually a system was developed employing lithium aluminum tetralkyls as cocatalyst and in an Al/Ti molar ratio of 1:1. The polymer possessed characteristics very similar to that produced by a VOCl₃ based catalyst. Surprisingly both systems polymerized propylene to a crystallinity content of about 10 to 15% (X-ray). Further work on the copolymer using NMR has shown the vanadium synthesized material to possess head to head structure whereas the titanium ones do not.[3]

Although at first these catalyst systems appeared to be completely soluble they are now considered to be highly dispersed. Catalyst efficiencies were in the order of 100 to 200 g of polymer per gram of transition metal halide. The low efficiency is due to the unavailability of all of the transition metal.

### Soluble Catalyst Systems

Certain advantages could be foreseen for developing a soluble type of catalyst for EPM. These were: more homogeneous copolymers; more efficient use of the transition metal halide (resulting in a cleaner polymerization and polymer) and better incorporation of termonomer. As a result, work was initiated along such lines.

Work in our laboratories concentrated on the use of aluminum alkyl halides with vanadium oxytrichloride or tetrachloride (Figure 9).

## SOLUBLE CATALYST SYSTEMS*

### for

### EPM/EPDM

| Cocatalyst | Catalyst |
|---|---|
| $R\,Al\,Cl_2$ | $VOCl_3$ |
| $R_3\,Al_2\,Cl_3$ | $VCl_4$ |
| $R_2\,Al\,Cl$ | $VO\,(OR)_3$ |
| | $V(Ac\,Ac)_3$ |

* Catalyst efficiencies of ca. 1200 to 1500 gms. polymer/gms. transition metal halide.

Figure 9.    Soluble catalysts systems for EPM/EPDM*

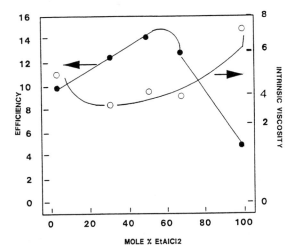

Variation of efficiency (grams EPM/gram
VOCl$_3$ x 10$^{-2}$) and intrinsic viscosity (cyclohexane
at 30° C.) with EtAlCl$_2$ /Et$_2$AlCl molar ratio

20 to 1 Al/V molar ratio, 0.05 mmole of VOCl$_3$,
350 ml. of n-heptane, 1 to 1 molar feed of
C$_2$H$_4$/C$_3$H$_6$ at 2 liters per minute, 30-minute
reaction time

Figure 10.    Catalyst efficiency.

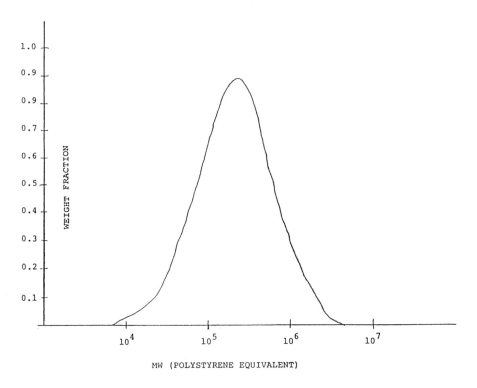

Figure 11.   GPC of polymer prepared with a single-species catalyst system.

$Et_2AlX$

or

$Et\ AlX_2$     $+\ VOCl_3 + VCl_4$

$Et_3\ Al + AlX_3$

This work[4] indicated that optimum yields and quality could be obtained with a catalyst system based on $Et_3\ Al_2\ Cl_3$ and $VOCl_3$.

In Figure 10 the variation of efficiency and intrinsic viscosity with $Et_2$ $Al\ Cl$ /$Et\ Al\ Cl_2$ molar ratio is illustrated.  A maximum in efficiency is found using a mixture consisting of 40 to 60 mol % $Et_2\ Al\ Cl$.  This efficiency is about 10 times greater  (1000-1500g/g $VOCl_3$) than the highly dispersed catalyst system.  A minimum in intrinsic viscosity is also observed in this range.  As shown in Figure 11 a monomodal molecular weight distribution is obtained, indicating single catalyst species.  Ethyl aluminum sesquichloride (EASC) is considered a 50:50 mixture of $Et_2\ Al\ Cl$ and $Et\ Al\ Cl_2$.

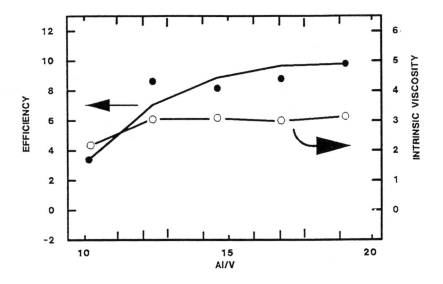

Variation of efficiency (grams EPM/gram VOCl$_3$
X 10-$^2$) and intrinsic viscosity (cyclohexane at
30$^\circ$ C.) with Al/V molar ratio


Catalyst: Et$_3$ Al$_2$Cl$_3$/VOCl$_3$
0.05 mmole of VOCl$_3$, 350 ml of n-heptane,
1 to 1 molar feed of C$_2$ H$_4$ /C$_3$H$_6$
at 2 liters per minute, 30-minute reaction time

Figure 12.    Catalyst efficiency.

In addition to the correct ratio of alkyl and halide groups in the aluminum alkyl it is also necessary to have a minimum Al/V molar ratio to have the optimum catalyst. This critical ratio is above 10:1 as shown in Figure 12. At values below this ratio, a certain portion of the catalyst becomes insoluble. This insoluble catalyst is longer-lived and gives rise to higher molecular weight and higher ethylene containing polymer. Examination of a GPC data of polymer produced with such a catalyst indicates a bimodal molecular weight distribution (Figure 13).

It can be speculated that the critical ratio (> 10/1) of EASC/VOCl$_3$ is related to having sufficient aluminum alkyl present to reduce the vanadium halide to the alkylated vanadium plus three species and to sufficiently complex the active site keeping it soluble in the polymerization medium. (See Figure 14). Sufficient chloride groups are necessary to complex with the vanadium catalyst system but also enough alkyl (ethyl) groups to keep the species soluble in hydrocarbon solvent. As a result, with the right combination one will achieve a soluble, single-species catalyst. With such a system a homogeneous monomodal copolymer will result. This type of catalyst does not homopolymerize propylene.

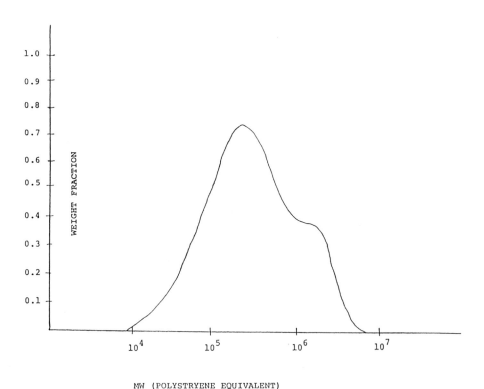

Figure 13.      GPC of polymer prepared with multi-species catalyst system.

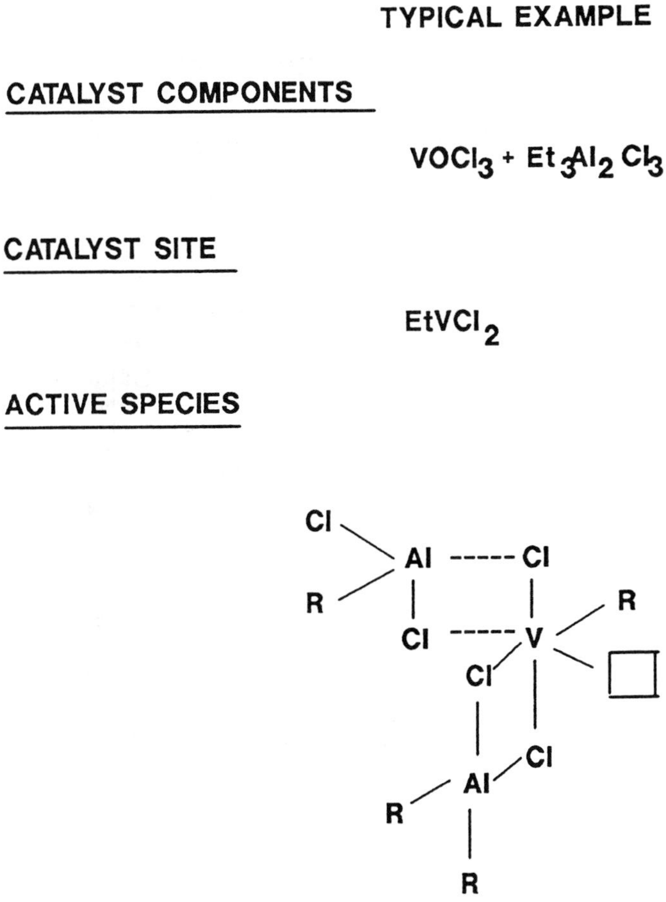
TYPICAL EXAMPLE
CATALYST COMPONENTS
$VOCl_3 + Et_3Al_2Cl_3$
CATALYST SITE
$EtVCl_2$
ACTIVE SPECIES
Cl
Al
Cl
R
R
Cl
V
Cl
Cl
R
Al
R
R

Figure 14. Homogeneous catalysts for ethylene/propolene copolymers.

In the case of $EtAlCl_2$ and $Et_2AlCl$ there are either not enough alkyl groups to keep the catalyst species soluble or not enough chloride groups to complex with the species. In the case of $Et_2$ Al Cl over reduction to inactive, insoluble $V^{2+}$ can occur. As a result, insoluble species are formed resulting in lower catalyst efficiency and higher intrinsic viscosity (molecular weight).

**Modified Soluble Catalyst Systems**

With the introduction of EPDM and a much sharper increase of usage of this kind of rubber it became necessary to fine-tune the catalyst systems. This came about due to a greater demand for improved processing and curability. These properties could be controlled primarily through molecular weight, molecular weight distribution and termonomer type.

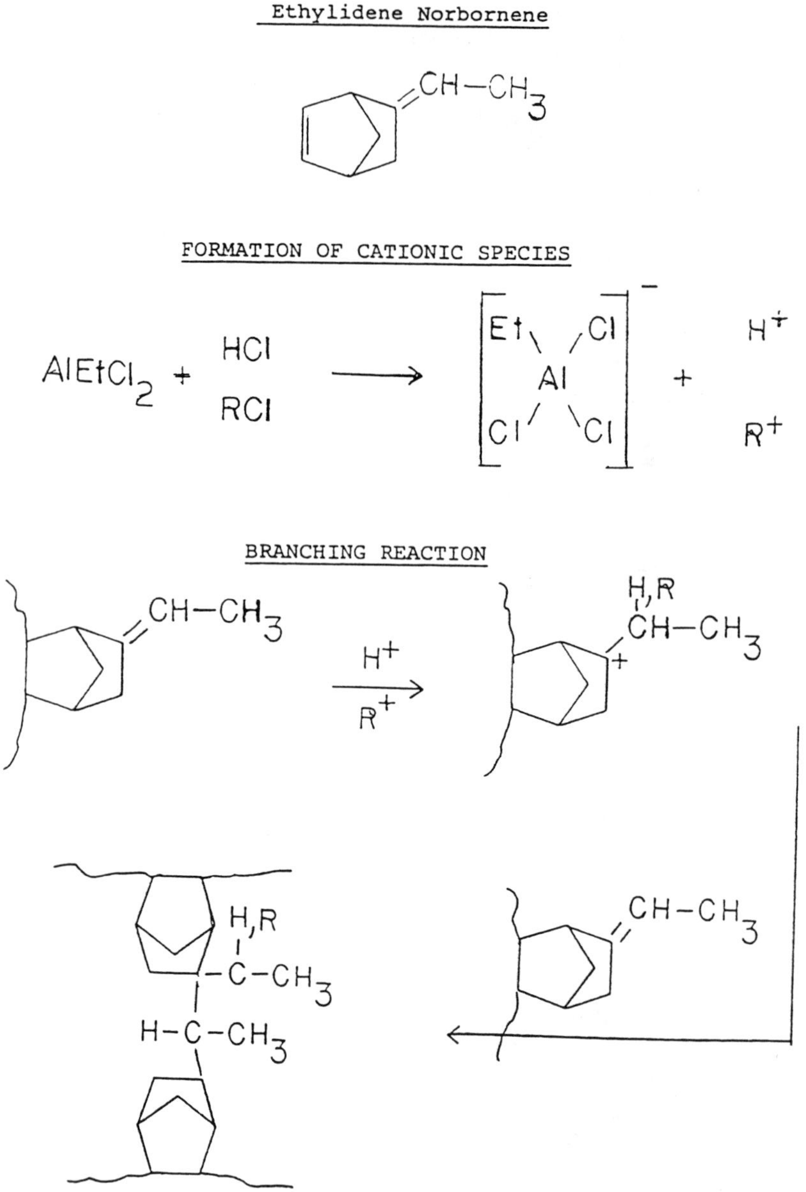

Ethylidene Norbornene

FORMATION OF CATIONIC SPECIES

BRANCHING REACTION

Figure 15.    Branching reactions in the presence of Lewis acids.

Ethylidene norbornene (ENB) became the termonomer of choice because of its ease of incorporation and ability to impart a fast cure rate. However, with this came sensitivity to a variability in branching. Although the second double bond of ENB is an internal double bond and not susceptible to polymerization, it does possess a nucleophilic character. As a result, because of the presence of strong Lewis acids emanating from the catalyst system, side reactions can occur leading to branching[5-6] (Figure 15). This branching can lead to molecular weight broadening as well as an increase in the elastomeric nature of the polymer, both of which can affect processing.

As a result, the extent of branching had to be controlled. Two different approaches have been employed to accomplish this. The first involved the use of Lewis bases to react with the generated Lewis acids. Compounds such as pyridine, aniline, ammonia, esters and polyethers have been employed[5]. This technique works to a certain extent but eventually leads to poorer monomer conversion. A second approach is to reduce the amount of chloride in the catalyst system. This can be accomplished by employing vanadates or vanadium acetylacetonate in place of $VOCl_3/VCl_4$. A better and more efficient way is through the use of dialkyl aluminum halides. Completely linear ENB terpolymers can thus be synthesized (Table 1).

Table 1
Effect of Cocatalyst on EPDM Branching

| Run No. | 1 | 2 | 3 | 4 | 5 |
|---|---|---|---|---|---|
| Cocatalyst | DEAC | DEAC/EASC | DEAC/EASC | DEAC/EASC | EASC |
|  | - | 3/1 | 1/1 | 1/3 | - |
| % Solids | 7.4 | 6.7 | 7.9 | 8.5 | 7.9 |
| ML-4 at 257° F. | 41.5 | 49 | 55 | 92.5 | 96 |
| E/P, wt. | 56/44 | 53/47 | 48/52 | 51/49 | 51/49 |
| Iodine Number, ENB | 12 | 12 | 9 | 8.7 | 8 |
| I.V. at 135° C. | 1.8 | 1.85 | 1.85 | 2.2 | 2.2 |
| % R.T. Insoluble | 4 | 1 | 2 | 2 | 2 |
| Branching Index | 1.0 | 1.1 | 1.4 | 1.9 | 2.2 |
| GPC Data: |  |  |  |  |  |
| $\bar{M}w/\bar{M}n$ | 4.1 | 5.7 | 4.4 | 5.4 | 5.8 |

The most convenient technique for controlling molecular weight and molecular weight distribution is through the use of hydrogen. However, at times this does not suffice and more drastic steps are required. As mentioned previously the combination of EASC and VOCl₃ gives rise to a monomodal type of distribution. On the other hand, a catalyst system composed of DEAC and VOCl₃ gives rise to a bimodal distribution (See Figure 13). Intermediate distributions can result through the use of combinations of the two cocatalysts or through the use of hydrogen.

## High Activity Catalyst Systems

In order to reduce manufacturing costs the most likely approach was to produce even more efficient catalyst systems. Catalyst efficiency is primarily dependent upon the lifetime of the catalyst. This is so because in a soluble, catalyst system, single species implies that each and every (vanadium atom) is being employed in making polymer.

Catalyst deactivation occurs by both a monomolecular and a bimolecular reaction[7] (Figure 16). The ultimate product is an inactive vanadium plus two species. Considerable effort was expended in finding compounds that would efficiently reoxidize the vanadium to a higher oxidation state, where it could be used to generate active catalyst again.

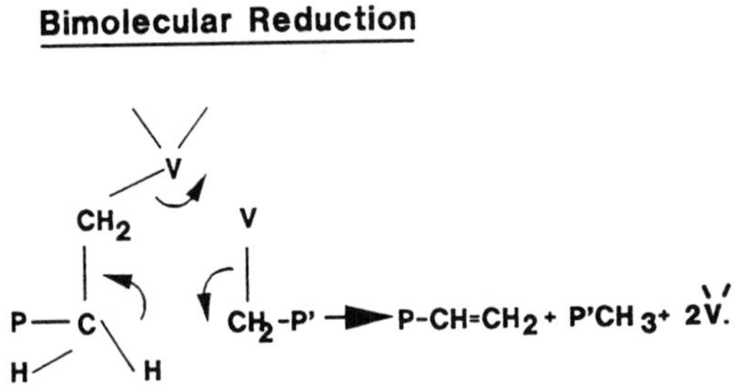

**Monomolecular Reduction**

$$\underset{Cl}{\overset{Cl}{\diagdown}} V - P \quad \blacktriangleright \quad \underset{Cl}{\overset{Cl}{\diagdown}} V \quad + CH_2 = \overset{\overset{CH_3}{|}}{C} - P + H$$

**Bimolecular Reduction**

$$P - C \quad \cdots \quad CH_2 - P' \quad \blacktriangleright \quad P - CH = CH_2 + P'CH_3 + 2\overset{\cdot}{V}.$$

Figure 16.    Catalyst deactivation.

Numerous activators were found. Examples are oxygen, nitro compounds, hydroquinones, phosphorus trichloride, sulfur and sulfuryl chloride[4]. The most effective and promising, however, were organic chlorides such as hexachlorocyclopentadiene, hexachloropropylene, ethyl trichloroacetate and butyl perchlorocrotonate.

These molecules apparently function by transfer of a chlorine[6] radical to the spent $V^{2+}$ species, thereby raising the valence (Figure 17). In this state the vanadium is available again for re-alkylation and formation of an active specie. Catalyst efficiencies (5000 g. polymer/g. $VOCl_3$), are increased 50-fold over the highly dispersedsystems or 5-fold over the normal soluble catalyst systems.

# ACTIVATORS*

## REOXIDIZE SPENT CATALYST TO ACTIVE STATE

### EXAMPLE: BUTYL PERCHLOROCROTONATE

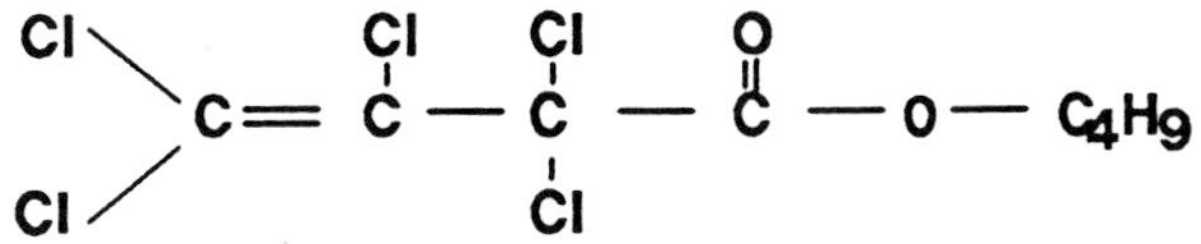

### MECHANISM:

$$R - Cl + Cl_2 \; V\bullet \longrightarrow R\bullet + VCl_3$$

OR

$$R - Cl + Cl_2 \; V-P \longrightarrow Cl_3 V + R-P$$

* Catalyst efficiencies of 5,000 to 6,000 gms. polymer/gm. transition metal halide.

Figure 17.    Catalyst activators.*

**Super High Efficient Catalyst System**

Because of the even more competitive nature of EPDM manufacture, it has become necessary to pursue even more efficient ways of producing EPDM. Additional motivation for this arises from the advancement made by the polyolefin industry not only in process but also in catalyst development. Polyolefin catalysts have been developed with efficiencies of at least 100,000 to 1,000,000 g polymer per gram of transition metal halide. As a result, it has become unnecessary to remove catalyst residues from the polymer. Such a step would be most welcome to EPDM manufacturers also.

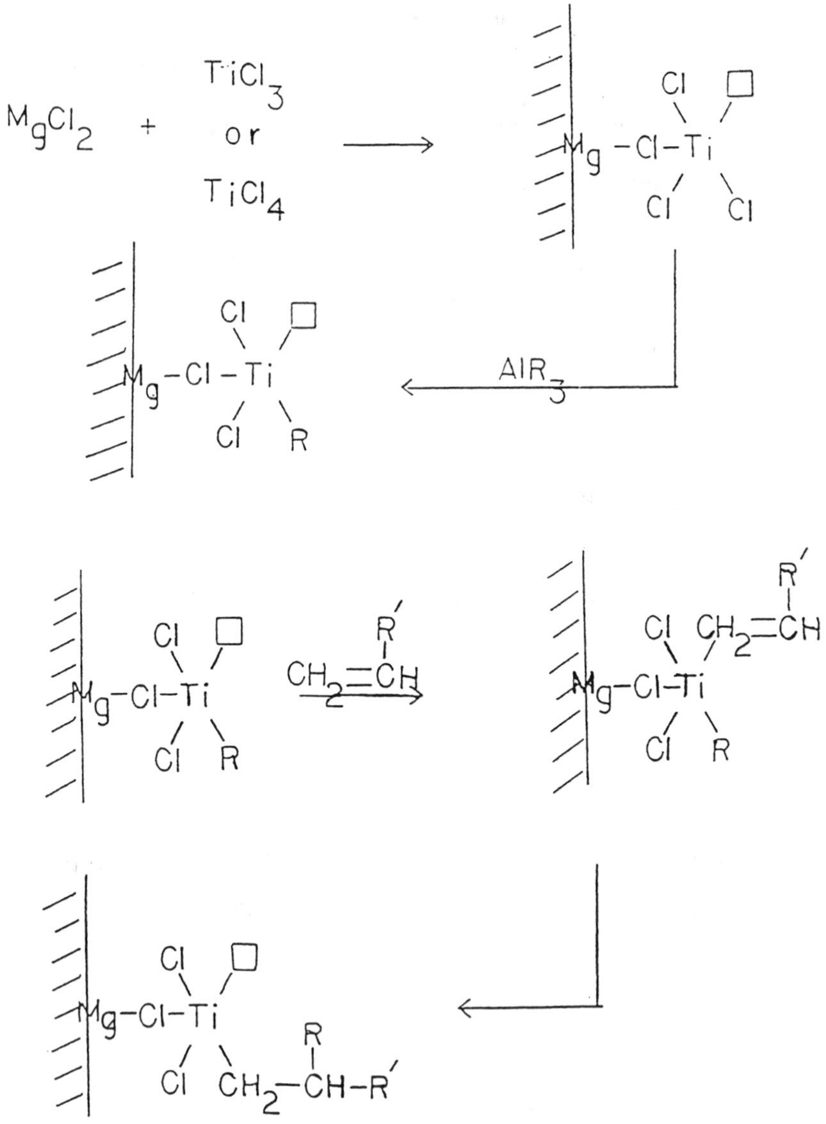

Figure 18.    MgCl$_2$ supported titanium catalyst.

The great advancement in these efficiencies has come about primarily through the supporting of titanium tetrachloride on magnesium chloride (Figure 18). Magnesium chloride is preferred in that it has a crystal structure very similar to that of titanium tetrachloride.[8] As a result it is possible to separate titanium catalyst centers sufficiently to maximize activity. In addition the life-time of such catalysts are hours rather than seconds as is the case with soluble vanadium catalysts. Dutral SA[9] has announced the commercialization of an EPM employing such a catalyst using their slurry process.

Although titanium-based catalysts have been the primary area of research other transition metals like vanadium and chromium[10] are being investigated. Another area of great interest are those based on zirconium halides as investigated by Professor Kaminsky. A combination of methyl aluminoxane and zirconecene results in a catalyst that can be considered a living catalyst with resulting efficiencies of 100,000 or greater (Figures 19 and 20). Although considered to be a soluble, single-species type of a catalyst, it readily polymerizes propylene to an atactic structure. In addition ethylene/propylene copolymers are readily obtained.[11] However chain transfer occurs readily, and high molecular weights are difficult to produce. Various modifications of this system are being evaluated, including supporting it on various substrates like magnesium chloride.

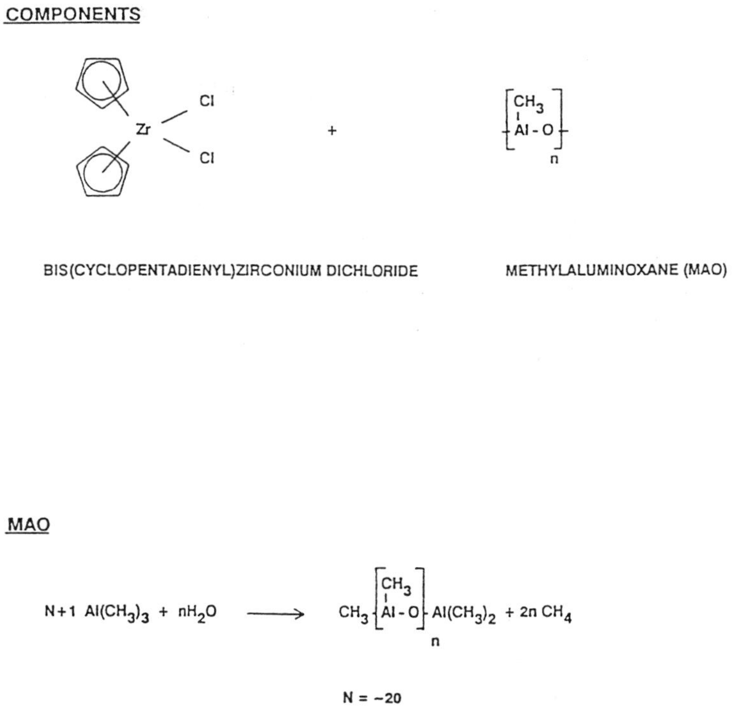

COMPONENTS

BIS(CYCLOPENTADIENYL)ZIRCONIUM DICHLORIDE          METHYLALUMINOXANE (MAO)

MAO

$$N+1 \ Al(CH_3)_3 + nH_2O \longrightarrow CH_3 \left[ \begin{matrix} CH_3 \\ | \\ Al-O \end{matrix} \right]_n Al(CH_3)_2 + 2n \ CH_4$$

N = ~20

Figure 19.     Kaminsky catalyst.

1. COMPLEXATION

$$Cp_2 \, Zr \, Cl_2 \; + \; MAO \qquad \longrightarrow \qquad Cp_2 \, Zr \, Cl_2 \cdot MAO$$

2. ALKYLATION

$$Cp_2 \, Zr \, Cl_2 \cdot MAO \qquad \longrightarrow \qquad Cp_2 \, Zr \, ClCH_3 \cdot MAO$$

3. INSERTION

$$Cp_2 \, Zr \, ClCH_3 \cdot MAO + n \, C_2H_4 \qquad \longrightarrow \qquad Cp_2 \, Zr \, Cl(C_2H_4)_n CH_3 \cdot MAO$$

Figure 20.     Possible mechanism of Kaminsky catalyst.

Another development is the preparation of long lived catalyst systems. Professor Doi et al has developed (Figure 21) living catalysts based on vanadium acetylacetonate. Since the system requires low temperature, (-78 C.); productivity is low. However, minor modifications of the catalyst structure[12] have allowed for polymerization at higher temperature (-40 to -20C.) resulting in higher productivity.

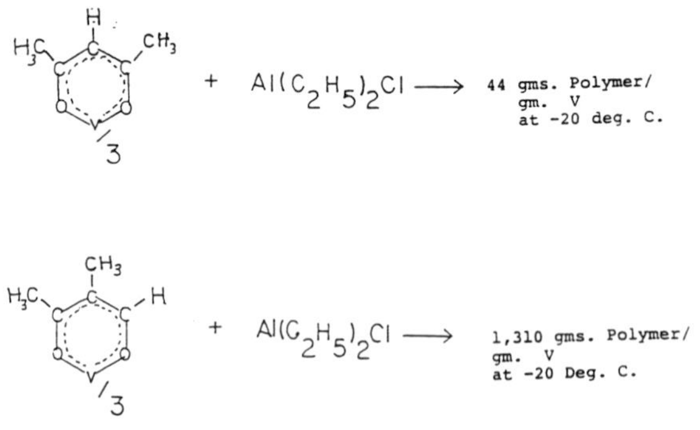

+ $Al(C_2H_5)_2Cl \longrightarrow$ 44 gms. Polymer/ gm. V at -20 deg. C.

+ $Al(C_2H_5)_2Cl \longrightarrow$ 1,310 gms. Polymer/ gm. V at -20 Deg. C.

Figure 21.     Living catalyst for block copolymers (Doi et al.[12]).

Table 2
Summary of catalyst type and efficiency

| EPM/EPDM Catalyst System | | Catalyst Efficiency gs. polymer/g transition metal compound |
|---|---|---|
| 1) Highly dispersed | AIR$_3$ /VOCl$_3$ | 100 to 200 |
| 2) Soluble | Et$_x$ AlCl$_{3-x}$/VOCl$_3$ | 1.200 to 1,500 |
| 3) Activated Soluble | Et$_x$ AlCl$_{3-x}$/VOCl$_3$ +Activator | 5,000 to 6,000 |
| 4) Supported & Single site catalyst (Zr) | MAO/Cp$_2$ ZrCl$_2$ | $\geq$ 100,000 |

# SUMMARY

All in all, research in catalyst development for EPM/EPDM has been an interesting and challenging field (Table 2). The future could be even more exciting as the search goes on for superactive catalyst systems for EPM/EPDM. They will make the present solution processes even more economical or will lead to new, even more economical processes as have occurred with the poly-$\alpha$-olefins.

# ACKNOWLEDGMENT

My career with Ziegler/Natta polymerization started through the efforts of Dick. Originally Dick and I were a group of two engaged in solid rubber research for Uniroyal Chemical Division just after the great flood of 1955. This effort was in free radical polymerization, primarily in the preparation of BDE/ACN copolymers. It was Dick, however, through his efforts in pushing back the frontier of science that got us started in Ziegler-Natta polymerization. Dick convinced management that an effort in Ziegler-Natta polymerization would be rewarding. Since I was the only member of the group at the time, I was the one who had the opportunity to initiate this work. What an opportunity this turned out to be.

Ziegler and Natta's discoveries led to a revolution in polymer chemistry. Through Dick's effort I had the good fortune of (playing) being a part of this great revolution. At first I did not realize whether this was good fortune or not. Here Dick had me grinding up 100 gram quantities of LiAlH with the potential of a spontaneous fire and also working with a World War II smoke screen chemical, $TiCl_4$. The aluminum alkyls also were capable not only of bursting into flames but also producing a burn much worse than one would get from bromine with severe pain lasting for days. This work was also occurring in the presence of volumes of flammable gaseous monomers, solvents and nonsolvents. Luckily for all of us, nothing serious happened and great strides were made in the Ziegler-Natta polymerization of ethylene and propylene.

## REFERENCES

1.  E. K. Easterbrook & R. D. Allen, *Ethylene-Propylene Rubber, Rubber Technology 3rd Ed.*, Maurice Morton, Chapter 9, 260-243 (1987).

2.  G. Natta et al, "Elastomeric Copolymers of Ethylene and Propylene", U.S.Patent 3,300,459 (January 24, 1967).

3.  W. N. Baxter & N. Merckling, U.S.Patent 4,371,680, February 1, 1983.

4.  R. J. Kelly, H. K. Garner, H. E. Haxo & W. R. Bingham, "Ethylene-Propylene Copolymer Produced with Soluble Catalysts", *Ind. and Eng. Chem., Product Research and Development, 1*,210 (1962).

5.  E. K. Easterbrook, T. J. Brett, Jr., F. C. Loveless & D. N. Matthews, "A Discussion of Some Polymerization Parameters in the Synthesis of EPDM Elastomers", *XXIII International Congress of Pure and Applied Chemistry, Macromolecular Preprint, II*, 712 (1971).

6.  E. N. Kresge et al, "Long Chain Branching and Gel in EPDM" *Rubber Division ACS Symposium*, May 8, 1984.

7.  F. P. Baldwin & G. VerStrate, "Ethylene-Propylene Elastomers", *Rubber Chemistry and Technology, 45*,718 (1972).

8.  J.C.W. Chien, *Recent Advances in Supported High Mileage Catalysts for Olefin Polymerization, 55* (1989), Transition Metal Catalyzed Polymerization Ziegler-Natta and Metathesis Polymerization, R. P. Quirk, Editor, Cambridge University Press.

9.  G. Foschini & F. Milani, *New EP Elastomers, Advances in Polyolefins*, R. Seymour & T. Cheng, Editors, 1987, Plenum Press.

10. F. J. Karol et al, New Catalysis & Process for Ethylene Polymerization, *Advances in Polyolefins*, R. Seymour & T. Cheng, Editors, 1987, Plenum Press, New York.

11. W. Kaminsky & M. Miri, Ethylene-Propylene Diene Terpolymer Produced with a Homogeneous and Highly Active Zirconium Catalyst, *J. Polymer Science, Polymer Chemistry Edition, 23*,2151-2164 (1985).

12. Y. Doi et al, Living Coordination Polymerization of Propene with a Highly Active Vanadium Base Catalyst, *Macromolecules, 19*,2896-2900 (1986).